Proceedings in Life Sciences

Modern Methods in Forest Genetics

Edited by J. P. Miksche

With Contributions by
F. Bergmann · G. P. Berlyn · W. Bücking
R. A. Cecich · L. S. Dochinger · F. H. Evers
P. P. Feret · R. B. Hall · K. M. Hansen · O. Huhtinen
K. F. Jensen · E. G. Kirby · J. Lunderstädt
J. P. Miksche · R. P. Pharis · B. R. Roberts
A. E. Squillace · R. G. Stanley · A. M. Townsend
R. B. Walker · K. von Weissenberg · L. Winton
W. Żelawski

With 38 Figures

Springer-Verlag
Berlin Heidelberg New York 1976

Professor Jerome P. Miksche
North Central Forest Experiment Station
U.S.Department of Agriculture, Forest Service
Institute of Forest Genetics, P.O.Box 898
Rhinelander, WI 54501, USA

ISBN 3-540-07708-1 Springer-Verlag Berlin Heidelberg New York
ISBN 0-387-07708-1 Springer-Verlag New York Heidelberg Berlin

Library of Congress Cataloging in Publication Data. Main entry under title: Modern methods in forest genetics. (Proceedings in life sciences). Papers developed from courses given at the International Union of Forest Research Organizations Biochemical Genetics Workshop, held at the University of Göttingen, Germany, July 5–28, 1973. 1. Forest genetics–Technique–Congresses. 2. Tree breeding–Technique–Congresses. I. Miksche, Jerome P. 1930– II. Bergmann, F. III. International Union of Forest Research Organizations Biochemical Genetics Workshop, University of Göttingen, 1973. SD399.5.M6. 582'.1601'5. 76-8828.

Offsetprinting: Beltz Offsetdruck, Hemsbach/Bergstr., Bookbinding: Brühlsche Universitätsdruckerei, Gießen.

Preface

The present volume contains papers developed from courses given at the International Union of Forest Research Organizations (IUFRO) Biochemical Genetics Workshop (Working Party S.04-5) held at the University of Göttingen, Germany on July 5 through 28, 1973. The workshop was organized by Professor Robert G. Stanley and was held in memory of Professor Klaus Stern. Unfortunately, both met with untimely deaths. Professor Stanley was also instrumental in initiating the process of having the workshop proceedings published. I was asked by the workshop participants to complete this task, and I wish to acknowledge their cooperation, advice and encouragement.

In addition to the courses and subsequent papers resulting from the above workshop, we have included some papers by colleagues who were unable to attend the meeting. The contents of this text may, therefore, be considered a working-manual of generally "modern" techniques that are applicable to forest genetics and breeding programs.

The chapters are placed in five major categories. The first three categories follow according to classes of chemical constituents inherent to plants which are nucleic acids (DNA, RNA), primary gene products (amino acids, proteins and enzymes) and primary and secondary metabolites (carbohydrate polymers, resins, phenolics, pigments, etc.). The fourth category is concerned with the interaction of environment and gene systems. Indirect selection, crossing and protoplasmic and flowering manipulation are factors covered in the fifth category.

Chapter 1 by Berlyn and Cecich offers a method of precisely determining DNA quantity with relative ease. This technique has particular relevance because recent findings indicate that the amount of DNA per cell may not be as invariant or as constant per species as previously thought and DNA variation is another genetic variable that can be utilized by the forest geneticist. In addition to DNA quantity other qualitative and quantitative factors characteristic of DNA are pertinent to forest genetics programs. Chapter 2 by Hall, Miksche and Hansen presents protocols for DNA separation, purification and characterization. Two DNA characterization techniques are, for example, C_ot analysis and nucleic acid hybridization. C_ot analysis demonstrates that tree DNA like the DNA of other eukaryotes consists of two kinds, highly repeated and single copy or unique DNA. The hybridization technique may serve as a useful tool for the study of phylogeny and genetic relatedness, which are two valuable areas of research for supporting tree breeding programs. The modes of genetic relationships as manifested by proteins can be further explored with electrophoretic methods provided by Feret and Bergmann in Chapter 3. They also present a rather thorough literature coverage of the application of electrophoresis to the plant sciences. Amino acids are the basic chemical constituents of proteins and Lunderstädt in Chapter 4 outlines methods of amino acid extraction and evaluation, i.e., the determination of types and amounts of amino acids present within the organism under study.

The physico-chemical processes and structures related to the conversion of light energy to chemical energy are intrinsic factors of the genomic composition of trees and these factors display genetic variation. Zelawski and Walker in Chapter 5 present techniques in measuring rates of photosynthesis and amounts of primary plant metabolites, but only after base-line terms are defined. Secondary plant metabolites, resin products and phenolics are discussed by Squillace and Lunderstädt in Chapters 6 and 7, respectively. The characterization and analysis of the diversity of resin components was greatly augmented with the advent of gas-liquid chromatography. Squillace presents the general use of GLC instrumentation in conjunction with monoterpene analysis in tree populations. In addition, a species cross reference list pertaining to monoterpene composition of conifers is included, and should prove useful to those individuals starting and those presently working in the area of GLC applications. Lunderstädt's Chapter (7) furnishes methods for the separation of phenolics, their characterization and quantitative analysis. The presence or absence of certain phenolic compounds in trees is related to resistance patterns of pathogenic microorganisms and the compounds may also serve as protectants against leaf-eating insects.

The above chapters are concerned with the detection and measurement of intrinsic genetic substances and their derived genetic and metabolic products. The following two chapters discuss some of the interaction factors of the soil and air of the milieu external to the tree. Genetic considerations of nutrient uptake and methods of mineral analyses are presented by Evers and Bücking in Chapter 8. Chapter 9 by Jensen, Dochinger, Roberts and Townsend presents a rather extensive discussion of air pollution as related to forest trees.

The remaining chapters deal with the manipulation of plant material and/or the application of exogenous treatments for the ultimate purpose of tree improvement. The method of application of indirect selection is presented by von Weissenberg in Chapter 10. Kirby and Stanley in Chapter 11 discuss pollen handling methods with a pertinent slant towards incompatibility. Protoplasmic fusion, a relatively new technique, that is gaining considerable favor in the planning of breeding programs is presented in Chapter 12 by Winton and Huhtinen. Flowering studies are an asset to tree breeding plans and the present "state of the art" is presented by Pharis in the final chapter.

I wish to thank G. Berlyn, M. Berlyn, M. Conkle, R. Dickson, D. Durzan, J. Hanover, R. Jeffers, G. Namkoong, H. Nienstaedt, A. Schipper and J. Zavitkovski for reviewing manuscripts. Mrs. Carol Brouchoud's typing of the manuscripts is appreciated. I wish also to extend my thanks to all the colleagues who unanimously concurred with the dedications. The appreciation of Klaus Stern's and Robert Stanley's dedication statements by H.H. Hattemer and F. Bergmann and E.G. Kirby and A.E. Squillace, respectively, is gratefully acknowledged.

Rhinelander, March 1976 JEROME P. MIKSCHE

Contents

Contributors

BERGMANN, F.	University of Göttingen, Büsgenweg 2, 3400 Göttingen-Weende, FRG
BERLYN, G.P.	School of Forestry and Environmental Studies, Greeley Memorial Laboratory, Yale University, 370 Prospect St., New Haven, CT 06511, USA
BÜCKING, W.	Department of Botany and Site Location Study, Fasanengarten, 7000 Stuttgart 31 (Weilimdorf), FRG
CECICH, R.A.	USDA Forest Service, North Central Forest Experiment Station, Institute of Forest Genetics, P.O.B. 898, Rhinelander, WI 54501, USA
DOCHINGER, L.S.	USDA Forest Service, Northeastern Forest Experiment Station, Forest Insect and Disease Laboratory, Delaware, OH 43015, USA
EVERS, F.H.	Forest Research Institute Baden-Württemberg, Fasanengarten, 7000 Stuttgart 31 (Weilimdorf), FRG
FERET, P.P.	Department of Forestry and Forest Products, Virginia Polytechnic Institute and State University, Blacksburg, VA 24061, USA
HALL, R.B.	Department of Forestry, Bessey Hall, Iowa State University, Ames, IA 50010, USA
HANSEN, K.M.	USDA Forest Service, North Central Forest Experiment Station, Institute of Forest Genetics, P.O.B. 898, Rhinelander, WI 54501, USA
HUHTINEN, O.	Tree Breeding of the Forestry Faculty of University of Göttingen, Büsgenweg 2, 3400 Göttingen-Weende, FRG
JENSEN, K.F.	USDA Forest Service, Northeastern Forest Experiment Station, Forest Insect and Disease Laboratory, Delaware, OH 43015, USA
KIRBY, E.G.	Tree Genetics Laboratory, Oregon Graduate Center, 19600 N.W. Walker Road, Beaverton, OR 97005, USA
LUNDERSTÄDT, J.	Institute of Forest Zoology, University of Göttingen, 3400 Göttingen, FRG
MIKSCHE, J.P.	USDA Forest Service, North Central Forest Experiment Station, Institute of Forest Genetics, P.O.B. 898, Rhinelander, WI 54501, USA

PHARIS, R.P.	Department of Biology, University of Calgary, Calgary, Alberta, Canada
ROBERTS, B.R.	USDA Agricultural Research Service, Shade Tree and Ornamental Plants Laboratory, Delaware, OH 43015, USA
SQUILLACE, A.E.	USDA Forest Service, Southeastern Forest Experiment Station, Naval Stores and Timber Production Laboratory, Olustee, FL 32072, USA
STANLEY, R.G.†	School of Forest Resources and Conservation, Forest Physiology-Genetics Laboratory, Institute of Food and Agricultural Sciences, University of Florida, Gainesville, FL 32611, USA
TOWNSEND, A.M.	USDA Agricultural Research Service, Shade Tree and Ornamental Plants Laboratory, Delaware, OH 43015, USA
WALKER, R.B.	Department of Botany, University of Washington, Seattle, WA 98195, USA
VON WEISSENBERG, K.	Finnish Forest Research Institute, Experiment Station of Reforestation, 77800 Iisvesi, Finland
WINTON, L.	Division of Natural Materials and Systems, The Institute of Paper Chemistry, P.O.B. 1039, Appleton, WI 54911, USA
ŻELAWSKI, W.	Institute of Silviculture, Warsaw Agricultural University, Warsaw, Poland

Dedication to ROBERT G. STANLEY

Professor of Forestry, University of Florida, Gainesville, FL, USA

Professor Robert G. Stanley died on April 15, 1974. For those who knew him, Dr. Stanley will long be remembered for his indefatigable enthusiasm and unyielding commitment to both science and the arts. He is survived by his wife, Patricia Pachler Stanley.

Dr. Stanley was born in New York in 1926. He graduated *cum laude* from Michigan State University in 1948. He then attended the University of California at Berkeley where he was awarded the Doctor of Philosophy degree in Botany in 1955. From 1955 to 1965 he was Principal Plant Physiologist with the United States Forest Service in Berkeley, California. In 1958-1959 he took leave from the Forest Service to serve as Research Fellow at the California Institute of Technology in Pasadena and in 1961-1963 he worked at University College in London on a Guggenheim Fellowship. Part of that time was shared with the Department of Botany, University of Nijmegen, Holland, where he served as a research professor. In 1965 he joined the School of Forest Resources and Conservation at the University of Florida, Gainesville. Since coming to Gainesville, Dr. Stanley served on the Editorial Board of the International Review of Forestry Research. He was designated by the United States Academy of Sciences as a visiting lecture-fellow to the Polish Academy of Sciences in 1968 and in 1972 as a visiting lecturer to the Romanian Academy of Sciences. He was also honored with the Charles L. Pack Lectureship at Yale University in 1971. In 1973, Dr. Stanley was chosen as a NATO senior research fellow. He also belonged to many societies and service organizations including the American Society of Plant Physiologists, Xi Sigma Pi, the American Chemical Society, the Scandinavian Society of Plant Physiologists, Sigma Xi, Freedom from Hunger-Food for Peace and the American Association for the United Nations.

Dr. Stanley's research interests spanned the broad field of plant biochemistry with specific interests in pollen physiology and forest tree reproduction. He is most recently remembered for his work on boron and nucleic acid metabolism in pollen. Just prior to his death he completed a book "Pollen-Biology, Biochemistry and Management", written together with Professor H.F. Linskens.

Of greater significance to the scientific community than his research contributions was his commitment to furthering international cooperation in science. Dr. Stanley was active in International Union of Forest Research Organizations (IUFRO) and served as chairman of the Biochemical Genetics Working Party. Another of his concerns was for the application of technology to world problems. He was the former California Associate Director of the Freedom from Hunger-Food for Peace campaign. In 1973 his significant book entitled "Food for Peace-Hope and Reality of U.S. Food Aid" was published.

There is yet another side to the man that was Dr. Stanley: the art patron and aficianado. Art, specifically graphic art, was his constant past time. The author of several papers in recognized art journals and an avid collector, he was constantly supporting and encouraging new artists. He was president of the Southern Graphic Arts Circle, based in Micanopy, Florida. In 1973 he initiated an exhibition of Romanian graphic art at the University Gallery in Gainesville.

The loss of Dr. Stanley is sorely felt by those who were infected by his particular kinetic enthusiasm for science and art and his devotion to his fellow man. We hope that his energy and excitement will survive in those that follow.

E.G. KIRBY and A.E. SQUILLACE

Dedication to Klaus Stern

Professor of Forest Genetics and Forest Tree Improvement, University of Göttingen, Göttingen, FRG

Professor Klaus Stern died in an automobile accident on May 7, 1973, when he was on the way from his home to his Göttingen office. His untimely death meant a profound loss to all who admired his enthusiasm for science and his ability to enjoy life. He is survived by his wife, Irmgard Stern, and son and daughter, Matthias and Claudia Stern. He left behind an institution, the existence of which was due almost entirely to his dedicated efforts, and a group of young scientists, all deeply devoted to him.

Klaus Stern was born in Hasenberg (then in Germany) on December 29, 1923. After World War II, and after recovering from injuries, he studied biology, agriculture and forestry at the Universities of Hamburg, Gießen and Berlin. In 1949 he was appointed Research Assistant at the forestry department of the former Erwin-Baur-Institut at Müncheberg. This was later integrated into the Akademie der Landwirtschaftswissenschaften, and in 1952 Klaus Stern was awarded the Doctor of Philosophy degree in Forestry at the Eberswalde Institute of that academy. In 1953 he took a one-year leave to work at the Brunsberg Institute of the Swedish Foundation for Forest Tree Breeding. This Institute was headed at that time by Professor Enar Andersson with whom he often collaborated professionally, and maintained a close friendship.

In 1954 Klaus Stern joined the research group at the Schmalenbeck Institute of Forest Genetics. He stayed there until 1966 and was among the founders of the journal which enjoys worldwide circulation under its new name, Silvae Genetica. During this 12-year period he did intensive work on both theoretical and practical problems in genetics. His name became closely connected with the development of forest tree breeding since he created a theory of forest genetics based on population-genetic concepts. Those who had the good fortune to work with him during those years will always remember his intellectual way of coping with scientific problems, both of his own and his coworkers', and also with personal problems of anyone who asked his advice.

His large number of publications reveal the scope of his work. In 1960 he became Associate Professor of Forest Genetics at the University of Hamburg. His second thesis was published as a book "Plusbäume und Samenplantagen". By then he had gained an international reputation. He spent half a year of study and research at North Carolina State University, Raleigh. He also was a Visiting Professor at the Royal College of Forestry, Stockholm, for one year.

Since he combined high quality of research with excellent teaching he was appointed Professor of Forest Genetics and Forest Tree Breeding at the University of Göttingen in 1966. Now as before in Schmalenbeck, he aggregated a small group of interested people and conveyed to them the enthusiasm to pursue challenging goals in teaching and research. At this time he passed away.

Two significant books were completed by dedicated fellow teachers: "Genetics of Forest Ecosystems" (published together with Lawrence Roche), and "Oekologische Genetik" (published together with Peter M.A. Tigerstedt). Besides widespread research activities, Klaus Stern served in many organizations. He was a member of Arbeitsgemeinschaft für Forstpflanzenzüchtung, Kuratorium für Forstpflanzenzüchtung, Biometry Society (GR), Gesellschaft für Umwelt-Mutationsforschung, und Genetische Gesellschaft. He was active in the International Union of Forest Research Organization (IUFRO) and served as joint leader (together with R. Toda) of Subject Group S2.04 on genetics. During the academic year, 1969-1970, he was elected Dean of the School of Forestry of the University of Göttingen. Many people have lost a friend.

H.H. HATTEMER and F. BERGMANN

CHAPTER 1
Optical Techniques for Measuring DNA Quantity

G. P. BERLYN and R. A. CECICH

Introduction

Not long ago Feulgen microspectrophotometry was considered to be such an advanced technique that it belonged only in the hands of specialists in microscopy. This situation has changed dramatically in the past few years because more and more scientists have come to appreciate the vast potential of the technique. This realization created a demand for commercial instruments, whereas previously, the microscopist of modest means had to assemble and, in some cases, construct his own instrument. Fully functional instruments are now commercially available over a wide price range and many of these are capable of providing accurate and precise data in the hands of most research scientists.

Microspectrophotometry, when applied to the cell, is often termed cytophotometry, especially if a single wavelength is employed. We shall use cytophotometry as a general term for cytological microspectrophotometry. The technique has many applications in forest genetics. For example, it can reveal: 1. the ploidy level of plant cells both within and between individuals; 2. the effect of aging, herbicides, pesticides, and growth regulators on this ploidy level; 3. the variation in the diploid genome within (Miksche, 1971) and between (Miksche, 1967) species; 4. the DNA content of individual chromosomes in development, heredity, and plant breeding; and 5. whether cell fusion has occurred in the new tissue culture genetics. In addition, the instrumentation for Feulgen cytophotometry, once acquired, can be used for a wide variety of other analytical techniques (cf. Rasch and Woodard, 1959; Wied, 1966; Berlyn, 1969; Wied and Bahr, 1970; Cecich et al., 1972; Berlyn and Miksche, 1976).

The Feulgen technique has been under almost continuous refinement for over 50 years since it was first proposed by Feulgen and Rossenbeck (1924) and it now ranks as the most quantitative of cytochemical techniques (e.g. Leuchtenberger, 1958; Kasten, 1960, 1964; Hardonk and van Duijn, 1964; Deitch, 1966; Hale, 1966; Murgatroyd, 1967; Greenwood and Berlyn, 1968; Garcia, 1962, 1970; Jacqmard and Miksche, 1971). Thus, Feulgen cytophotometry has great utility, proper instrumentation is readily available, and the preparative techniques for biological tissues are well developed. Consequently the technique has evolved to the point where it can now be effectively employed in applied fields and it is the purpose of this chapter to provide a manual for the use of Feulgen cytophotometry in forest genetics.

Several cytophotometric techniques will be presented in "cook book" fashion for use in the laboratory. However, when a new technique is first attempted it seldom seems as easy or precise as described in the manual. It is first necessary to gain experience and build up confidence in your capacity to perform the technique. This is critically important for the practitioner of cytophotometry and, in addition, the investigator must learn to evaluate the data and to identify possible sources of error along with their relative importance. Each technique has been explored in depth in the literature and, therefore, we do not intend to treat them in detail, but we do urge the investigator to study the original citations to become familiar with both the potentialities and limitations of cytophotometry. We do not wish to promote a false sense of security in the techniques or the reliability. Any technique has certain pitfalls and it is important to know what these are and how to avoid them. Therefore, we will first consider the primary processes involved in cytophotometry and briefly discuss the possible errors inherent in the method.

Light Absorption

The main physical event in any photometric procedure is light absorption and this means the chemical being detected must contain a chromophore with a selective pattern of light absorption. This pattern or absorption curve provides a sort of finger-print by which the molecule can be identified and quantitatively measured (Fig. 1). The absorption curve consists of a plot of absorption or concentration against wavelength. The absorption peak for Feulgen-stained DNA from various sources lies around 550 to 570 nm. DNA does not have a natural chromophore in the visible range of light and therefore an exogenous dye (Feulgen staining) must be introduced into the molecule in order for it to display absorption in visible light. DNA does have a native absorption in the UV range with a peak absorption of ca. 260 nm, but utilizing this chromophore requires special and very expensive equipment that is subject to a number of additional difficulties such as non-specific light loss. Therefore, the more practical and convenient alternative to determine nuclear DNA content is to use visible light Feulgen cytophotometry.

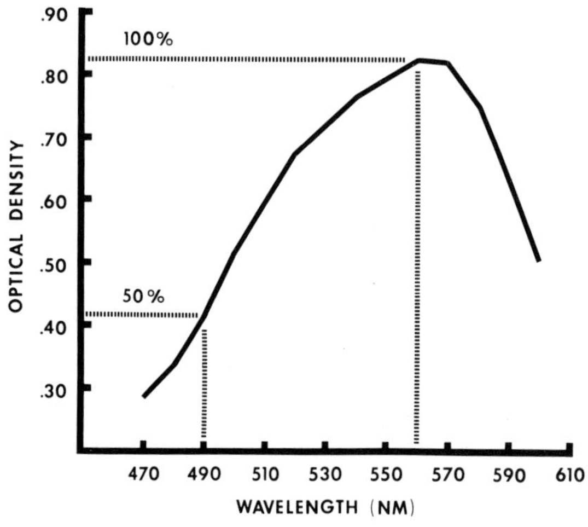

Fig. 1. A Feulgen absorption curve. Dash lines indicate procedure for determining choice of wavelengths for the two-wavelength technique

Molecules consist of atoms that are composed of nuclei with orbiting electrons. Some of the electrons move in spherically symmetrical orbits around their nucleus (s-electrons) while others move in asymmetrical orbitals (p-electrons). In molecules some of the electrons are shared between atoms. This sharing of electrons (bonding or Π electrons) holds the molecule together. These "delocalized" bonding electrons are the ones primarily responsible for light absorption.

For most microscopical phenomena we think of light as electro-magnetic radiation consisting of rapidly alternating electric and magnetic fields. But we need to remember that light can also be conceived of as consisting of a shower of particles. Each particle, called a photon, contains a quantum of energy $E = h\ c/\lambda$ where h is Planck's constant, c is the velocity of light, and λ is the wavelength of the light. Thus, wavelength is inversely correlated with the amount of energy per unit of light. The smaller the wavelength, the greater the energy per photon.

Absorption of light leads to a more energetic state of the absorbing species; it moves from a ground state to a higher energy level. Only certain levels are "allowed" for a given molecule and to reach the allowed levels whole quanta must be absorbed, thus only photons containing exactly the correct amount of energy can be absorbed by that particular absorbing species (Fig. 2). That is, for light absorption to occur, the energy (quantum) of the incident photon must exactly equal the difference in energy between the ground state and one of the sub-levels of the excited state.

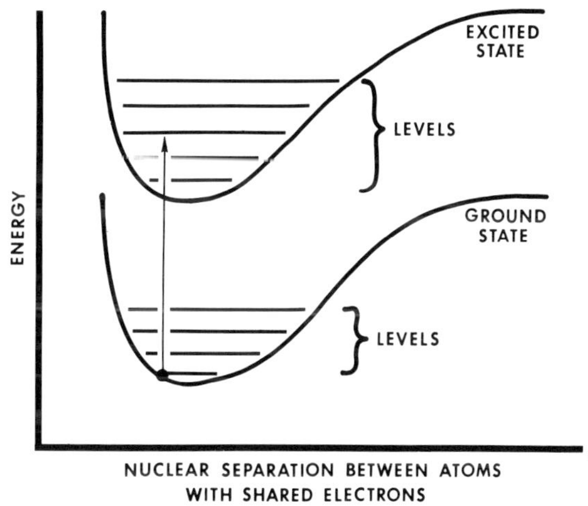

Fig. 2. Potential energy curves showing energy differences between the vibrational sublevels of the ground and excited states. The vertical arrow represents a transition from a ground state to a sublevel of the excited state. For this transition to occur a photon containing exactly the correct amount of energy must be absorbed

When a suitable photon is absorbed its oscillating electric and magnetic fields interact with the negatively charged electron causing it to oscillate and creating an electric dipole. The orientation of the potential electric dipoles in the molecule will also influence the specificity of light absorption. In other words, the direction and magnitude of the induced electric dipole will depend on the resisting forces provided by the structure and qualities of the rest of the molecule which is another reason for the distinctive form of absorption curves. Therefore, the probability that a photon will be absorbed depends on both wavelength (and consequently energy per photon), the direction of the electric field vector of the photon, and the molecular structure of the absorbent. We can think of the absorption curve as a probability curve, the peak representing the most probable transition and the off-peak portions representing energy levels less likely to be attained with the given molecule.

Upon absorption of a photon some of the light energy is transferred to the nucleus and some remains with the excited electron. Consequently light absorption accelerates both nuclear and electronic vibration. However, the excess vibrational energy is rapidly dissipated as heat and in some cases by luminescence (either fluorescence or phosphorescence). Luminescence, of course, greatly complicates cytophotometric analysis and must usually be eliminated by barrier filters. Fortunately this is not usually a problem in Feulgen cytophotometry. Of course, sometimes one is specifically interested in the fluorescent light and in this case only the monochromatic background light is filtered off and fluorescence intensity is measured.

Our discussion to this point has considered the absorption of single photons, but in cytophotometry we are dealing with absorption of trillions of photons and at this macroscopic level certain empirical equations have been shown to summarize the molecular events and therefore predict, rather precisely, the absorption of light. This approach was developed by Bouger in 1729, and Lambert in 1768 and extended by Beer in 1852. The mathematical arguments for the final equations (shown below) can be found in Berlyn and Miksche (1976).

$$\log \frac{Io}{I} = k\ c\ l$$

where log Io/I = optical density (OD) = extinction (E) = absorbance (A).

Io = background light coming through a blank portion of the microscope preparation

I = light being transmitted by the specimen

k = extinction coefficient

c = concentration of the chromophore

l = thickness of the specimen

Consequently, transmittance (T) = I/Io and absorption = (I-T).

Also by rules of logarithms

$$\frac{Io}{I} = 10^{kcl}$$

and

$$T = 10^{-kcl}.$$

The above terms are the fundamental relationships to know for cytophotometry. These so-called absorption laws were developed for dilute solutions but experience has shown that they work well under proper circumstances in cytophotometry (Wied, 1966; Wied and Bahr, 1970; Berlyn and Miksche, 1976).

The quantity actually measured in cytophotometry is transmittance T. To convert this quantity to mass, M, the following relationships are used:

$$M = A \times l \times C = \frac{A \log 1/T}{k}$$

where A = area of specimen.

Note that the extinction coefficient k, is a function of wavelength and two optical density values are only comparable at the same wavelength. In cytophotometry it is often difficult to calculate k and therefore when comparing nuclei a term called relative mass, RM, is often used.

$$RM = kM.$$

However, because of the inclusion of the extinction coefficient relative masses can only be compared when they are calculated from optical density values taken at the same wavelength.

Sources of Error

Feulgen cytophotometry is at least 90% accurate and perhaps somewhat better under ideal conditions. Today, instrument error is usually within 1% and seldom more than 5% where the absorption laws hold (see Garcia, 1970; Mayall and Mendelsohn, 1970; for further discussion of this). Berlyn (1969) provides methods of testing the absorption laws and this should be done for all new material. An internal standard on all slides is also recommended (see Methods for details).

There are essentially two types of errors operating in cytophotometry, viz. random (stochastic) errors and mean-biasing errors. The random or stochastic errors are those that are randomly distributed about the mean values and therefore do not affect the accuracy of the mean. They do affect the sensitivity of the method but can be tuned out to any desired degree by simply increasing sample size. These random errors can be due to a variety of causes such as dirt flecks on the slide, random fluctuations in light intensity or light detectors, positioning of the cell, focus, manual error in wavelength setting, and variation in measuring spot size.

Mean-biasing errors alter the value of the mean, degrading the accuracy of the method although sensitivity may remain high. These errors can be divided into specimen errors and absorption errors. Specimen errors include: 1. loss of DNA during specimen preparation and/or storage; 2. staining errors (masking, chromatin compaction and orientation, pH interactions and steric interference of chromophore molecules and, competing absorbents in the preparation); and 3. optical effects (excessive refractive index divergence between specimen and mounting medium, polarization, scattering, etc.).

Absorption errors may in turn be separated into optical and distribution error. Optical absorption errors are usually due to conical light (high condenser aperture and Köhler illumination), steep changes in optical density with small changes in wavelength, and glare. These errors can be minimized by: 1. using low condenser numerical aperture so that the rays of light passing the specimen are parallel (limiting the illumination beam to the central paraxial rays); 2. making measurements on a flat portion or gently sloping portion of the absorption curve; and 3. eliminating glare by working in a darkened room and keeping diaphragms partially closed (a tube diaphragm in the trinolcular measuring tube is very helpful in reducing glare).

Distribution error is probably the most widely discussed and serious error in cytophotometry and yet, according to Swift (1966) and our own experience, it is not often a factor. Fortunately, the use of the two wavelength procedure completely obviates this error and has the additional virtue of permitting accurate measurements of any shaped nuclei including metaphases and telophase chromosome masses.

Distribution error is caused by an unequal distribution (heterogeneity) of DNA (the chromophore) in the nucleus. It is a so-called negative error because it inappropriately reduces optical density from the true value although distribution error itself increases as optical density increases. The reason distribution error is negative is that the absorption law is a logarithmic equation. Distribution error means that each transmitting field is an average of two or more unequal transmissions (partially clear areas in the field). The log of the mean of the whole field is always less than the sum of the logs of the individual means.

$$\log \left[\frac{n}{\Sigma Ti}\right] < \Sigma\left[\frac{n}{\log Ti}\right] \quad \text{(see Mendelsohn, 1966).}$$

However, this type of error is not a factor in the two-wavelength procedure since the whole chromophore (nucleus or chromosome mass) is measured. Distribution error is also avoided in the more sophisticated scanning microspectrophotometers which scan over the entire specimen and feed the data into a computer.

Stain Specificity

As mentioned previously, the Feulgen staining procedure has been studied for over 50 years and numerous comprehensive reviews are available (Kasten, 1960; Wied, 1966; Wied and Bahr, 1970). Prior to staining, the tissue is first hydrolyzed in 1N HCl at 60°C or 5N HCl at room temperature for a period of time that will produce the deepest stain intensity. Depending upon the species and type of fixation, this time is usually between 8 and 16 min for 1N HCl and 15 to 40 min for 5N HCl. The treatment selectively removes the purines from the deoxyribose, freeing functional aldehyde groups.

The stain itself is prepared from the dye basic fuchsin whose main constituent is pararosaniline. In acid solution with excess SO_2, basic fuchsin is converted to the Schiff reagent (bis-N-aminosulfinic acid of pararosaniline leucosulfonic acid). The proposed reaction sequence for this conversion is presented in Fig. 3. The hydrolyzed tissue (apurinic DNA) is then incubated with the Schiff reagent (see Methods for details) whereupon two aldehyde groups condense with each molecule of Schiff's reagent. According to the most widely accepted model, the sulfonic acid group is released from the central carbon of the reagent and the dye complex is converted to a relatively stable, magenta-colored quinoid form (Fig. 3). The above mechanism is essentially that of Wieland and Scheuing (1921) and it has also been proposed that the products of the aldehyde condensation with Schiff's reagent are amino-alkylsulfonic acid derivatives (see Hörmann et al., 1958; Kasten, 1960, 1964; Hardonk and van Duijn, 1964; Deitch, 1966).

Materials and Techniques

We would now like to discuss briefly basic instrument components, considerations for tissue handling, and the various cytophotometric techniques, including a procedure for making an absorption curve for Feulgen-stained material. An appreciation of these topics will reduce errors, and consequently, provide more accurate data.

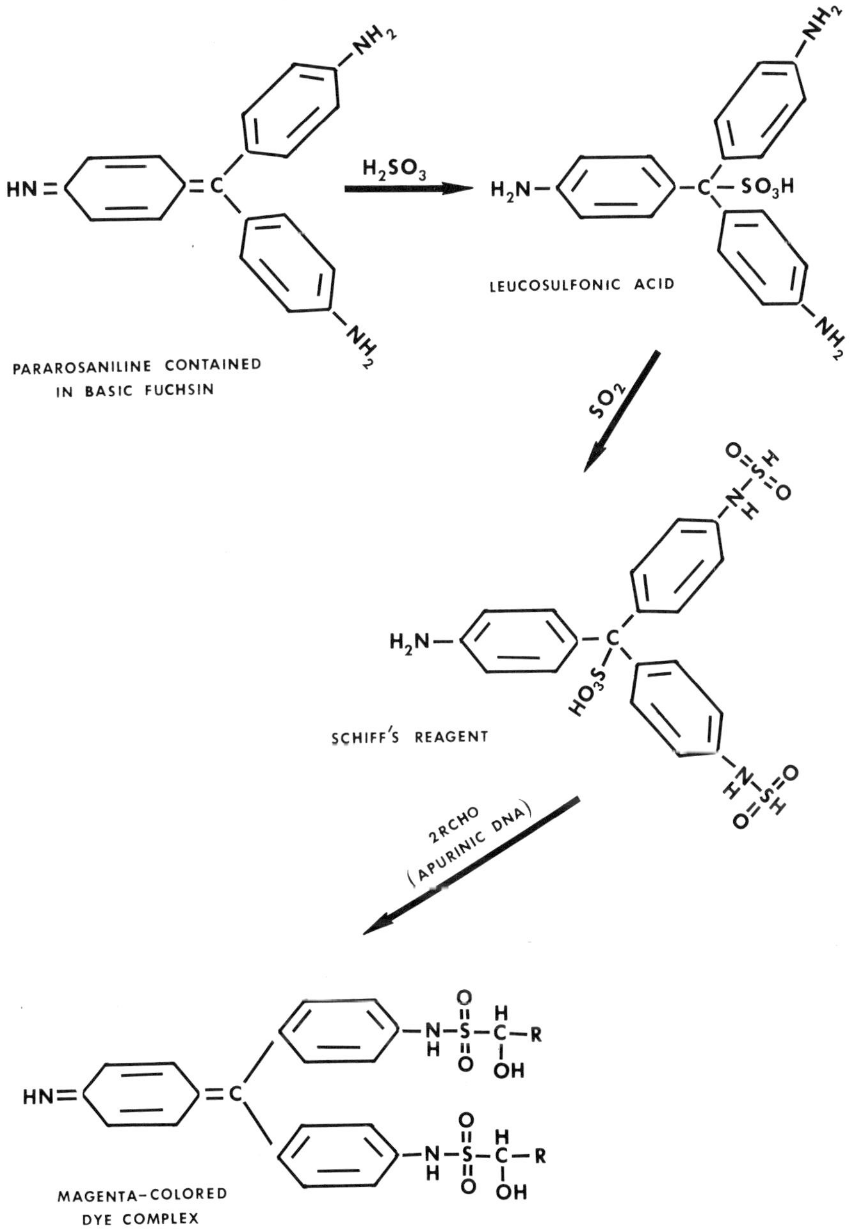

Fig. 3. Proposed reaction pathway in the Feulgen staining process

Cytophotometer Components

A simple microspectrophotometer, utilizing visible wavelengths of light, consists of three major components: 1. a microscope with good-quality glass optics; 2. a light source, with a monochromator; and 3. a photomultiplier-galvanometer device (Fig. 4). The light source directs white light into the monochromator, which separates the white light into the visible color spectrum. The monochromator emits visible wavelengths which are selected by the operator. The monochromatic light is collected by the microscope condenser system of low numerical aperture and is directed through the specimen, with which it interacts as previously described. The transmitted light is collected by the objective and ocular lenses and passed through a diaphragm into the photomultiplier device, which converts the light energy into an electrical signal. This electrical signal is amplified and read out on the galvanometer.

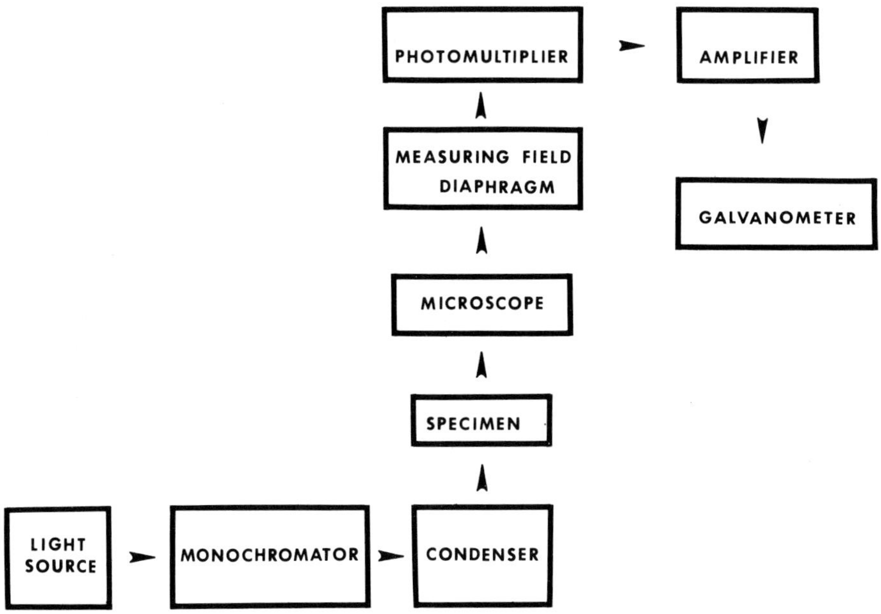

Fig. 4. A schematic representation of a basic microspectrophotometer. Arrows indicate the path of light and electrical signal

Cytophotometer Alignment

1. Turn on light source and amplifier for the galvanometer and allow to equilibrate as suggested by manufacturer.

2. Place a microscope slide with a specimen on the microscope stage. Find the specimen with the 10x objective and switch to the objective lens with which you will be measuring. Refocus and adjust condenser elevation so that the image of the field diaphragm is in the same plane of focus as the specimen.

3. The width of the exit slit on the monochromator is kept constant during an experiment.

Specimen Handling

This section is not intended to be an in-depth presentation of microtechnique but, a brief statement to remind or make the investigator aware that specimen preparation for quantitative histochemistry requires care, cleanliness and attention to procedures. In addition, we also want to stress that all specimens within an experiment must receive the same treatment so that meaningful comparisons can be made.

These approaches to histochemical microtechnique begin with fixation. Fixatives must be reproduced exactly if the fixative must be freshly made when sample collection does not permit fixation at one time. Time of fixation must be the same for all samples, or there may be a differential coagulation of the substrate within a cell.

Subsequent handling after fixation is important. If the experimental materials are not to be squashed or embedded immediately after fixation, they are generally dehydrated to 70% ethanol and held in the refrigerator until further processing. If the material is subsequently to be squashed (Conger and Fairchild, 1953), it is hydrated to water and broken into isolated cells with the aid of acid hydrolysis or pectinase. With the acid treatment cell isolation is accomplished by tapping with a glass rod and pressure on a coverslip. The number of taps with the glass rod is exactly reproducible from sample to sample, but thumb-pressure on the coverslip can only be approximated. After spreading the cells, the time of freezing with dry ice or gaseous CO_2 should be the same. Pectinase has the advantage that, with proper concentration and incubation time, the cells will separate without the application of external force or by gentle shaking of the incubation tube.

For microtomy, different samples of tissues within the experiment must be embedded in the same manner. This includes types of embeddment, times in dehydration fluids and transition fluids (e.g., xylene, n-butanol, benzene).

Within an experiment different types of adhesives should not be used to mount paraffin ribbons. Formalin-containing adhesives should be avoided since the aldehyde may interfere with the Feulgen reaction and give spurious values for the nuclei. A suitable adhesive consists of 2.5 - 5.0 g gelatin dissolved in 500 ml warm distilled water to which is added 0.25 g of chromium potassium sulfate. This may be used as a one-solution adhesive whereby the ribbon is floated on a drop of the adhesive, flattened over a flame or hot plate, excess moisture removed with filter or bibulous paper, and the slides dried on a warming pan. A more uniform procedure consists of "subbing" the slides by immersing them in the adhesive and allowing them to dry in a vertical position so that the excess adhesive drains off leaving a thin, uniform film of adhesive on the glass slide. The ribbon can then be floated on a drop of water and processed as above.

Staining of either squashed or microtomed specimens for quantitative histochemistry requires close control over temperature and pH of staining solution. It, therefore, differs from a "standard" staining technique, such as a safranin-fast green, which does not depend upon critical staining or differentiating times, or pH of the solutions.

Feulgen Staining Procedure

1. Hydrolyze in either 1N HCl at 60°C or 5N HCl at room temperature for the required time.

2. Ice-cold distilled water for 5 min to stop hydrolysis.

3. Schiff's reagent (see below) for 1 h in dark at room temperature.

4. Three changes of 5% sodium metabisulfite (NaS_2O_5), 10 min each.

5. Distilled water for 10 min.

6. Dehydrate and air-dry.

7. Mount specimen in refractive index oil or a synthetic resin.

Schiff's Reagent (Lillie, 1951). Mix 2.0 g basic fuchsin and 3.8 g sodium metabisulfite ($Na_2S_2O_5$) in 200 ml of 0.15 N HCl. Shake on a mechanical shaker for 2 h. Add 1.0 g fresh activated charcoal, shake for 2-5 min, and filter through Whatman No. 1 filter paper in a Buchner funnel. Repeat the charcoal treatment until the Schiff reagent is perfectly clear. Store in the refrigerator until ready for use. The reagent lasts 2-3 weeks under refrigeration. Discard when white precipitate forms or if reagent turns pink.

Obtaining an Absorption Curve for the Feulgen Reaction

1. Select a phototube diaphragm (measuring field) whose image is just smaller than that of a nucleus. Set the monochromator at 470 nm.

2. Move the nucleus out of the measuring field so that a clear area next to the nucleus fills the field. Open the shutter to the phototube and zero the galvanometer at 0.0 optical density (100% transmittance).

3. The operator has a choice of 2 procedures for obtaining a spectral curve. The first procedure requires fewer replications, but an additional arithmetic step. The second choice omits the arithmetic part, but has a larger error and, therefore, requires more replications.

a) Procedure 1

1. After zeroing at 470 nm, close shutter and set monochromator at 480 nm. Open shutter, record O.D., close shutter, move to 490 nm. Repeat sequence in 10 nm steps to 600 nm. You will have to re-zero occasionally; note the wavelengths where you do this.

2. Move nucleus into measuring field and record O.D. at 600 nm. Close shutter, change monochromator to 590 nm, open shutter, record. Repeat sequence to 470 nm, re-zeroing in clear area at the wavelengths noted in step 1.

3. Substract O.D. of clear area from O.D. of nucleus at each wavelength. Repeat the entire sequence for several nuclei and plot the means of the O.D.s against the wavelengths from 470 nm to 600 nm.

b) Procedure 2

1. Zero the galvanometer at 470 nm on a clear area next to the nucleus, close shutter, move nucleus into measuring field so that no clear area shows; open shutter, record O.D. of nucleus, and close shutter.

2. Change monochromator to 480 nm. Move nucleus out of field, zero galvanometer, close shutter, move nucleus into field so that same region of nucleus fills the measuring field, open shutter, record O.D., close shutter.

3. Repeat sequence to 600 nm.

4. This procedure requires more replications because the measuring field does not always contain the same exact portion of the nucleus at each wavelength. Thus, there is a difference in depth of the curved nucleus and corresponding error in O.D. because the nucleus O.D. at each wavelength is based upon zero for each blank.

Plug Method (Swift and Rasch, 1956)

When the researcher wants to measure differences on a per cell basis between tissues, he must do so on a microtomed section to maintain spatial orientation. If using a squash or smear preparation, this orientation is lost and recognition of cells from different tissues is usually impossible. In return for maintaining spatial orientation of the tissues, the researcher accepts a larger error of reproducibility (10-15%) by using the plug method, the technique usually employed with sectioned materials. As will soon be evident, the plug technique is very similar to the procedure for determining an absorption curve.

1. Select a phototube diaphragm smaller than the nucleus. Determine its diameter with a stage or ocular micrometer and calculate its area. Multiply the area (μm^2) of the diaphragm image by the thickness of the section (as estimated from the microtome setting). This volume, which approximates a cylinder, is called the plug.

2. From the absorption curve, select the peak wavelength (between 550 and 570 nm) for Feulgen-DNA complex. Set monochromator to peak wavelength.

3. Move nucleus away from measuring field; zero galvanometer in clear area adjacent to nucleus.

4. Move stage so that entire measuring field is filled with nucleus. Open shutter to phototube, record optical density value, close shutter.

5. With ocular micrometer, measure length and width of the nucleus if it is oblong; if circular, just measure diameter. Record these values.

6. If the nucleolus is large, measure its diameter and record this value. The nucleolus volume should be subtracted from the volume of a nucleus to give a more accurate estimate of chromophore volume.

7. Calculate Feulgen-DNA amount in relative units from:

$$M = \frac{\Pi E C^2}{F}$$

where M = relative unit of chromophore.

E = optical density of plug having radius C.

C = radius of plug (in micrometers).

F = ratio of plug volume/nucleus volume, where

$$\text{Nucleus Volume} = \frac{\Pi d^3}{6}$$

$$\text{Plug Volume} = \Pi C^2 h$$

h = section thickness.

Π can be eliminated as a constant in the ratio, therefore:

$$F = C^2 h \Big/ \frac{d^3}{6}$$

e.g., Calculate relative units of DNA in a sectioned nucleus where:

E = optical density of plug = 0.60

C = radius of plug = 2 μm

d = diameter of nucleus = 10 μm

h = section thickness = 5 μm

n = radius of nucleolus = 1 μm

nv = nucleolar volume = $\frac{4}{3}\Pi r^3$ = 4.19

$$M = \frac{EC^2}{F} = \frac{(0.60)\ (2^2)}{\dfrac{(2^2)\ (5)}{\dfrac{(10^3)}{6} - (4.19)}} = \frac{2.4\ (162.51)}{20} = 19.50$$

If M is calculated without subtracting the nucleolus volume, the formula can be reduced, so that

$$M = \frac{Ed^3}{6h}$$

In this case, M = 20.0, a difference of 2.5%.

Two-Wavelength Technique (Ornstein, 1952; Patau, 1952)

When the researcher needs the most precision in his measurements, the two-wavelength technique is chosen. This technique deletes the distributional error, reducing the total error to about 3-5%. The technique is especially useful on smear or squash preparations which isolate cells from each other.

From the absorption curve (Fig. 1), note the wavelength (approximately 560 nm) which gives maximum optical density for the Feulgen-DNA complex. Now find the wavelength which corresponds to 50% of the maximum O.D. For instance, if O.D. at 560 nm = 0.82; 50% of 0.82 = 0.41 corresponding to a wavelength of 490 nm (Fig. 1).

1. Select, measure, and record the diameter of a phototube diaphragm whose image is slightly larger than the nucleus.

2. Move nucleus out of measuring field, open shutter to phototube and zero the galvanometer in the clear area to give 100% transmittance at 560 nm. Close shutter.

3. Move nucleus into measuring field, open shutter, record % transmittance, and close shutter.

4. Change wavelength to 490 nm and follow procedures in steps 2 and 3.

5. The derivation of the final equation for the technique has been presented by many authors and will not be redone at this time. The final equation:

$$M = \frac{BL_1C}{k\ln 10}$$

is required to calculate the relative amount of chromophore. The denominator kln 10 is the extinction coefficient of the chromophore complex and is not required to calculate relative amounts of M.

B = area of measuring field. B can be omitted from calculations if the measuring area is kept constant throughout the experiment.

$$L_1 = (1 - T_1) \qquad L_2 = (1 - T_2)$$

$$Q = \frac{L_2}{L_1}$$

$$C = \frac{1}{2-Q} \ln \frac{1}{Q-1}$$

An example is given below where the transmittance at 560 nm (T_2) = 0.40 and transmittance at 490 nm (T_1) = 0.60. We will exclude B from the calculation.

$$L_1 = (1-0.60) = 0.40 \qquad L_2 = (1-0.40) = 0.60$$

$$Q = \frac{0.60}{0.40} = 1.50$$

$$C = \frac{1}{(2-1.50)} \ln \frac{1}{(1.50-1)} = \frac{1}{.5} \ln \frac{1}{.5} = 2 \ln 2 = 1.386$$

$M = BL_1C$ B is omitted, therefore

$$M = L_1C$$

$$= (0.40)(1.386)$$

$$= 0.554$$

To omit these cumbersome calculations, the reader is referred to a set of tables for finding M in relative units by finding the intersect of the two transmittance values (Mendelsohn, 1958). These tables cover the range of T values that are likely to be encountered. The value of M, based upon the T values above, found in the Mendelsohn's tables is 555 ($M = L_1C \times 10^3$).

One-Wavelength/2-Area Technique (Garcia, 1965)

If changing wavelengths is cumbersome for the operator, another method is available which is subject to less error than the plug technique. See Garcia and Iorio (1966) for an excellent discussion of the technique, its uses, and errors.

1. Select the peak wavelength from the absorption curve.

2. Select a diaphragm size to give a plug of the nucleus. Move nucleus out of the way, open shutter to phototube, zero galvanometer to 100% transmittance, and close shutter.

3. Move nucleus into field, open shutter to phototube, record transmittance (T_1) and close shutter. If your galvanometer is so constructed, also record the optical density (O.D. = $\log \frac{1}{T}$).

4. Open measuring diaphragm to include whole nucleus and some surrounding, clear background. Move nucleus out of the field and zero the galvanometer to 100%.

5. Move nucleus back into field (now having Area A_i), open shutter to phototube, record transmittance (T_i), and close shutter.

6. The arbitrary amount of chromophore in the nucleus is calculated from:

$$E_1A_1 = (\log \frac{1}{T}) \quad \frac{(1-T_i)\ A_i}{(1-T_1)}$$

where E_1A_1 = extinction of nucleus having area A_1.

Values of T_1, T_i, and A_i are substituted into the equation.

$$\text{Log} \frac{1}{T_1} = \text{optical density of plug reading.}$$

The quantity $\frac{(1-T_i)\ A_i}{1-T_1}$ is the area of the chromophore.

Multiply this by the O.D. reading in step 3 to get total extinction of the nucleus.

Integrating Techniques

Several commercially-available, very expensive, cytophotometers provide a combination of speed and ease of operation. The specimen is scanned in a variety of modes, depending upon the manufacturer, yielding many "plug values" which are accumulated and integrated into one value for the specimen.

Internal Standard

A source of error in microspectrophotometry is slide to slide variation in staining. Two specimens in the same stage of development, but on different microscope slides, may indicate a difference in the amount of stain. Omission of this problem is thus desirable. A technique to omit this variability in Feulgen cytophotometry is to in-

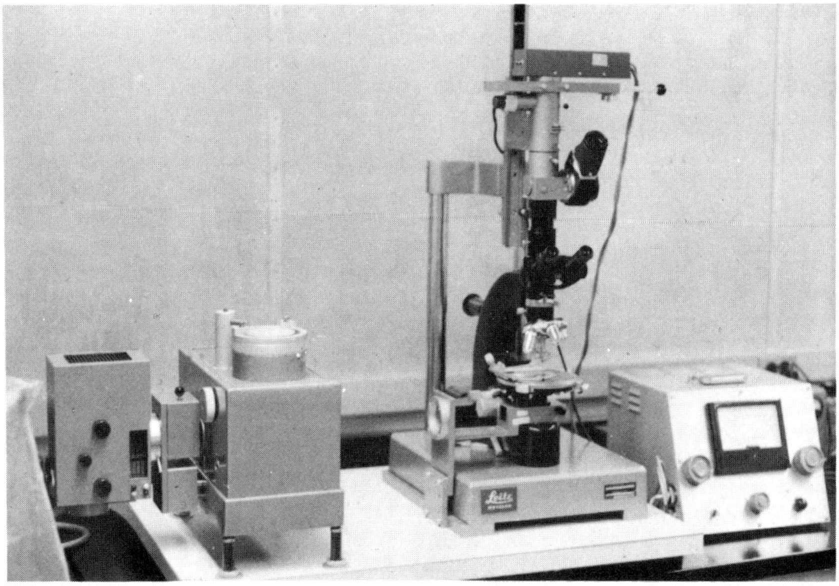

Fig. 5. A commercially-available microspectrophotometer

clude an internal standard with the specimen on the microscope slide. Chicken erythrocytes are a commonly used standard because they are nucleated and their DNA is non-replicated (2c) at maturity.

Slides are smeared with the chicken blood, air-dried, and stored. When needed, the blood-smeared slides are fixed in the same manner as the experimental material. If the experimental material has been bulk-stained, the slides with erythrocytes must be Feulgen-stained prior to squashing. If you squash unstained experimental material on the slides or mount paraffin ribbons, the erythrocytes and experimental material are stained simultaneously. Either procedure is required to convert the relative Feulgen value to absolute amounts, especially when comparisons among experiments or investigators are desirable. The first procedure cannot be used for determining slide to slide variation because the experimental material and erythrocytes were stained independently of each other. Their inclusion on the same slide is a random event. In the second procedure, the experimental material and the erythrocytes can be considered as paired comparisons, since they were stained together. Therefore, after conversion to absolute amounts, slide to slide variation can be omitted.

Interference microscopy, a quantitative measuring technique, is used to measure the nuclear dry mass of a sample of the erythrocytes before and after extraction of the DNA. This difference, in picograms (10^{-12}g), is the amount of DNA per erythrocyte. With an interference microscope the researcher can take advantage of the refraction of light as it passes from one medium (water) to another, the erythrocyte nucleus. The amount of refraction can be numerically described in degrees $^{\circ}$ and is termed optical path difference ($\Delta\delta$), where

$$\Delta\delta = \frac{\Delta\rho GAx2}{360} \text{ x } \lambda \text{ (cm)}$$

ρGA is the maximal darkness of the background or the object (in degrees ^{o}), as determined by the analyzer of the microscope. The background maximal darkness is set with the analyzer at 0°. Therefore, $\Delta\rho GA$ is the maximal darkness (extinction) of the object; that is, the difference between background and object maximal darkness. With the extinction transfer method of interference microscopy, the multiplication of $\Delta\rho GA$ by a factor of 2 is necessary because the phase difference is only measured in half wavelengths ($\lambda/2$) and not whole wavelengths (one $\lambda = 360^{o}$).

λ = wavelength of light used (546×10^{-7} cm)

$$\Delta\delta = \frac{\Delta\rho GA \times 2}{360} \times 546 \times 10^{-7} \text{ cm}$$

$$\Delta\delta = \Delta\rho GA \times 3.03 \times 10^{-7} \text{ cm}$$

Therefore, to obtain $\Delta\delta$ one simply multiplies the extinction value, $\Delta\rho GA$, by 3.03×10^{-7} cm. Determine the area, F, of the nucleus, using an ocular micrometer. Repeat these two steps on at least 10 nuclei per slide. Remove the coverslip and extract the DNA with DNAase. Mount a new coverslip in water and repeat the extinction measurements for the same nuclei. The difference between the two extinction values per nucleus is substituted into the equation

$$M = \frac{\Delta\delta_w \cdot F}{100 \cdot a} \quad \text{where}$$

$\Delta\delta_w = \Delta\delta_1 - \Delta\delta_2$

$\Delta\delta_1$ = optical path difference of whole nucleus in water

$\Delta\delta_2$ = optical path difference after DNAase

F = area of the nucleus ($\times 10^{-8}$ cm^2)

a = the material constant (a specific refraction increment for protein) for dissolved substances

a = 0.0018 (cm^3/g/100)

M is, therefore, the amount of DNA, in picograms (10^{-12} g), per erythrocyte.

In addition to determining the relative values for the experimental materials, cytophotometer measurements are also made of the chicken erythrocytes. A mean for the erythrocyte Feulgen values on each slide is determined and is used as the basis for standardizing all Feulgen measurements of experimental materials on that slide. For instance, let the mean erythrocyte Feulgen value for slide number one be 44 and the mean dry mass of DNA for an erythrocyte nucleus be 2.5×10^{-12} g. If the relative Feulgen value for an experimental nucleus in telophase is 550, a ratio can be made to determine the picograms of DNA for that nucleus.

$$\text{DNA} = \frac{2.5 \times 10^{-12} \text{ g}}{44} (550) = 31.25 \times 10^{-12} \text{ g.}$$

In slide number two, the erythrocyte Feulgen value is 40, the erythrocyte DNA dry mass is still 2.5×10^{-12} g but the telophase figure on this slide has a Feulgen value of 500. This 10% difference in relative Feulgen values for the telophase figures disappears when the relative amounts are converted to absolute amounts. By substituting into the above equations, we find that both telophase figures have 31.25 picograms of DNA.

References

Berlyn, G.P.: Microspectrophotometric investigations of free space in plant cell walls. Amer. J. Botany 56, 498-506 (1969).

Berlyn, G.P., Miksche, J.P.: Botanical microtechnique and cytochemistry. Ames, Iowa: Iowa State University Press 1976.

Cecich, R.A., Lersten, N.R., Miksche, J.P.: A cytophotometric study of nucleic acids and proteins in the shoot apex of white spruce. Amer. J. Botany 59, 442-449 (1972).

Conger, A., Fairchild, L.M.: A quick-freeze method for making smear slides permanent. Stain Technol. 28, 281-283 (1953).

Deitch, A.: Cytophotometry of nucleic acids. In: Introduction to Quantitative Cytochemistry (ed. G. Wied), pp. 327-354. New York: Academic Press 1966.

Feulgen, R., Rossenbeck, H.: Mikroskopische-chemischer Nachweis einer Nucleinsäure von Typus der Thymonucleinsäure und die darauf beruhende elektive Färbung von Zellkerzen in mikroskopischen Präparaten. Z. Physiol. Chem. 135, 203-248 (1924).

Garcia, A.M.: Studies on DNA in leucocytes and related cells of mammals. II. On the Feulgen reaction and two-wavelength microspectrophotometry. Histochemie 3, 178-194 (1962).

Garcia, A.M.: A one-wavelength, two-area method in microspectrophotometry for pure amplitude objects. J. Histochem. Cytochem. 13, 161-167 (1965).

Garcia, A.M.: Stoichiometry of dye binding versus degree of chromatin coiling. In: Introduction to Quantitative Cytochemistry II (eds. G. Wied, G. Bahr), pp. 153-170. New York: Academic Press 1970.

Garcia, A.M., Iorio, R.: A one-wavelength, two-area method in cytophotometry for cells in smears or prints. In: Introduction to Quantitative Cytochemistry (ed. G. Wied), pp. 239-255. New York, London: Academic Press 1966.

Greenwood, M.S., Berlyn, G.P.: Feulgen cytophotometry of pine nuclei; effects of fixation, role of formalin. Stain Technol. 43, 111-117 (1968).

Hale, A.J.: Feulgen microspectrophotometry and its correlation with other cytochemical methods. In: Introduction to Quantitative Cytochemistry (ed. G. Wied), pp. 183-199. New York: Academic Press 1966.

Hardonk, M.G., van Duijn, P.: The mechanism of the Schiff reaction as studied with histochemical model systems. J. Histochem. Cytochem. 12, 748-751, 758-767 (1964).

Hörmann, H., Grassmann, W., Fries, G.: Über den Mechanismus der Schiffschen Reaktion. Liebigs Ann. Chem. 616, 125-147 (1958).

Jacqmard, A., Miksche, J.P.: Cell population and quantitative changes of DNA in the shoot apex of *Sinapsis alba* during floral induction. Botan. Gaz. 132, 364-367 (1971).

Kasten, F.H.: The chemistry of Schiff's reagent. Intern. Rev. Cytol. 10, 1-100 (1960).

Kasten, F.H.: The Feulgen reaction - An enigma in cytochemistry. Acta Histochem. 17, 88-99 (1964).

Leuchtenberger, C.: Quantitative determination of DNA in cells by Feulgen microspectrophotometry. In: General Cytochemical Methods (ed. J.F. Danielli), pp. 219-278. New York: Academic Press 1958.

Lillie, R.D.: Simplication of the manufacture of Schiff reagent for use in histochemical procedures. Stain Technol. 26, 163-165 (1951).

Mayall, B.H., Mendelsohn, M.L.: Errors in absorption cytophotometry; some theoretical and practical considerations. In: Introduction to Quantitative Cytochemistry (eds. G. Wied, G. Bahr), pp. 171-197. New York: Academic Press 1970.

Mendelsohn, M.L.: The two-wavelength method of microspectrophotometry. II. A set of tables to facilitate the calculations. J. Biophys. Biochem. Cytol. 4, 415-424 (1958).

Mendelsohn, M.L.: Absorption cytophotometry: comparative methodology for heterogeneous objects, and the two-wavelength method. In: Introduction to Quantitative Cytochemistry (ed. G. Wied), pp. 201-214. New York: Academic Press 1966.

Miksche, J.P.: Variation in DNA content of several gymnosperms. Can. J. Genet. Cytol. 9, 717-722 (1967).

Miksche, J.P.: Intraspecific variation of DNA per cell between *Picea sitchensis* (Bong.) Carr. provenances. Chromosoma 32, 343-352 (1971).

Murgatroyd, L.D.: A quantitative investigation into the effect of fixation, temperature, and acid strength upon the Feulgen reaction. J. Roy. Microscop. Soc. 88, 133-139 (1967).

Ornstein, L.: The distributional error in microspectrophotometry. Lab. Invest. 1, 250-265 (1952).

Patau, K.: Absorption microphotometry of irregular-shaped objects. Chromosoma 5, 341-362 (1952).

Rasch, E., Woodard, J.W.: Basic proteins of plant nuclei during normal and pathological cell growth. J. Biophys. Biochem. Cytol. 6, 263-276 (1959).

Swift, H.: Analytical microscopy of biological materials. In: Introduction to Quantitative Cytochemistry (ed. G. Wied), pp. 1-39. New York: Academic Press 1966.

Swift, H., Rasch, E.: Microphotometry with visible light. In: Physical Techniques in Biological Research, Vol. III (eds. G. Oster, A.W. Pollister), pp. 353-400. New York, London: Academic Press 1956.

Wied, G.L. (ed.): Introduction to quantitative cytochemistry. New York: Academic Press 1966.

Wied, G.L., Bahr, G.F. (eds.): Introduction to quantitative cytochemistry II. New York: Academic Press 1970.

Wieland, H., Scheuing, G.: Die Fuchsin-Schweflige Säure und ihre Farbreaktion mit Aldehyden. Ber. Dtsch. Chem. Ges. 54, 2527-2555 (1921).

CHAPTER 2

Nucleic Acid Extraction, Purification, Reannealing, and Hybridization Methods

R. B. HALL, J. P. MIKSCHE, and K. M. HANSEN

Introduction

The study of nucleic acids is an important area of research in biology and particularly in genetics. DNA is "the molecule of heredity", and RNA is the keystone in transferring genetic information into cellular structure and function. The techniques developed to study nucleic acids are providing tools to understanding such questions as the organization of the eukaryotic genome, species taxonomic relationships, chromosome structure, and potential productivity. This chapter describes certain nucleic acid techniques that have been applied to forest genetics research.

The body of publications related to nucleic acid research is too voluminous for total inclusion in this chapter. However, the references in the citation section are both general and specific enough to give the reader a starting base in nucleic acid research.

Details on the preparation of the chemical reagents used in the various techniques are given at the end of the appropriate sections.

Material and Methods

DNA Extraction and Purification

Modifications and variations of extraction procedures to accommodate tissue consistency, pH, DNA content per cell and bacterial and/or fungal contamination must be devised in order to obtain a pure product (Stern, 1968; Pearson and Ingle, 1972). Basically, three general processes are followed in DNA extraction and subsequent purification: 1. the material (tissue, cells and organelles) must be physically disrupted without major DNA shearing. 2. The deoxyribonucleoprotein (DNP) is extracted from the broken cells by detergent solution. 3. Enzymatic degradation of the DNA by endogenous and exogenous nucleases must be prevented by the use of high salt concentrations, detergents, chelating agents, chemical denaturants (e.g., phenol, chloroform) and deoxyribonuclease free ribonuclease and pronase.

If the above general procedures are not followed, the quantity and quality of the DNA yield will not be high. Therefore, rectification of a poor product yield can be accomplished by checking disruption, solubilization of DNP and enzymatic digestion procedures.

DNA extraction procedures from conifer whole seeds (sphorophytic and gametophytic tissue), seedlings or needles (Fig. 1) and young dicotyledonous leaves (Fig. 2) are given below.

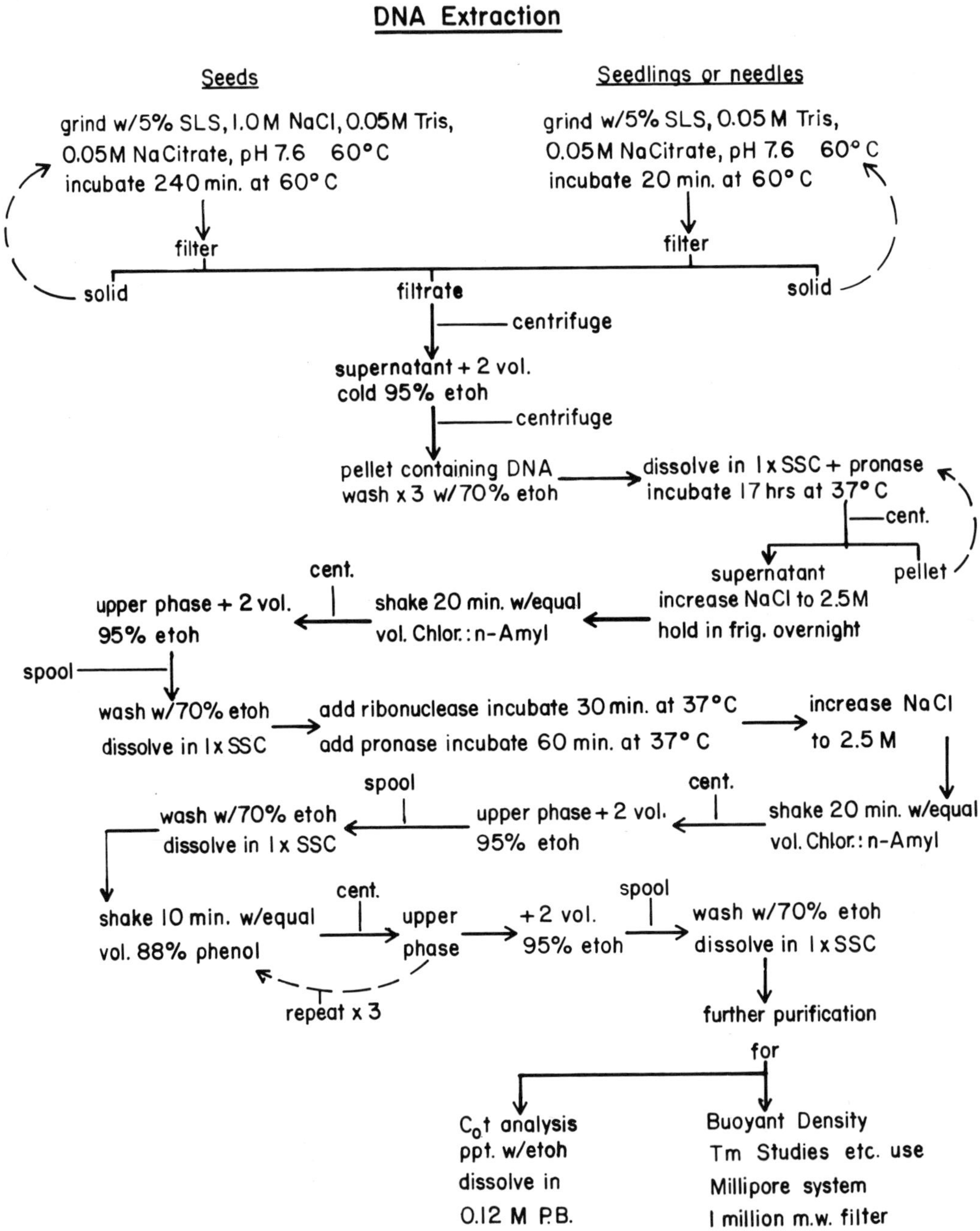

Fig. 1. DNA extraction and purification procedure for conifer seeds, seedlings and needles

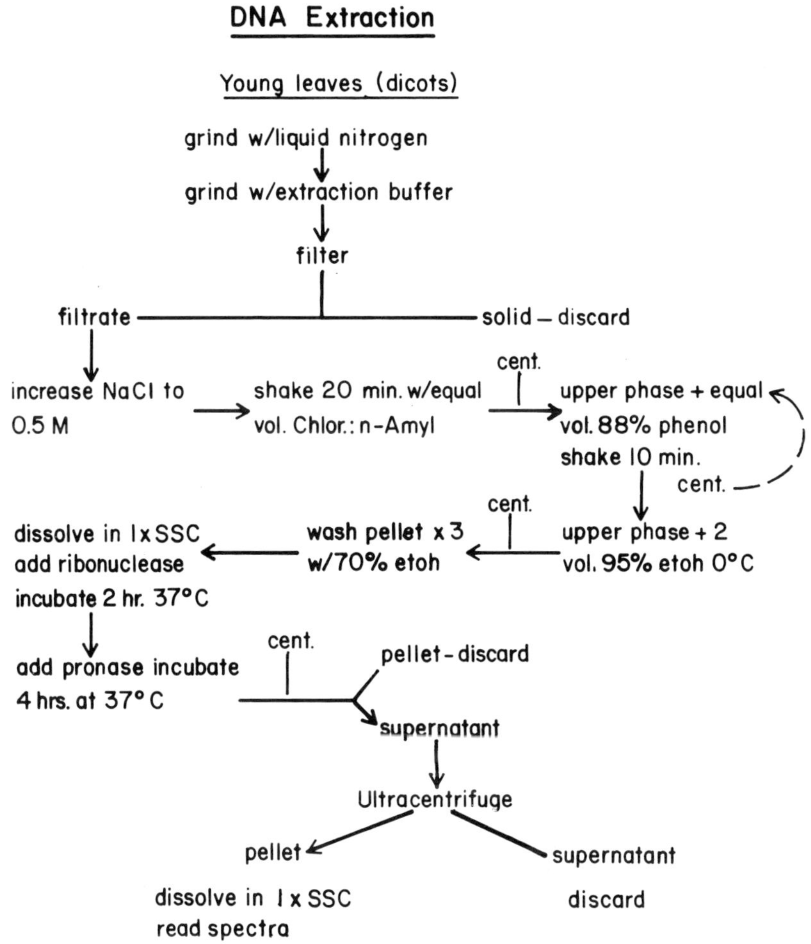

Fig. 2. Procedure for extraction and purification of DNA from young dicotyledonous leaves

Whole Seed, Seedlings or Needles

The conifer seeds are washed with 10 volumes of 10% sodium hypochlorite ($NaClO \cdot 5H_2O$) in a beaker or similar container covered with cheesecloth that is held in place with a rubber band.

The seed and sodium hypochlorite mixture is stirred vigorously for 15 to 20 min. After stirring, the antiseptic is decanted and the seeds are rinsed in running cold tap water for 30 to 45 min. The seeds are placed on a flat surface and air dried with a hair dryer or blower.

Seeds or mature gametophytic tissue are ground with sand (acid washed and ignited) in a mortar and pestle until the mixture becomes a fine powder. Ten volumes of 1M NaCl-SLS extraction buffer is added, and the mixture is incubated up to 4 h at 60°C with intermittant stirring. The mixture is filtered through two or three layers of cheesecloth and the filtrate is saved. The residual solid is reextracted with 5 volumes of extraction buffer, the material is ground in a mortar and pestle (preheated to 60°C prior to adding the buffer) and incubated 4 h.

The cheesecloth filtration step is repeated and the solid material is discarded. The filtrates are centrifuged 15 min at approximately 30,000 x g at 20°C and the pellet is discarded. The pellet contains cell wall fragments and other insoluble materials. The supernatant is overlain and mixed with two volumes of 95% ethyl alcohol and placed either in a refrigerator overnight or a freezer for 3 h. The DNA protein precipitates are collected using repeated five-minute 20,000 x g centrifugations at 4°C. The supernatant is discarded. The precipitate is washed three times with cold 70% ethyl alcohol by means of 15,000 x g, five-minute centrifugations.

To extract DNA from embryo axes, yound seedlings, or needles an alternative first step may be used. Ten to twenty grams of fresh tissue are submerged in a mortar filled with liquid nitrogen. Just as the last of the liquid nitrogen evaporates, the tissue is vigorously ground to a fine powder. The powder is transferred to the 60°C SLS (sodium-lauryl-sulfate) extraction buffer (10 ml for every gram fresh weight of tissue) and incubated for 10 min or longer. The suspension is filtered through cheesecloth and the solid material is ground briefly in a heated mortar using a small amount of solution. The ground tissue is recombined with the filtered solution and centrifuged to remove insoluble materials. The pellet collected in the centrifugation step can be reworked one or more times recovering significant amounts of DNA each time. The combined supernatants are then overlain with two volumes of 95% ethanol, and the precipitated DNA is collected as described previously. At this and following alcohol precipitation steps, the DNA can be covered with 70% ethanol and stored in a freezer at -20°C.

The deoxyribonucleoprotein precipitate is resuspended in 1 x SSC (3 vol.) and pronase solution (1 vol.) at room temperature with the aid of a magnetic stirrer and incubated at 37°C for 12 to 16 h. The mixture is then centrifuged at 25,000 x g for 50 min at 4°C. The supernatant is held under refrigeration while the pellet is resuspended in 1 x SSC and pronase and processed again to insure complete extraction of DNA. The supernatant from the reworked pellet is combined with the first supernatant and the NaCl concentration is adjusted to 2.5 M using solid NaCl.

An equal volume of chloroform:n-amyl alcohol is added to the solution and mixed by slow inversion shaking for 30 min. The mixture is centrifuged for 50 min at 20,000 x g and 4°C. The upper phase is pipetted off (discarding the interface and organic phase) and is precipitated by layering on 2 volumes of cold 95% ethyl alcohol. The fibrous precipitate is spooled on a thin glass stirring rod. The spooled fibers are transferred to and held in 70% ethyl alcohol.

If the DNA fails to spool around the glass rod and remains as strands in suspension, too great a quantity of 1 x SSC was used in the above deproteinization step. The proper amount of SSC to use must be determined empirically when extracting DNA from unfamiliar material. A concentration 0.5 mg/ml DNA would be ideal. If all of the DNA fails to spool properly, it can be collected by centrifugation and subsequently resuspended in a smaller volume of 1 x SSC.

The crude DNA precipitate is resuspended in 1 x SSC plus 1/10 volume of ribonuclease solution and incubated 60 min at 37°C. Stock pronase is added to make the solution 0.5 mg/ml pronase, and it is incubated overnight. The mixture is centrifuged for 50 min at 20,000 x g at 5°C, the molarity of NaCl in the supernatant is increased to 2.5 M, and it is stored in the refrigerator overnight as previously described.

An equal volume of chloroform:n-amyl alcohol is added and the mixture is shaken for 30 min at room temperature as previously described. Again, the solution is centrifuged for 50 min at 20,000 x g at 4°C and the aqueous phase is pipetted off and layered with two volumes of cold 95% ethyl alcohol. The DNA is then spooled with a glass rod and the DNA precipitate is stored in 70% ethyl alcohol. The phenol denaturation step in Fig. 1 can be excluded or included according to purity of sample.

The DNA precipitate is resuspended in 1 x SSC solution and centrifuged at 25,000 x g at 4°C for 50 min. A small aliquot is removed to determine the optical density 260/280 nm ratio. At this stage of purification, the ratio should be > 1.80. If not, the pronase step is repeated. The DNA solution is then further purified by passing it through a Pellicon 1 million molecular filter (Millipore filtration system). The effluent should be monitored until a spectral scan (230-300 nm) indicates that most of the small molecules have passed through the filter. A 150 ml solution of DNA in SSC can be processed through a filter in approximately 10 to 11 h. This filtration method removes the small pieces of DNA and protein contaminants while retaining the larger macromolecules of DNA above the filter. A 260/280 nm reading should be approximately 1.87 at this point. The DNA solution is centrifuged at 40,000 x g, 4°C for 60 min, the supernatant is decanted and DNA is precipitated with two volumes of 95% ethyl alcohol by spooling. The spooled DNA is held in cold 70% ethyl alcohol for subsequent use.

Additional purification procedures are diagrammed in Figs. 1 and 2. The protocol used will vary with the material and must be determined empirically.

Dicotyledonous Leaf Material (Jaworski and Key, 1974)

Seventy-five grams of young leaves are ground to a powder in a dry ice-sand (acid-washed and ignited) mixture using a precooled mortar and pestle. The finely ground powder is transferred to a fresh mortar and pestle (room temperature) and ground with 300 ml of PAS extraction buffer. The mixture is filtered through two or three layers of cheesecloth and the residue is reground with a small quantity of the original filtrate. The filtrate is then stirred with a magnetic stirrer at room temperature for 60 min.

The NaCl molarity is increased to 0.5 M and an equal volume of chloroform:n-amyl alcohol is added to the filtrate. The mixture is agitated gently on a shaker or a magnetic stirrer for 30 min. The mixture is centrifuged for 50 min at 25-30,000 x g at 4°C. The aqueous upper phase is collected by pipette, discarding the interface and organic phases, and an equal volume of redistilled, 88% phenol is added. The DNA-phenol mixture is placed on a magnetic stirrer or shaker for 30 min and agitated gently. The solution is centrifuged at 20,000 x g at 4°C for 50 min. The upper phase is saved and the phenol step is repeated. Again, the phenol-DNA solution is centrifuged and the DNA from the separated upper phase is precipitated by overlaying it with two volumes of cold 95% ethyl alcohol. The DNA is precipitated and stored 3-4 h in the refrigerator. The precipitate is collected by centrifugation and washed three times with 70% ethyl alcohol.

The DNA-containing precipitate is resuspended in 14 ml of 1 x SSC plus 2 ml of RNAse solution and incubated for 2 h at 37°C. Then 4.5 ml of the pronase solution is added and incubation is continued for 2 to 4 h at 37°C. After incubation, the mixture is centrifuged at 25,000 x g and 4°C for 50 min and the pellet is discarded. The supernatant is centrifuged at 110,000 x g (IEC, Rotor A-321 or equivalent) for 17 h at 4°C, and discarded. The translucent-like DNA pellet is resuspended in 6 ml of 1 x SSC. An aliquot of the solution should yield a 260/280 optical density ratio of approximately 1.85. The DNA may be precipitated from the 1 x SSC solution by spooling and stored in 70% ethyl alcohol.

DNA Extraction and Purification Solutions:

1. Sodium hypochlorite (10%): add 100 mls of Chlorox (approximately 5% Na hypochlorite) 900 mls of distilled water

2. NaCl-SLS extraction buffer (60°C):5% w/v sodium-lauryl-sulfate (Gallard and Schlesinger Corp., especially pure grade), 1.0 M NaCl, 0.05 M Tris, 0.05 M Na citrate, pH 7.6 adjusted with HCl. The extraction buffer must be heated to go into solution.

3. SLS extraction buffer (60°C):5% w/v sodium-lauryl-sulfate (Gallard and Schlesinger Corp., especially pure grade), 0.05 M Tris, 0.05 M Na citrate, pH 7.6 adjusted with HCl. The extraction buffer does not require heating.

4. Standard saline citrate (SSC)
   0.15 M NaCl, 0.015 M Na citrate
   0.1 x SSC, 20 x SSC, etc., refers to different strengths of the SSC. Buffered 1 x SSC contains 1.21 g/l Tris, pH 7, adjusted with HCl.

5. Pronase stock solution (Calbiochem)
   2.0 mg/ml in 1 x SSC, self-digested (incubated at 37°C for 60 min)
   Store this stock solution in a freezer

6. Chloroform:n-amyl alcohol:24:1 V/V

7. Ribonuclease:
   2 mg/ml in 0.15 M NaCl ph 5.0, incubate at 80°C for 10 min, cool and store in a freezer.

8. PAS extraction buffer used for young dicotyledonous leaves:
   7.7% sodium aminosalicylate acid
   1% Tri-isopropyl naphthalene sulfonic acid
   0.05 M NaCl, 0.05 M Tris, pH 7.5 adjusted with HCl

9. Redistilled phenol, 88% V/V.

RNA Extraction and Purification (Fig. 3):

Ten to 12 g of young conifer seedlings are blended with sand for 1 min at high speed in a Waring Blender with 100 ml of the extraction buffer plus 100 ml of 88% phenol. The blended material is then agitated with a magnetic stirrer for 60 min at room temperature.

The mixture is centrifuged for 10 min at 15,000 x g and the upper phase is collected. An equal volume of 88% phenol is added to the

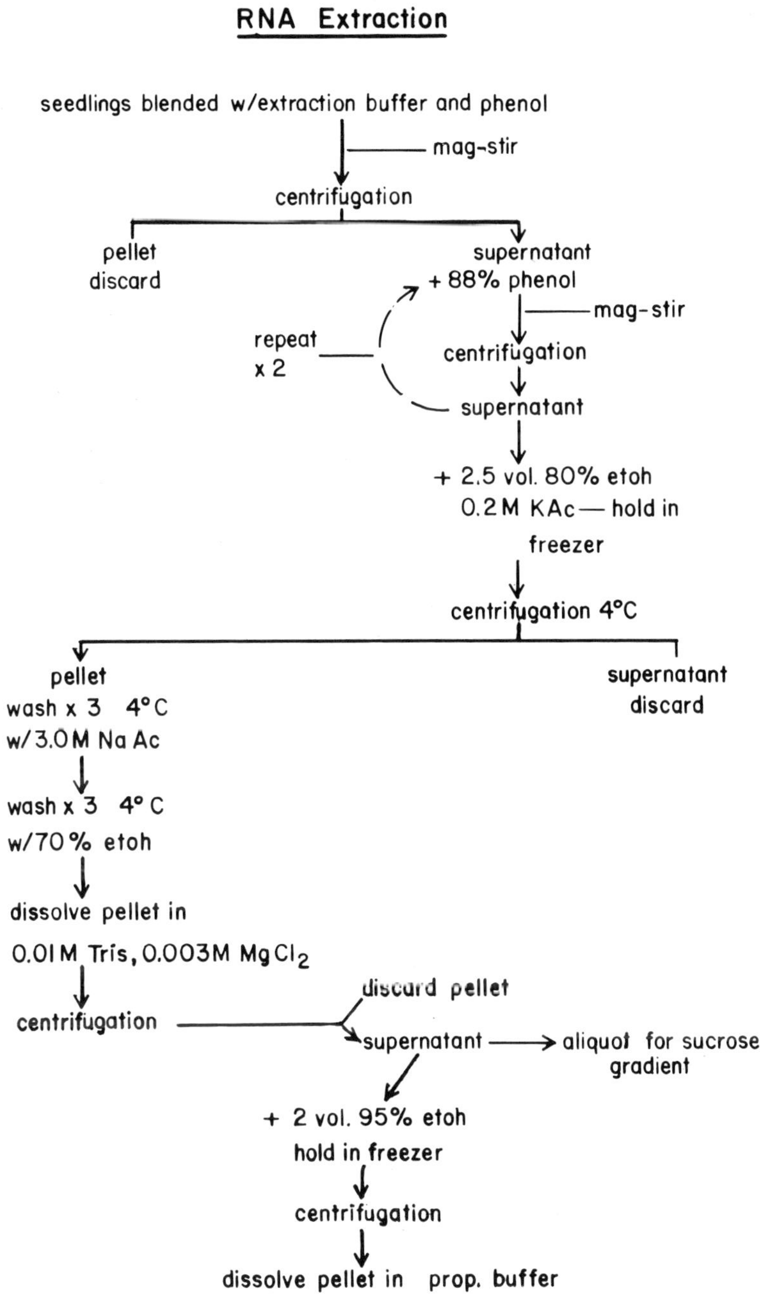

Fig. 3. Extraction flow diagram of RNA from conifer and dicotyledonous organisms

RNA solution, it is stirred on a magnetic stirrer for 20 min and the mixture is centrifuged for 10 min at 15,000 x g. The upper phase is saved and the above phenol step is repeated twice. The upper phase volume decreases after each phenol treatment yielding approximately 30 ml of RNA-containing solution after the final treatment.

The final supernatant is precipitated with 2.5 volumes of 80% ethyl alcohol-potassium acetate solution and placed in a freezer overnight to complete precipitation. The mixture is centrifuged at 10,000 x g for 10 min and the pellet is washed three times by stirring for 60 min with 3.0 sodium acetate at 4°C. The pellet is washed three additional times with 70% ethyl alcohol. The washed pellet is resuspended in 10 ml of 0.01 M Tris, 0.003 M $MgCl_2$ and centrifuged 10 min at 25,000 x g and the pellet is discarded. The supernatant is precipitated with two volumes of cold 95% ethyl alcohol and held in the freezer for further analysis (Fig. 3).

To procure radioactively labeled material, seeds are sown in a 1 liter beaker on four layers of filter paper soaked with 15 ml of H^3 uridine, the beaker is then covered with "Handiwrap" or similar covering material to contain moisture and radioactivity. The germinating seeds are harvested and the labeled RNA is extracted as described above. Germination time will vary with the species used.

RNA Extraction and Purification Solutions:

1. 0.2% sodium-lauryl-sulfate (Gallard and Schlesinger Corp., especially pure grade)
   0.03 M Tris, 0.003 M $MgCl_2$, pH 7.8, adjusted with HCl.
2. Redistilled phenol 88%, v/v.
3. 0.2 M potassium acetate in 80% ethyl alcohol (the potassium acetate is dissolved in water prior to dilution in alcohol).
   Stored in a freezer.
4. 3.0 M sodium acetate at pH 6.0 and stored in the refrigerator.
5. 0.01 M Tris with 0.003 M $MgCl_2$ at pH 7.8.

Organelle Nucleic Acid

The preceding nucleic acid extraction and purification procedures deal with whole seeds (dormant and post-dormant), seedlings, and organs, particularly young leaves. However, since organelles such as chloroplasts and mitochondria contain DNA and RNA there may be situations where nucleic acids from subcellular organelles may be desired. For example, one approach to the study of differential photosynthetic efficiency of provenances and clones would be to study differences between chloroplasts at the nucleic acid level. To provide the purified nucleic acids for such studies, techniques for "cell fractionation" or selective release of nucleic acids must be employed (Spencer and Wildman, 1964; Whitefield and Spencer, 1968).

Approximately 20 gm of tissue (young leaves 5-7 cm in length) are blended in 60 ml of STN or Honda medium using a Waring Blender or Omnimixer at 40 V for 30 sec at 4°C. In some species, the yield can be improved by mincing the leaves with a razor blade in 2 vol of Honda medium prior to the blending processes.

The ground tissue mixture is filtered through prewetted (STN or Honda media) flanellette, or a double layer of Miracloth (Chicopee Mills Inc., New York). The "cell free" filtrate is centrifuged at 2,500 x g for 10 min to yield a pellet containing chloroplasts and nuclei. The centrifugation force must be modified according to the material under study. The pellet is processed through 1. a discontinuous gradient (Tewari and Wildman, 1966) or 2. a selective chloroplast solubilization treatment (Whitefield and Spencer, 1968) or both treatments, depending on the material.

Discontinuous Gradient. The chloroplast and nuclei pellet is suspended in 3 ml of Honda medium per 10 g of the original leaf weight and the suspension is layered onto a 25 ml gradient that consists of 60% (w/v) sucrose in Honda medium (10 ml), 45% (w/v) sucrose in Honda medium (10 ml), and 20% (w/v) sucrose in Honda medium (5 ml). The samples are centrifuged for 1 to 2 h at 15,000 x g. The bulk of the nuclei are found pelleted at the bottom of the tube and the chloroplasts should be observed in the sucrose layers, which layer, however, will depend on the plant material. The various fractions are collected by pipetting off the layers of material. The subcellular composition of the layers can be checked microscopically by observing chloroplast fluorescence and treating the material with nuclear stains (Berlyn and Miksche, 1976). Chloroplast recovery is facilitated by diluting the fraction with an equal volume of 0.01 M Tris (pH 7.8), 0.01 M $MgCl_2$ and centrifuging at 8,000 x g for 15 to 20 min.

The nuclear pellet in the centrifuge tube is covered with a small amount of Honda medium. The tube is sealed with parafilm and stored in the refrigerator. Subsequent nuclear and chloroplast DNA extraction and purification is accomplished by previously described methods.

Selective Solubilization. The use of this method facilitates the recovery of intact nuclei and a soluble fraction from chloroplasts which serves as a source of chloroplast DNA. The pellet containing chloroplasts and nuclei is resuspended in 10 ml of STMMM solution and centrifuged for 10 min at 2,500 x g. The pellet is resuspended in Triton X-100 solution (4 ml per 10 gm of starting material) and the nuclei-chloroplast mixture is centrifuged at 4,000 x g for 10 min. The Triton solution solubilizes chloroplast membranes but not nuclear membranes, so the upper 85-90% of the supernatant may be used as a source of chloroplast DNA (Fig. 4).

Nuclei-containing pellets recovered from both the discontinuous gradient and the Triton solubilization methods can be combined for DNA extraction and purification.

The isolated nuclei are lysed with a one to three fold dilution (depends on DNA quantity) of 2% SLS in buffered 1 x SSC and incubated at 60°C for 10 min. The mixture is centrifuged at 10,000 x g for 25 min. The supernatant is processed through the NaCl, chloroform: iso-amyl alcohol, pronase, and RNAse treatments which have been previously described in this chapter.

Solutions Used for Cell Free Organelle Fractionation:

1. STN medium consists of 0.4 M sucrose, 0.05 M Tris (pH 7.9), 0.01 M NaCl.

2. Honda medium (Honda et al., 1966) consists of 2.5% Ficol, 5% dextran, 0.25 M sucrose, 25 M Tris (pH 7.8), 1 mM $MgCl_2$, 4 mM mercaptoethanol.

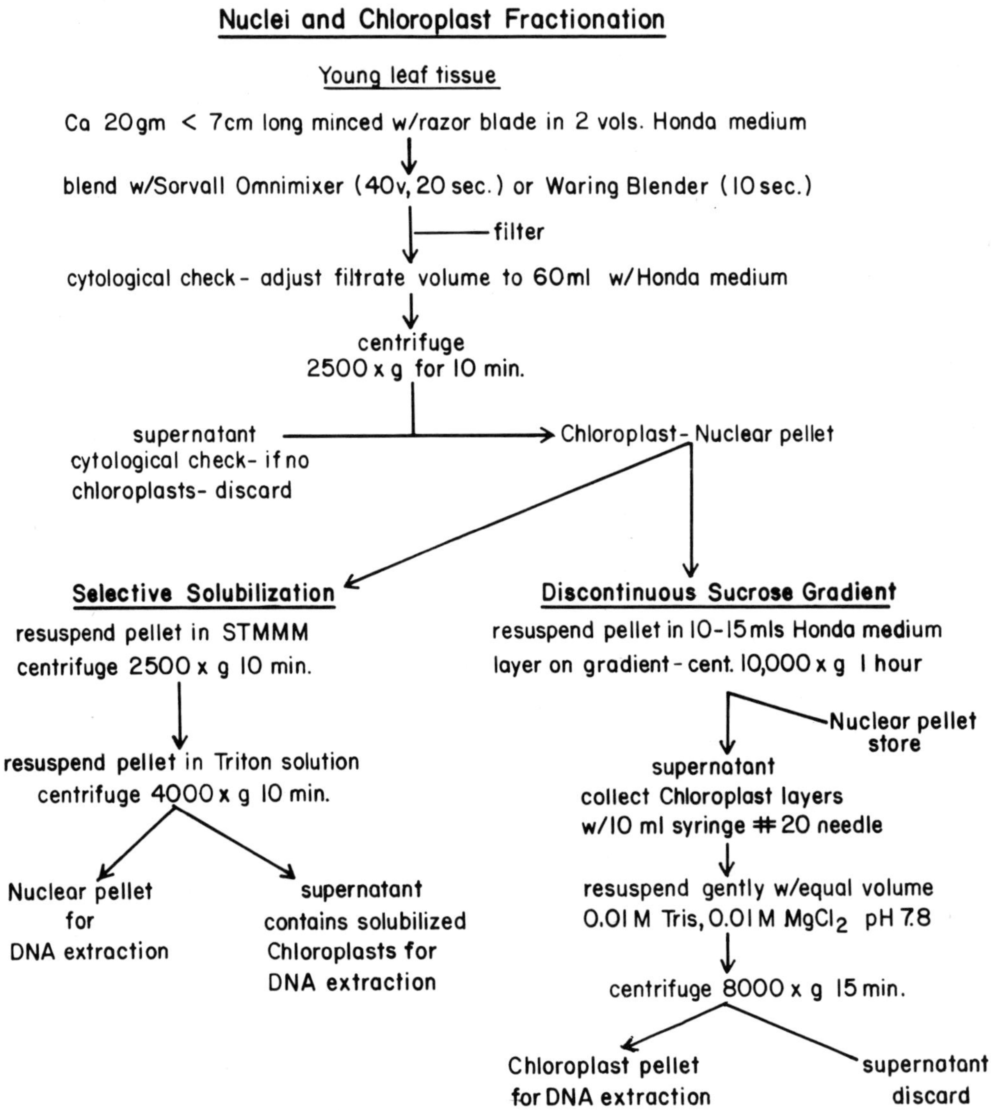

Fig. 4. A protocol for nuclei and chloroplast fractionation for subsequent DNA extraction from young dicotyledonous leaves

3. STMMM solution consists of 0.4 M sucrose, 0.025 M Tris (pH 7.8), 0.01 M KCl, 0.01 M $MgCl_2$, 0.004 M mercaptoethanol.

4. The Triton containing solution consists of 0.2 M sucrose, 0.01 M Tris (pH 7.8), 0.002 M $CaCl_2$ and 1 ml of 20% Triton.

5. 2% sodium-lauryl-sulfate, 0.15 M NaCl, 0.015 M Na citrate, 0.01 M Tris (pH 7.5).

Characterization and Analytical Methods for Nucleic Acids

Once isolated, nucleic acids can be analyzed by several means. This chapter outlines techniques for buoyant density and base composition determination; molecular weight estimation; and molecular and re-annealing hybridization. Cytological approaches to studying DNA are discussed by Berlyn and Cecich (Chap. 1).

Buoyant Density and Base Composition Determinations

Preparative and analytical ultracentrifugation techniques can be used for determining the buoyant density and the base composition (% G+C) of DNA samples. Thermal denaturation or Tm analysis is an alternative means of determining base composition. Dramatic changes in DNA buoyant density profiles or base compositions between or within species can be evidence of large scale changes in genomes during evolution. In some cases such changes reflect the addition or deletion of particular repetitious sequences of DNA. However, base composition has not proved to be a very meaningful taxonomic tool for the conifers (Miksche and Hotta, 1973). Base composition, through its effect on Tm, is an important variable in the DNA reannealing technique described in a later section.

Preparative Ultracentrifugation. The DNA is mixed with a cesium chloride solution and centrifuged at high speed. The CsCl solution forms a density gradient under these conditions and the DNA molecules migrate to an equilibrium density position which depends on their average % G+C (guanine plus cytosine) composition (Szybalski, 1968). Three basic pieces of equipment are needed for the technique: a preparative ultracentrifuge such as an IEC B-60 or Spinco Model L; a refractometer such as the Bausch and Lomb Abbe-3L; and a U-V spectrophotometer.

The refractometer is used to measure the density of the CsCl solution. A 15-20 µl drop of CsCl solution is placed on the prism, taking care not to scratch the glass surface. The prism is carefully closed and the refractive index reading is taken. The relationship between refractive index and the density of CsCl solutions is linear (Fig. 5).

Angiosperm and gymnosperm DNA's have densities of approximately 1.7 gm/ml at pH 8.5 (Kirk et al., 1971; Miksche and Hotta, 1973). Therefore, DNA samples for centrifugation should be adjusted to a density of about 1.7 or a CsCl refractive index of about 1.4. Solution preparation is accomplished by adding 70 to 80 µg of DNA in solution to a stock solution of CsCl that will yield 5 ml of final solution that is 0.02 M Tris-HCl buffer (pH 8.5) and 56% CsCl.

The 5 ml of DNA-CsCl solution is placed in a centrifuge tube and overlain with mineral oil (= paraffin oil). The centrifuge tubes are then run in a swinging bucket rotor (IEC SB-283 or Spinco SW39 head on the respective instrument) at 35 K for 72 h. After centrifugation, samples are collected by puncturing the bottom of each centrifuge tube and collecting 10-20 drop fractions (the same number of drops in each fraction) into calibrated test tubes. A 15-20 µl aliquot is taken from every 5th - 10th test tube and its refractive index is read. A straight line should result from a plot of these refractive index data. The volume of each test tube is brought to 2.0 ml with distilled water, the solution is mixed and the optical density at 260 nm is determined. Alternatively, labeled DNA can be precipitated with TCA and counted by scintillation methods.

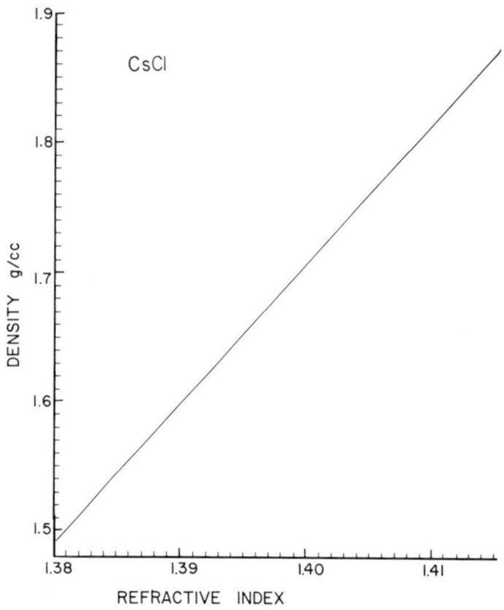

Fig. 5. Linear relationship between refractive index and density of CsCl solutions. The CsCl and refractive index values should be plotted on 10 x 10 to the centimeter paper to procure accurate readings of density

A plot of the refractive index and optical density data should resemble Fig. 6, the buoyant density determination of *Picea mariana* DNA. The refractive index at the bulk DNA peak is 1.3995 which is equal to a density of 1.6970 gm/ml. From this data the guanine plus cytosine content of the DNA sample is calculated as follows:

$$\% \text{ G+C} = \frac{\text{density} - 1.660 \text{ gm/ml}}{0.098}$$

$$= \frac{1.6970 - 1.660}{0.098}$$

$$= 37.8\%$$

Notice that the refractive index data points in Fig. 2 are not perfectly linear. The deviations are due mainly to evaporation of the CsCl solution. This emphasizes the need for rapidity and uniformity in taking readings. Also, a reference DNA of known density such as *Escherichia coli* DNA (density = 1.7114) should be included in the buoyant density runs as a check on their accuracy.

Analytical Centrifugation. This method is sometimes used for determining the buoyant density and base composition of DNA samples, but its most important use is in estimating the molecular weight of DNA fragments.

The solution used for a buoyant density determination is prepared by dissolving 5 µg (2 µg to 8 µg) of DNA in CsCl, buffered with Tris-HCl (pH 8.5). The refractive index is adjusted to 1.4000, and 2 µg of 1 mg/ml reference DNA is added. A 0.7 ml aliquot of the DNA mixture is placed in a centrifuge cell using a 22 gauge needle. Centrifugation is usually performed at 44,770 RPM and 25°C in an instrument such as the Beckman Model E ultracentrifuge. The equilibrium buoyant density of the unsheared DNA is established in 20-24 h and a UV absorption photograph is taken. The photograph may be analyzed directly, but preferably it should be precisely scanned on a microdensitometer. Fig. 7 illustrates a microdensitometer tracing from a buoyant

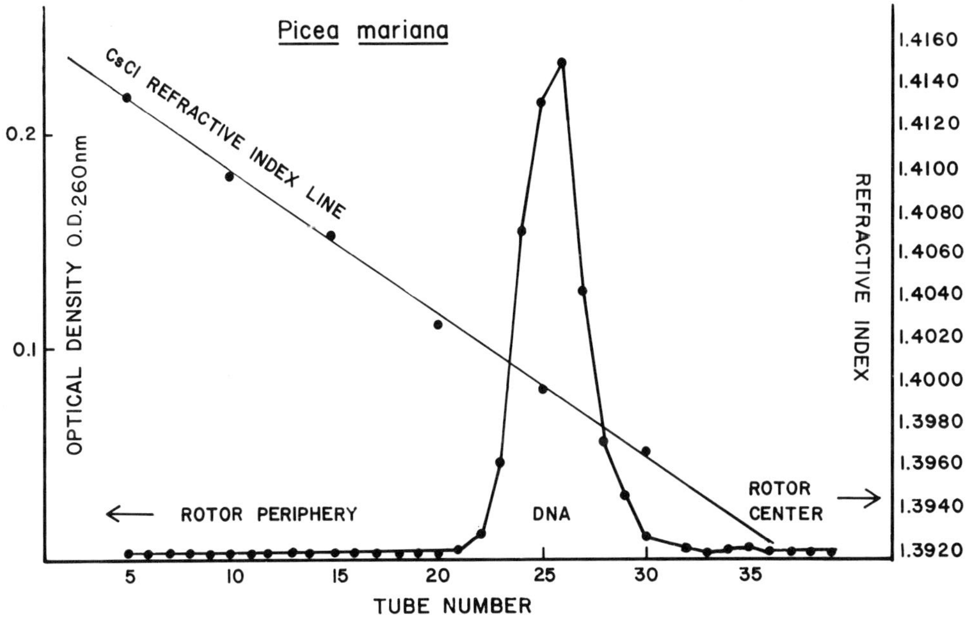

Fig. 6. Refractive index and optical density plot of *Picea mariana* DNA using the CsCl buoyant density method. See text for the formula used to calculate the percent guanine plus cytosine of a DNA sample

density run and demonstrates the mathematics involved in calculating % G+C (Mandel et al., 1968).

Tm Analysis. Guanine-cytosine base pairs are held together in the double helix by three hydrogen bonds while adenine-thymine base pairs are stabilized by only two hydrogen bonds. Hence, more energy, i.e., higher temperatures, are required to melt G+C rich DNA samples. Accordingly, the following technique has been developed (Mandel and Marmur, 1968) for determining the % G+C in a DNA sample based on the temperature at which melting is one-half complete. This temperature is referred to as the Tm, and it is determined from a plot of hyper chromicity vs. temperature.

The instrumentation consists of a spectrophotometer with a controlled temperature water jacket surrounding the cuvette compartment. The temperature variation is regulated by a waterbath. The temperature within the cuvette is read by a small thermistor probe fed through a tight fitting stopper to prevent evaporation (Fig. 8). The technique is becoming widely used and, therefore, many spectrophotometer manufacturers are producing controlled temperature apparatus for Tm profile determinations and related methods.

An unsheared DNA sample is diluted with 1 x SSC solvent to a concentration of 20 μg/ml yielding an OD_{260} around 0.500. This is the accurate portion of the optical density scale. The sample is placed in a cuvette and the temperature is brought to 30°C.

The optical density of the sample is recorded at this initial temperature. The cuvette temperature is increased up to 75°C and a second OD_{260} reading is taken. Then in a series of small steps or a programmed continuous rate of 1-2°C/min the cuvette temperature is raised to a

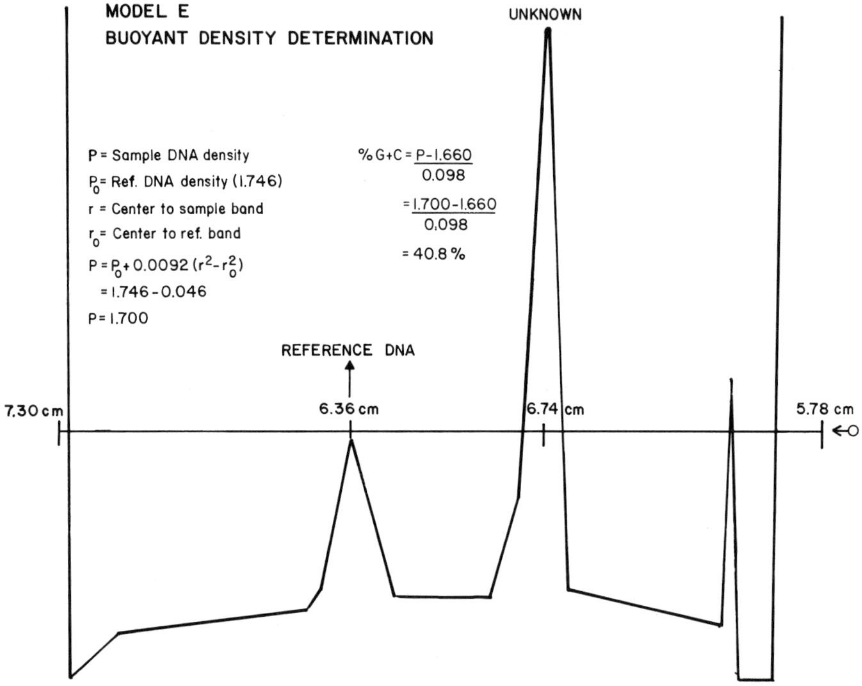

Fig. 7. A CsCl buoyant density tracing of DNA from a Model E analytical centrifuge. The tracing is derived from a densitometer scan of a UV photograph. See text for the formula used in calculation of percent guanine plus cytosine

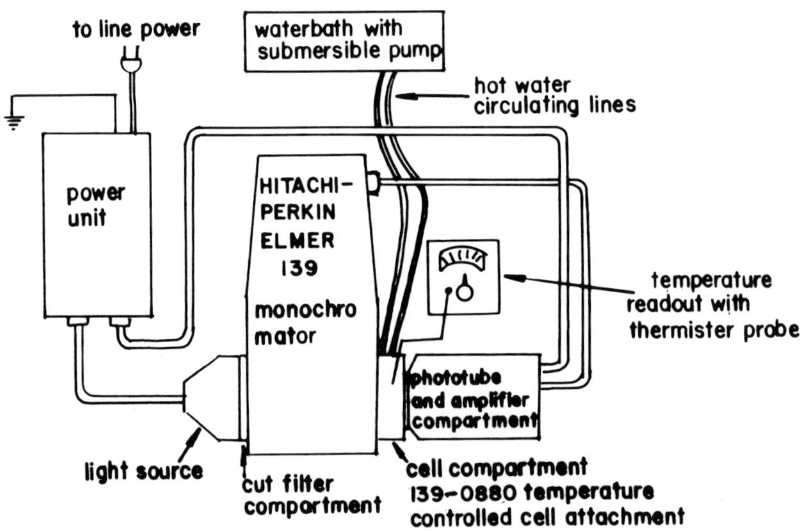

Fig. 8. Diagram of the instrumentation used for temperature profile studies

final thermistor reading of 98-105°C with OD_{260} readings being taken at every change of 1-2°C along the way. All these data are recorded in tabular form and the appropriate temperature correction factors are applied (Table 1) to each optical density reading to adjust for volume changes during heating.

$$\text{Corrected OD} = \text{Observed OD}\ \frac{V_T}{V_{25}}$$

Table 2 illustrates these steps using data from Sitka spruce.

Table 1. Temperature conversion factors for compensating changes in OD_{260} readings during solvent expansion in the cuvette resulting from high temperatures involved in thermal profile determinations (Mandel and Marmur, 1968)

T	V_T/V_{25}	T	V_T/V_{25}
25	1.0000	66	1.0174
26	1.0003	67	1.0180
27	1.0005	68	1.0185
28	1.0008	69	1.0191
29	1.0011	70	1.0197
30	1.0014		
31	1.0017	71	1.0203
32	1.0020	72	1.0209
33	1.0024	73	1.0215
34	1.0027	74	1.0221
35	1.0030	75	1.0228
36	1.0034	76	1.0234
37	1.0037	77	1.0240
38	1.0041	78	1.0247
39	1.0045	79	1.0253
40	1.0049	80	1.0260
41	1.0053	81	1.0266
42	1.0057	82	1.0273
43	1.0061	83	1.0280
44	1.0065	84	1.0287
45	1.0069	85	1.0293
46	1.0073	86	1.0300
47	1.0078	87	1.0308
48	1.0082	88	1.0314
49	1.0087	89	1.0321
50	1.0091	90	1.0329
51	1.0096	91	1.0336
52	1.0100	92	1.0343
53	1.0105	93	1.0351
54	1.0110	94	1.0358
55	1.0115	95	1.0365
56	1.0120	96	1.0373
57	1.0125	97	1.0380
58	1.0131	98	1.0388
59	1.0135	99	1.0396
60	1.0141	100	1.0404
61	1.0146	101	1.0411
62	1.0152	102	1.0419
63	1.0157	103	1.0426
64	1.0162	104	1.0433
65	1.0168	105	1.0441

Table 2. Illustration of data procurement for the thermal profile (Tm) of *Picea sitchensis* whole genome DNA

Temp.[a]	O.D.	Temperature expansion correction[b]	30° $A_t/A_{30°}$[c]	Percent absorption conversion[d]
30	0.342	0.342	1.000	
75	0.369	0.377	1.102	17.8
76	0.369	0.378	1.105	18.3
77	0.370	0.379	1.108	18.9
78	0.371	0.380	1.111	19.4
79	0.375	0.385	1.126	22.0
80	0.378	0.388	1.135	23.6
81	0.380	0.390	1.140	24.4
82	0.385	0.396	1.158	27.6
83	0.393	0.404	1.181	31.6
84	0.408	0.420	1.228	39.8
85	0.428	0.441	1.290	50.6
86	0.445	0.458	1.339	59.2
87	0.458	0.472	1.380	66.3
88	0.475	0.490	1.433	75.6
89	0.487	0.503	1.471	82.2
90	0.498	0.514	1.503	87.8
91	0.505	0.522	1.526	91.8
92	0.510	0.528	1.544	94.9
93	0.515	0.533	1.559	97.6
94	0.519	0.538	1.573	100.0

[a] OD_{260} changes are slight from 30° to 75° with eukaryotic DNA

[b] Corrected for concentration dilution caused by solvent expansion at higher temperatures. For calculation refer to attached temperature correction table (Mandel and Marmur, 1968); e.g., at 82° - (1.0273) (0.385) = 0.396

[c] 0.396/0.342 = 1.158....n, for remaining temperatures

[d] For percent calculation at 82° 0.158/0.573 = 27.6%

Temperature and A_t/A_{30} (260 nm absorbance at elevated temperature/absorbance at 30°C) readings are plotted on the abscissa and ordinate axes, respectively, on linear paper as seen in Fig. 9. The infection point of the sigmoid curve is at 86.6°C. From the medium point of the temperature melting profile curve, the G+C% content of the DNA sample can be calculated using Mandel and Marmur's (1968) equation:

$$\begin{aligned} \%\ G{+}C &= (Tm - 69.3)\ 2.44 \\ &= (86.6 - 69.3)(17.3) \\ &= 42.2\% \end{aligned}$$

Simplification of Tm determination is accomplished by plotting the data on probability paper. Percent of full absorption for the A_t/A_{30} readings is calculated by taking the upper plateau value, 1.573 in this case, and expressing the other A_t/A_{30} readings as percentages of full hyperchromicity, e.g., at 75°C 0.102/0.573 = 17.8%....n 100% The percent hyperchromicity values and temperatures are plotted as shown in Fig. 10. The Tm is then found at the 50% point on the curve.

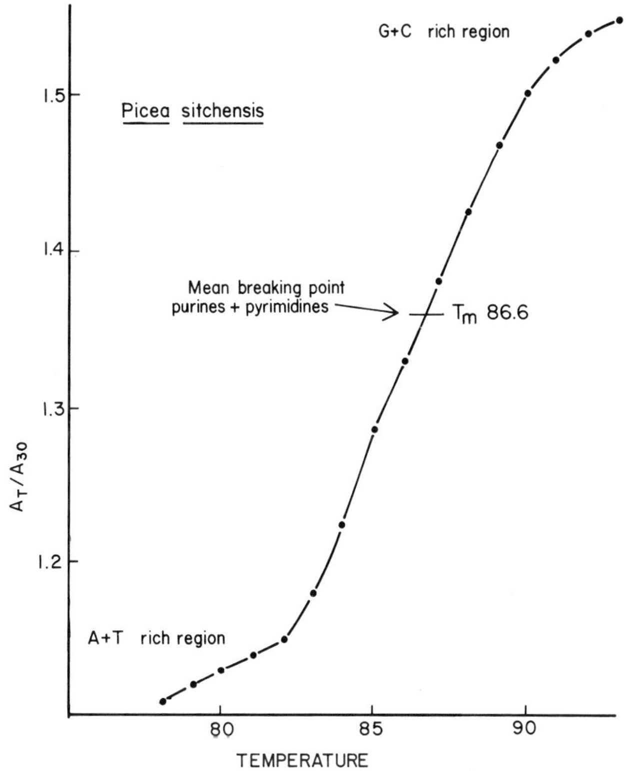

Fig. 9. Thermal profile of *Picea sitchensis* whole genome DNA showing a melting curve mid-point (Tm) of 86.6°C. See text for the procedure used to calculate percent guanine plus cytosine from the mean thermal denaturation temperature of the profile

See the Mandel and Marmur (1968) reference for further ramifications and embellishments of this useful technique. Readers interested in further chemical and physical aspects of DNA unwinding are referred to the report by Crothers (1969).

Molecular Weight Determination

As discussed later, DNA fragment size (molecular weight) is one important consideration in some types of experiments. It also is a means of characterizing different functional classes of RNA and of assessing the degradation of nucleic acid samples.

Analytical Centrifugation. The standard approach to determining the molecular weight of DNA preparations is analytical centrifugation as described by Studier (1965). Either single strand or double strand molecular weights can be determined depending on the choice of pretreatment and centrifugation medium.

The DNA sample should be prepared so that the buffer molarity is low (0.015 M for potassium phosphate buffer) and the DNA concentration is 50 μg/ml or greater (optical density at 260 nm is greater than 1.2 for double strand DNA and greater than 1.5 for single strand DNA). A 25 μl aliquot of this sample is then placed in the sample well of an ultracentrifuge cell. The cell is filled with about 1.7 ml of centrifugation medium. To run double strand DNA or single strand DNA at neutral pH a centrifugation medium of 1 M NaCl - 0.01 M tris

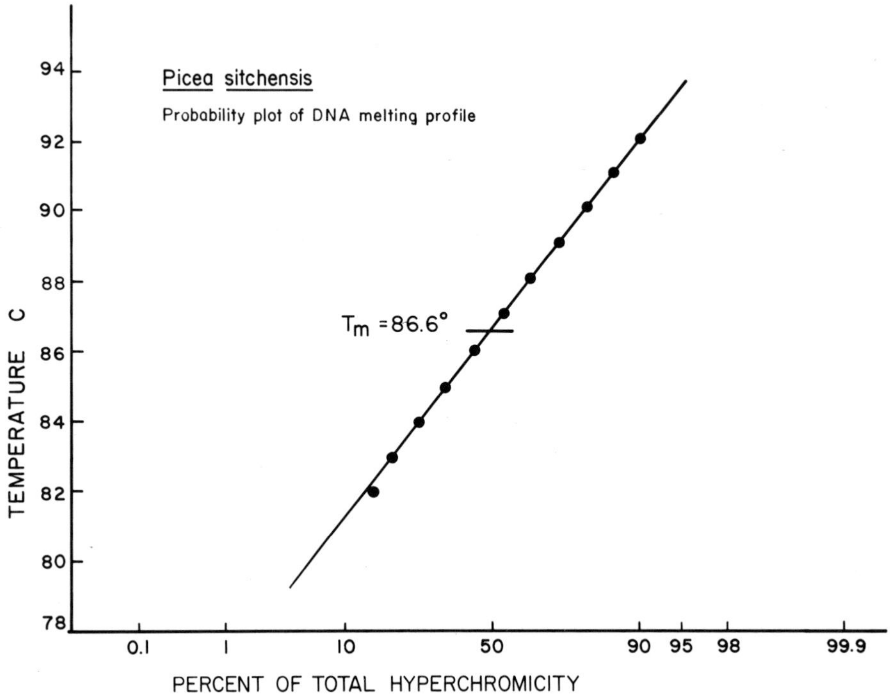

Fig. 10. Linear transformation of the *Picea sitchensis* sigmoid thermal profile curve illustrated in Fig. 9 using probability paper

buffer is used. To run single strand DNA at alkaline pH a medium of 0.9 M NaCl - 0.1 M NaOH is used (Studier, 1965).

The samples are centrifuged (Beckman Model E) at a temperature of 20°C with speeds from 26,000 up to 48,000 RPM depending on the expected size of molecule. The small DNA fragments used in reannealing experiments should be run at the higher speed. Using the special optics system of the ultracentrifuge, UV-absorption photographs of the migrating zone of DNA are taken at regular intervals such as 4 min. The photographic negatives are scanned on a microdensitometer and the distance from the DNA peak to the center of rotation (r) is determined for each photograph. From a plot of log r (Y axis) versus time in minutes (X axis) the DNA sedimentation coefficient, S, is read as the slope of the straight line drawn through the data points. The sedimentation coefficient (Studier, 1965) is converted to standard conditions of 20°C in water as follows:

$$S_{W,20} = 1.144\ S \quad \text{for 1 M NaCl - 0.01 M tris}$$

and

$$S_{W,20} = 1.160\ S \quad \text{for 0.9 M NaCl - 0.1 M NaOH}$$

Molecular weights (M) are then calculated from the following equations:

$$\text{Neutral single strand DNA} \quad M = \text{antilog} \frac{\log \frac{S_{W,20}}{0.0105}}{0.549}$$

Alkaline single strand DNA $M = \text{antilog} \dfrac{\log \dfrac{S_{W,20}}{0.0528}}{0.400}$

Double strand DNA $M = \text{antilog} \dfrac{\log \dfrac{S_{W,20}}{0.0882}}{0.346}$

The molecular weight as calculated by Studier's equation is in units of daltons. For some applications it is more convenient to express DNA molecular size in terms of number of nucleotides (single strand) or nucleotide pairs (double strand). To make this conversion:

Single strand:
Number of nucleotides $= \dfrac{\text{molecular weight}}{\text{330 daltons/nucleotide}}$

Double strand:
Number of nucleotide pairs $= \dfrac{\text{molecular weight}}{\text{660 daltons/nucleotide pair}}$

Electrophoresis. Analytical centrifugation is an expensive and time consuming technique. Hence, electrophoresis is a desirable alternative technique for the routine determination of RNA and DNA molecular weights.

The electrophoretic mobility of nucleic acids is inversely proportional to the logarithm of their molecular weight under appropriate conditions (Peacock and Dingman, 1968). Thus, measurements of DNA or RNA peak Rfs on electrophoresis gels can be converted to estimates of molecular weight.

General procedures for electrophoresis are described by Feret and Bergmann (Chap. 3): the following paragraphs describe the pertinent details for the electrophoresis of DNA samples used in reannealing experiments.

A 0.1 ml aliquot of DNA sample is mixed with an equal volume of sucrose-bromphenol blue solution (25 mg sucrose, 1 mg bromphenol blue, and distilled H_2O to 100 ml). One or two drops of this solution are layered on a 7.5% acrylamide disk gel and electrophoresis is carried out.

The position of the DNA peak in the gel can be determined directly by scanning at 260 nm, or by staining with toluidine blue, methylene blue, or Stains-all (Peacock and Dingman, 1968; Dingman et al., 1972).

Because gel formulations and other electrophoretic procedures vary between laboratories the relationship between DNA molecular weight and electrophoretic Rf should be independently established under each new set of conditions. DNA samples with different molecular weights can be prepared using different sonication intensities. By performing both analytical centrifugation and electrophoresis on several samples the regression coefficients can be determined for the equation:

$$\log(\text{molecular weight}) = a + b\,(\text{electrophoretic Rf})$$

which is then used for the routine determination of the fragment size of DNA samples.

DNA C_0t Analysis or DNA Reassociation Using Hydroxyapatite

Only within the last decade has the phenomenon of repetitious DNA been recognized, but this discovery and its continued study is likely to lead to considerable new knowledge on the structure and function of higher organism genomes. Repetitious DNA or repeated sequences refers to the presence of multiple copies of certain DNA sequences being present in each nucleus. The best general explanation and review of the topic has been made by Britten and Kohne (1968). Operationally, repetitious DNA is defined by and studied through DNA reannealing and hydroxyapatite chromatography experiments. In these experiments the double strand nature of the DNA molecules is disrupted by heat or chemical treatment. The single strand fragments are then allowed to incubate and molecular motion causes collisions between strands. If two colliding strands have dissimilar sequences, nothing results from the collision. If, however, there is similarity (complementarity) between the strands a stable duplex molecule can be reformed. The rate at which two complementary strands will reanneal depends on their respective concentrations in solution. When DNA is extracted from an organism, those DNA sequences which are repeated in the genome will have higher relative concentrations and will reanneal faster.

Some of the possible implications that repetitious DNA has for forest genetics and tree improvement have been previously discussed (Hall et al., 1974). It is likely that there is redundancy of genes which code for products needed in large amounts by the cell. In fact, a many thousand-fold redundancy of ribosomal RNA genes has been demonstrated for several tree species (Hotta and Miksche, 1974). In such a gene system, if independent mutation and recombination of the many copies of a single gene are not prevented by some mechanism, the gene may with time evolve to have more than two alleles per diploid organism. Such gene systems may be partially responsible for what we know as quantitative genetic inheritance.

A second function which has been attributed to redundant DNA is the control and coordination of the many gene systems present in a genome. The habitats of many tree species cover a wide ecological breadth. Rehfeldt and Lester (1969) have suggested that the particularly variable environments to which pioneer tree species are subjected have led to the evolution of genetic systems with a high degree of developmental homeostasis or a fairly constant physiological and morphological response to a range of environments. Such homeostasis could be due in part to redundant DNA control systems which activate the same developmental gene systems in response to varied stimuli.

A final category of speculated functions for redundant DNA is the control of chromosome structure and behavior. The constancy of metaphase chromosome morphology suggests an integration of a recognition system in chromatin which directs the same set of folding and "packaging" steps at each nuclear division. This might depend on certain DNA sequences that are repeated at critical points along the DNA strand (Comings and Riggs, 1971; Sutton, 1972). Redundant sequences have also been suggested as being important in mechanisms for the meiotic pairing of chromosomes (King, 1970; Comings and Riggs, 1971) and the attachment of spindle fibers to the centromeres (Walker, 1969).

Hydroxyapatite (HAP) chromatography is one means by which single strand DNA molecules can be separated from double strand molecules to assess the amount of reannealing that has occurred. The results

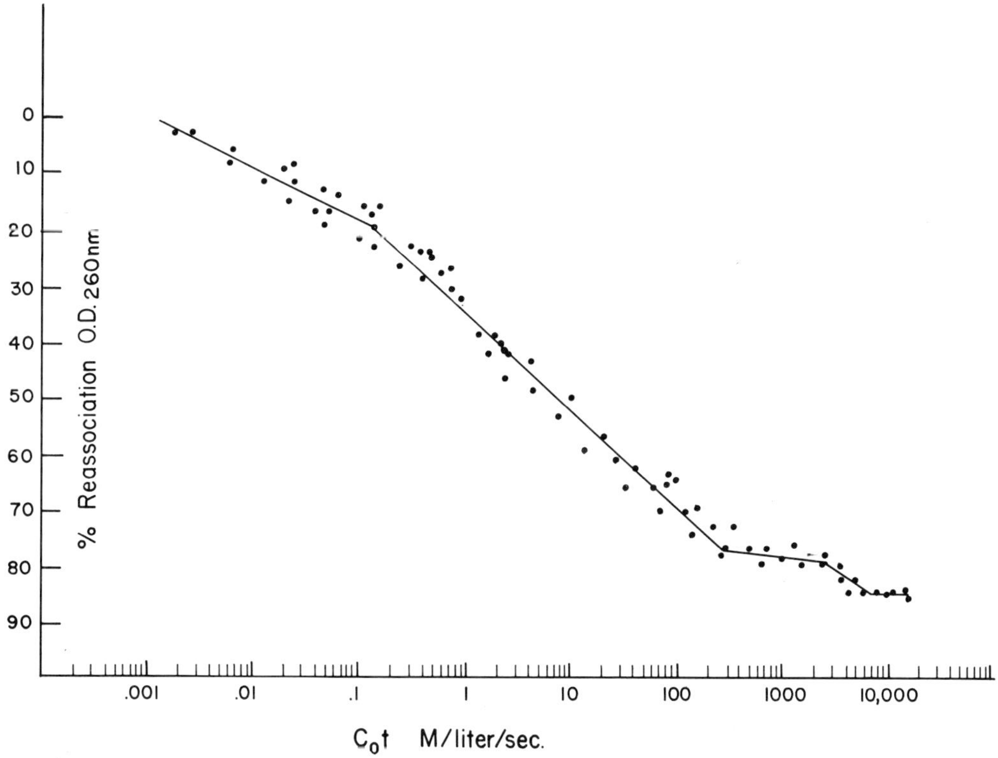

Fig. 11. Whole genome DNA reannealing C_ot plot of *Pinus banksiana*

of reannealing studies are expressed in terms of percent reassociation at different C_ot values. The latter term represents the product of the initial concentration and the time of incubation, and has the units moles, per liter, per second. Fig. 11 shows a plot of data from a reannealing experiment with *Pinus banksiana*.

Hydroxyapatite (HAP) Chromatography. HAP chromatography can be used to separate the single and double strand fractions of DNA because the single strand molecules of DNA are washed off a HAP column by low concentrations of phosphate buffer (0.12 M) while double stranded DNA is removed with a second, high concentration phosphate buffer (0.5 M) elution. Hydroxyapatite is commercially available from Clarkson Chemical Company or can be made by the procedures described by Bernardi (1971). Columns are commercially available from Biorad, or they may be fabricated to meet specific requirements. Fig. 12 illustrates a HAP column apparatus that is convenient for large scale experiments.

DNA Reannealing. Purified DNA samples are dissolved in 0.12 M phosphate buffer (PB) and are subjected to some form of treatment to reduce the molecular weight or fragment size of the DNA to the range of 300 to 500 nucleotides (see molecular weight determination procedures). The two most common fragmentation procedures are sonication and high pressure shear. In sonication, ice cooled samples of DNA are sonicated for several short pulses with a high power setting on an instrument

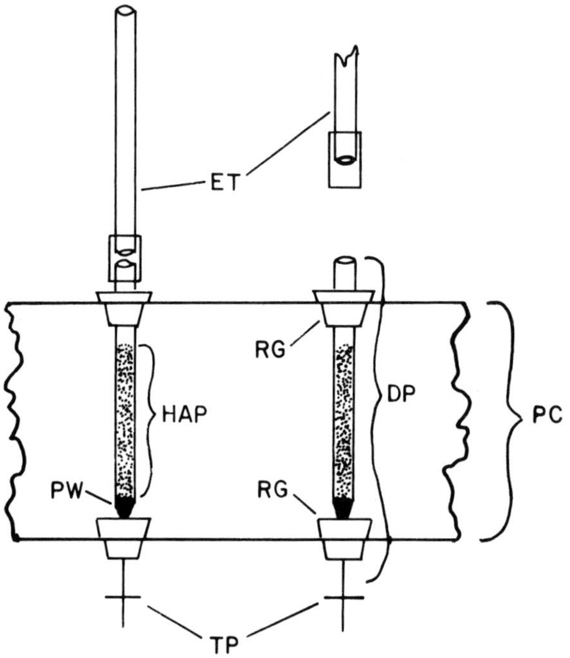

Fig. 12. Schematic diagram of hydroxyapatite column apparatus. Water is circulated through a plexiglass chamber (PC) to maintain a constant 60°C temperature. Six disposable Pasteur pipettes (DP) are used as columns and are sealed into the chamber by means of rubber gaskets (RG). A small plug of Pyrex wool (PW) supports the hydroxyapatite crystals (HAP) in the column, and a piece of flexible tubing with a pinch cock (TP) controls flow out of the column. After a sample is applied to a column an extension tube (ET) is connected to the column to hold the elution buffers

like the Biosonik II. For high pressure shear the DNA solution is forced through a French Pressure Cell at 12,000 p.s.i.

DNA is further purified after sonication by loading the sheared DNA on a hydroxyapatite (HAP) column at 60°C and washing with 0.12 M PB. The double stranded DNA fragments are then eluted with 0.24 M PB. A 1000 mw millipore filter can be used to reconcentrate the sample and reduce PB to 0.12 M.

To study DNA reannealing across a range of C_0t values a variety of DNA samples with different concentrations are needed. Very low concentration DNA samples are obtained by dilution. The easiest procedure to make high concentration (4 to 8 mg/ml) samples is to concentrate DNA solutions with a microconcentrator (e.g., Minicon-A25 by Amicon). Individual DNA samples are then prepared for the reannealing experiments by sealing them into incubation containers. For high concentration, low volume samples, a technique that has worked well has been to use 50-100 µl disposable micropipettes that are flame sealed.

The DNA samples are heated at 100°C for 10 min to denature. Following this treatment all DNA should be in the single strand form. The samples are transferred to a controlled temperature waterbath and the reannealing incubation period is started. The proper temperature for incubation is Tm - 25°C or 25°C below the melting temperature of the DNA in 0.12 M phosphate buffer (see section on melting determinations). The temperature most commonly used is 60°C.

When the appropriate time of incubation for a sample has elapsed it is removed from the batch and the reannealing reaction is stopped by freezing the sample. The samples may be analyzed immediately or stored frozen.

HAP columns are maintained at 60°C to achieve proper binding between the crystals and the DNA. Since this is also the optimal temperature for reannealing, samples should be kept cold and be added to the column as small aliquots to minimize exposure of the DNA in free solution to the elevated temperature. When the entire sample has been loaded onto the column the two step elution of single and double strand DNA is begun. The volume of eluting buffers needed will vary with the column and sample size. In most cases a volume of 0.12 M KPB (potassium phosphate buffer) that is 2-3 times the column volume will suffice to wash off the single strand component of DNA. An equal volume of 0.5 M KPB is then used to elute the double strand component. A UV analyzer (260 nm) can be used to monitor column outflow to insure that both DNA peaks are properly eluted from the column.

Quantification of the DNA in the two fractions is by measurement of optical density at 260 nm. Note, however, that both samples should be heat denatured before the optical density readings are taken to avoid errors due to hyperchromicity.

There is an alternative procedure for following the progress of reannealing experiments. The DNA solution can be incubated directly in a spectrophotometer cuvette and the OD_{260} monitored periodically or continuously. As the conversion of single strand to double strand molecules occurs the OD_{260} declines, and this trend can be used to quantitate the degree of reannealing. However, it should be realized that the two methods of quantifying reannealing give different results. This is because absorbance is a measure of the number of nucleotides which have paired while HAP chromatography quantifies the number of DNA fragments which are sufficiently reannealed to act like double stranded DNA on the HAP column. Reaction rates assayed on HAP appear to be as much as two-fold more rapid than optical density rates.

The general method for the use of hydroxyapatite in reassociation of DNA is outlined below:

1. DNA is dissolved in .12 M phosphate buffer (pH 6.8). The DNA concentration is determined by measuring the OD_{260}. For native DNA, 1 OD unit = 40 μg/ml; denatured DNA, 1 OD unit = 33 μg/ml.

2. The DNA is broken into smaller segments by means of a French press or sonicator.

3. The DNA samples are placed in ampules, sealed and heated at 100°C for 10-20 min to denature. Denaturation is checked by reading the OD_{260} after heating, it should register a 35-40% increase in OD.

4. The ampules are placed in a 60°C waterbath and incubated for appropriate lengths of time.

5. The sealed ampules are removed from incubation and the contents loaded on the HAP column maintained at 60°C. The column is washed with an appropriate volume of 0.12 PB to elute single stranded DNA and then with another volume of .5 M PB to elute double-stranded DNA. The collected volumes of the eluted DNA samples are registered.

6. The OD_{260} of the eluted DNA is read and the percent reassociation is calculated.

Phosphate buffer solutions for reannealing experiments:

1. 0.12 M phosphate buffer solution (solutions A and B are mixed just prior to use).

A. 0.24 M solution of monobasic potassium phosphate KH_2PO_4 (136)(.24) = 32.64 g/l.

B. 0.24 M solution of dibasic potassium phosphate K_2HPO_4 (174)(.24) = 41.80 g/l.
51 ml A. soln. + 49 ml B soln. diluted to a final volume of 200 ml.

2. 0.5 M phosphate buffer solution.

A. 0.5 M solution of monobasic potassium phosphate 68 g/l.

B. 0.5 M solution of dibasic potassium phosphate 87 g/l.
51 ms. A. plus 49 mls. B.

The calculations involved in a reannealing experiment are illustrated by the following example:

A 1 ml sample containing 216 µg/ml DNA was incubated for 5400 sec.

$$C_ot \text{ (moles/sec/liter)} = \left(\frac{\mu g/ml\ DNA}{3.3} \times 10^{-5}\right)(t_{sec})$$

$$= \left(\frac{216}{3.3} \times 10^{-5}\right)(5400) = 3.53$$

The sample was then separated into single and double strand (reassociated) components on a HAP column.

Buffer	Fraction eluted	Eluted volume	OD_{260} in single strand state	OD Units	µg DNA
0.12 M	single strand	17.2 mls	0.211	3.64x33	120
0.5 M	reassociated	18.6 mls	0.156	2.91x33	96
					216

$$\% \text{ DNA renatured} = \frac{96}{216} = 44.4\%$$

Once the reannealing data are collected they should be plotted in a form similar to Fig. 11 or Fig. 13. The percentage of DNA reassociated or its converse, percentage remaining single stranded, is plotted on the ordinate. The calculated C_ot values are entered on a logarithmic scale on the abscissa.

To analyze the data and interpret it in terms of genome characteristics a computer technique has been developed (Hall, 1974) and is currently being refined. As previously discussed, reannealing is dependent on the interactions between complementary single strand molecules of DNA. For this reason the rate at which reannealing occurs follows second order kinetics. For a genome that contains no repetitious DNA a plot of the percentage of DNA remaining in the single strand state versus the logarithm of the C_ot value yields a simple sigmoid curve. This curve is described by the equation:

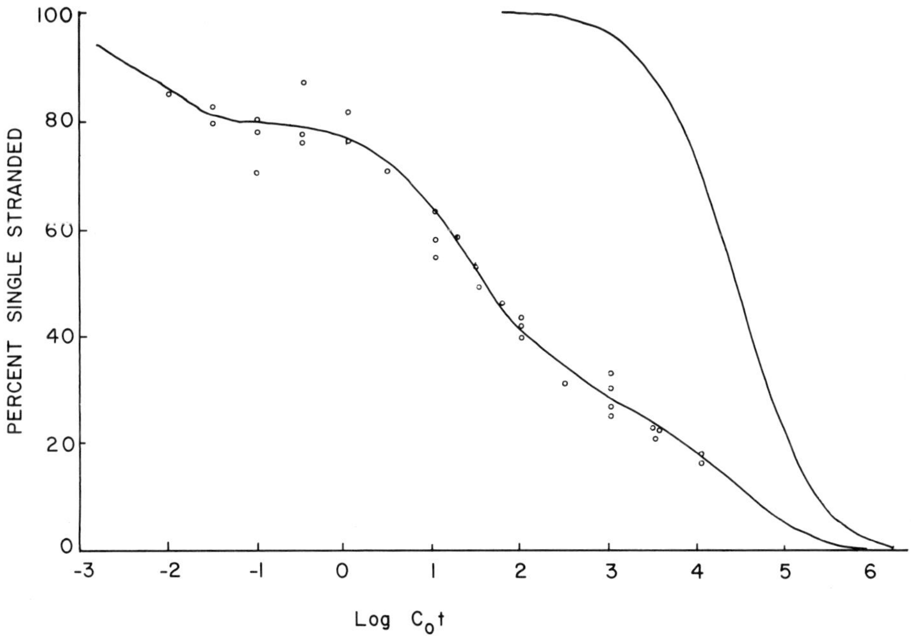

Fig. 13. Fitting computer-calculated reannealing curves to observed data for *Pinus banksiana*. Significance of the two curves is explained in the text

$$\frac{C}{C_o} = \frac{1}{1 + k_2 C_o t}$$

Where C/C_0 is equal to the percent of DNA remaining single stranded at time t and k_2 is a rate constant dependent on two genome characteristics.

The first of these characteristics is genome size, or DNA content per cell. The larger the genome size, the more dilute the concentration of individual sequences becomes, and the slower the reannealing reaction will proceed.

The second genome factor that affects reannealing rate is the presence of repetitious DNA. Any DNA sequence which is repeated in a genome will have a correspondingly greater concentration in the reannealing solution, and it will reanneal more rapidly. If all DNA sequences are repeated the same number of times (e.g., polyploidy), reannealing will still follow a simple sigmoid curve. If, however, all DNA sequences in the genome do not have the same frequency, a complex reannealing cuve will be observed. In these cases the genome can be considered to be composed of two or more "fractions" each of which can be identified on the basis of its average reiteration frequency. Each fraction will reanneal according to the above equation. By summing across all fractions of the genome the overall reannealing rate can be expressed by the equation:

$$\frac{C}{C_0} = \sum_{i+1}^{P} \frac{C_{0i}}{C_0 + k_{2i}C_{0i}C_0t}$$

Where p equals the number of genome fractions.

This second equation forms the basis for the computer analysis technique. In its present form, a series of simulations are run. For each simulation it is necessary to specify the genome size (pg. of DNA per cell as determined from other studies), the number of genome fractions, the percentage of DNA in each fraction, and a reiteration factor for each fraction. The simulated reannealing curve that is produced is then compared to the observed data. By an empirical, iterative process different sets of genome parameters can be used in the simulation until a calculated curve is arrived at that agrees closely with the observed data.

The use of the technique is illustrated in Fig. 13 where two computer-calculated curves are superimposed on observed reannealing data for *P. banksiana* (Hall, 1974). The simple curve that falls far to the right of the data was calculated on the assumption that the jack pine genome has no repetitious DNA. The curve which fits the observed data rather closely was based on the following genome parameters for jack pine: genome size of 27.6 pg.; five genome fractions containing 20, 15, 29, 12, and 24% of the total DNA, and having reiteration factors of 7 million, 3500, 900, 40, and 1, respectively.

It should be kept in mind that these estimates of genome parameters were arrived at by an empirical, non-statistical approach, and that curve shape and position are not highly sensitive to some changes in genome parameters. Nevertheless, the technique provides order of magnitude estimates which give a better insight into genome organization than any other available procedure. Copies of the computer program and further details on its use are available from R.B. Hall.

Nucleic Acid Hybridization

The method is potentially valuable in estimating relationships between organisms and in determining the percentage of a genome that produces a particular RNA. The technique consists of combining strands of DNA and RNA and quantifying the degree to which the nucleic acid strands reassociate or are complementary. Two "alike" or complementary strands should ideally, form a perfect double helix and the degree to which the strands do not reassociate should indicate the degree to which the two strands differ in base pair composition and genetic relatedness. Therefore, it should be possible to ascertain the amount of difference in the genetic message coded by two species under study or the percentage of DNA that is complementary to a certain RNA. Preliminary experiments using the hybridization technique yielded promising results in molecular hybridization among *Picea glauca*, *Picea sitchensis*, and *Picea abies*. Greater complementarity of the DNA bases was demonstrated between *P. glauca* and *P. sitchensis* than between *P. glauca* and *P. abies* (Miksche and Hotta, 1974).

DNA-RNA Hybridization. Several methods have been used to form and assay DNA-RNA hybrid molecules (Kennel, 1971). The formation of duplex hybrid molecules is accomplished by exposure of denatured DNA to an RNA solution under annealing conditions. The DNA may be immobilized on a filter (Gillespie and Spiegelman, 1965); in an agar column (Bolton and McCarthy, 1962); or free in solution (Marmur et al.,

1963). The general conditions that facilitate annealing are: 1. high cation concentration (0.01 M Na+ or higher), 2. elevated temperatures (approximately 25°C below the Tm of the DNA) to weaken intrastrand secondary structure, and 3. sufficient concentrations and incubating times to permit collisions for hybridization to occur.

Any method of analysis that distinguishes the bihelical structure from a single-stranded structure may be utilized to assay hybrid formation. In the case of agar, the amount of RNA bound to the column is measured (Bolton and McCarthy, 1962). Filter formed hybrids (Gillespie and Spiegelman, 1965) are generally assayed by radioactivity scintillation methods. Hybrids formed in solutions can be assayed by CsCl density gradients (Hall and Spiegelman, 1961), MAK chromatography (Hayashi et al., 1965) and filtration on nitrocellulose filters (Nygaard and Hall, 1963).

CsCl density gradients are too costly in time and equipment for hybrid assay; however, Hotta and Miksche (1974) utilized this method to ascertain that for several conifer species the 28S and 18S RNA genes are located next to each other on a relatively short DNA strand. The agar column technique requires separate columns for each assay and considerable experience before reproducible results are yielded. Nitrocellulose filtration techniques are rapid, relatively inexpensive and generally require a minimum of initial experience.

The use of the filter (Gillespie and Spiegelman, 1965) or the solution (Nygaard and Hall, 1963) techniques rests upon experimental conditions. For example, at high RNA/DNA ratios where a small fraction of the RNA will hybridize with DNA, free solution annealing should be used to minimize the "sticking" of unannealed RNA to the filters. On the other hand, at low RNA/DNA ratios, DNA-DNA reannealing is less of a problem when hybrids are formed directly on a filter.

DNA-DNA Hybridization. The nitrocellulose filter technique may be used here also, but methods must be used that facilitate the removal of added DNA which has not annealed with the initially bound DNA. Two methods have proven to be successful in remedying this excess DNA problem: 1. preincubation of filters loaded with DNA to prevent binding of added single-stranded DNA (Denhardt, 1966) and, 2. removal of single-stranded DNA by washing filters with a buffer of low ionic strength and a high pH after hybridization is complete (Warnaar and Cohen, 1966).

The details of the second method are as follows: DNA in 6 x SSC is denatured by heating to 100°C for 5 to 10 min and then rapidly cooled in an ice bath. The denatured DNA solution is passed through a B6 filter (Schleicher and Schuel, Keene, N.H.) prewashed with 6 x SSC. The loaded DNA-filters are washed with 5 ml of 6 x SSC and dried thoroughly at room temperature overnight and are then placed in a vacuum oven at 80°C, 73 cm Hg for 2 h. Following this treatment, about 95% of the DNA is retained by the filter, and less than 1% is lost during a 12-24 h incubation period. Radioactively labeled DNA from a second species (in 1.25 x SSC, 0.01 M Tris-HCl buffer, pH 7.0) is sonicated and denatured. The DNA-containing filters are incubated with the comparison DNA by suspending them in 3.2 ml of the labeled DNA solution for 12-50 h at 60°C.

The filters are removed from the incubation vials and rinsed briefly in 0.003 M Tris-Cl, pH 9.4, and the nonhybridized DNA is removed by washing the filters by suction on each side with 100 ml 0.003 M Tris-HCl, pH 9.4. The filters are then dried with either a hair dryer or lamps.

The amount of labeled DNA that has annealed to the DNA on the filter is determined by liquid scintillation counting. Hybridization is quantified as the percentage of membrane-bound DNA that has become associated with labeled DNA.

Hybridization experiments can also be performed at the chromosome level. For a discussion of those methods the reader is referred to Wimber and Steffensen (1970, 1974) and Berlyn and Miksche (1976).

Conclusion

Nucleic acids can be extracted and purified in good yield from a variety of tree tissues. Specific treatments must be modified to fit the characteristics of the tissue used. However, by monitoring the nucleic acid quantity (OD_{260}) and quality (OD_{260}/OD_{280} ratio) during the purification procedure the need for adjustments in technique can be identified.

Purified nucleic acid samples can be used to study a number of important questions about tree genomes. Reannealing studies provide information on genome organization, the amount of repetitious DNA and the degree of its reiteration. Hybridization studies can be aimed at determining the reiteration of specific DNA sequences and at elucidating species similarities at the molecular level. Other techniques described in this chapter can answer questions that are prerequisite or related to such studies.

Because the study of nucleic acids is both a very broad and a relatively new field, this chapter is presented as a starting point for techniques and as a stimulus to further research on tree nucleic acids.

Acknowledgement. We wish to thank Dr. Mary Berlyn of the Connecticut Agriculture Experiment Station for reviewing the manuscript.

References

Berlyn, G.P., Miksche, J.P.: Botanical microtechnique and cytochemistry. Ames, Iowa: Iowa State Univ. Press 1976.

Bernardi, G.: Chromatography of nucleic acids on hydroxyapatite columns. In: Methods in Enzymology, Vol. 21 (eds. Grossmann and Moldave), pp. 95-139. New York: Academic Press 1971.

Bolton, E.T., McCarthy, B.J.: A general method for the isolation of RNA complementary to DNA. Proc. Nat'l Acad. Sci. (USA) 48, 1390-1397 (1962).

Britten, R.J., Kohne, D.E.: Repeated sequences in DNA. Science 161, 529-540 (1968).

Comings, D.E., Riggs, A.D.: Molecular mechanism of chromosome pairing, folding and function. Nature 233, 48-50 (1971).

Crothers, D.M.: On the mechanism of deoxyribonucleic acid unwinding. Accts Chem. Res. 2, 225-232 (1969).

Denhardt, D.T.: A membrane filter technique for the detection of complementary DNA. Biochem. Biophys. Res. Comm. 23, 641-646 (1966).

Dingman, C.W., Kakefuda, T., Fisher, M.P.: Electrophoretic properties of low molecular weight DNA fragments in agarose-acrylamide gels. Anal. Biochem. 50, 519-528 (1972).

Gillespie, D., Spiegelman, S.: A quantitative assay for DNA-RNA hybrids with DNA immobilized on a membrane. J. Mol. Biol. 12, 829-842 (1965).

Hall, B.D., Spiegelman, S.: Sequence complementarity of T2-DNA and T2 specific RNA. Proc. Nat'l. Acad. Sci. (USA) 47, 137-146 (1961).

Hall, R.B.: Variation in DNA content per cell and repetitious DNA of selected populations of *Pinus banksiana* Lamb. Ph.D. Thesis University of Wisconsin, Madison, 209 pp. (1974).

Hall, R.B., Stairs, G.R., Miksche, J.P.: Variation in DNA content and redundancy - possible significance to forest tree breeding. Proc. 8th Cent. State Forest Tree Improv. Conf., School of Forestry, Fisheries and Wildlife. Columbia, Missouri: Univ. of Missouri 1974.

Hayashi, M., Hayashi, M.N., Spiegelman, S.: Chromatographic separation of annealed and enzymatically synthesized RNA-DNA hybrids. Biophys. J. 5 (2), 231-246 (1965).

Honda, S.I., Hongladarom, T., Laties, G.G.: A new isolation medium for plant organelles. J. Exp. Botany 17, 460-472 (1966).

Hotta, Y., Miksche, J.P.: Molecular hybridization of coniferous ribosomal RNA to DNA. Proc. 8th Cent. State Forest Tree Improv. Conf., School of Forestry, Fisheries and Wildlife. Columbia, Missouri: Univ. of Missouri 1974.

Jaworski, A., Key, J.L.: Distribution of ribosomal deoxyribonucleic acid in subcellular fractions of higher plants. Plant Physiol. 53, 366-369 (1974).

Kennel, D.E.: Principles and practices of nucleic acid hybridization. In: Progress in Nucleic Acid Research and Molecular Biology (eds. J.N. Davidson and W.E. Cohn). New York: Academic Press 1971.

King, R.C.: The meiotic behavior of the Drosophila oocyte. Intern. Rev. Cytology 28, 125-168 (1970).

Kirk, J.T.O., Reese, H., Evans, F.: Base composition of nuclear DNA within the genus Allium. Heredity 25, 507-512 (1971).

Mandel, M., Marmur, J.: Use of ultraviolet absorbance-temperature profile for determining the quanine plus cytosine content of DNA. In: Methods in Enzymology (eds. L. Grossman and K. Moldave), XII Part B, pp. 195-206. New York: Academic Press 1968.

Mandel, M., Schildkraut, C.L., Marmur, J.: Use of CsCl density gradient analysis for determining the quanine plus cytosine content of DNA. In: Methods in Enzymology (eds. Grossman and Moldave), XII Part B, pp. 184-194. New York: Academic Press 1968.

Marmur, J., Rowand, R., Schildkraut, C.L.: Denaturation and renaturation of deoxyribonucleic acid. Prog. Nucleic Acid Research 1, 231-300 (1963).

Miksche, J.P., Hotta, Y.: DNA base compositions and repetitious DNA in several conifers. Chromosoma 41, 29-36 (1973).

Miksche, J.P., Hotta, Y.: Preliminary evaluation of biochemical hybridization methods. 3rd North Amer. For. Biology Workshop 361 (eds. C.P.P. Reid and G.H. Fechner). Fort Collins, Colorado: Colorado State Univ. 1974.

Nygaard, A.P., Hall, B.D.: A method for the detection of RNA-DNA complexes. Biochem. Biophys. Res. Comm. 12, 98-104 (1963).

Peacock, A.C., Dingman, C.W.: Molecular weight estimation and separation of ribonucleic acid by electrophoresis in agarose-acrylamide composite gels. Biochemistry 7, 668-674 (1968).

Pearson, G.G., Ingle, J.: The origin of stress induced satellite DNA in plant tissues. Cell Differentiation 1, 43-52 (1972).

Rehfeldt, G.E., Lester, D.T.: Specialization and flexibility in genetic systems of forest trees. Silvae Genetica 18, 118-123 (1969).

Spencer, D., Wildman, S.G.: The incorporation of amino acids into protein by cell-free extracts from toba leaves. Biochemistry 3, 954-959 (1964).

Stern, H.: Isolation and purification of plant nucleic acids from whole tissues and from isolated nuclei. In: Methods in Enzymology (eds. L. Grossman and K. Moldave), XII Part B, pp. 100-112. New York: Academic Press 1968.

Studier, F.W.: Sedimentation studies of the size and shape of DNA. J. Mol. Biol. 11, 373-390 (1965).

Sutton, W.D.: Chromatin packing, repeated DNA sequences and gene control. Nature New Biology 237, 70-71 (1972).

Szybalski, W.: Equilibrium sedimentation of viruses, nucleic acids, and other macromolecules in density gradients, pp. 1-15. Palo Alto, Calif.: Beckman Instruments 1968.

Tewari, K.K., Wildman, S.G.: Chloroplast DNA from tobacco leaves. Science 153, 1269-1271 (1966).

Walker, P.M.B.: The specificity of molecular hybridization in relation to studies on higher organisms. Prog. Nucl. Acid Res. and Mol. Biol. 9, 301-326 (1969).

Warnaar, S.O., Cohen, J.A.: A quantitative assay for DNA-DNA hybrids using membrane filters. Biochem. Biophys. Res. Comm. 24, 554-558 (1966).

Whitefield, P.R., Spencer, D.: Buoyant density of tobacco and spinach chloroplast DNA. Biochim. et Biophys. Acta 157, 333-343 (1968).

Wimber, D.E., Steffensen, D.M.: Localization of 5S RNA genes on drosophila chromosomes by RNA-DNA hybridization. Science 170, 639-641 (1970).

Wimber, D.E., Steffensen, D.M.: Localization of gene function. Annual Review of Genetics 7, 205-223 (1974).

CHAPTER 3

Gel Electrophoresis of Proteins and Enzymes

P. P. FERET and F. BERGMANN

Introduction

Multiple Molecular Forms of Proteins and Enzymes

Molecular Structure of Proteins. Proteins are defined as large molecular weight substances (10^4 - 10^5) composed of amino acids coupled by the α carboxyl group of one acid and the α amino group of another forming a polypeptide:

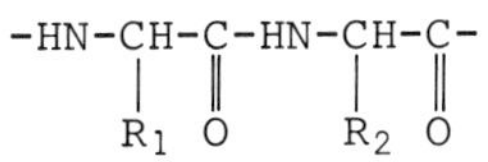

Any given protein may be composed of one or more polypeptide chains. Each polypeptide has a specific sequence of most or all of the 20 amino acids. For any given polypeptide, there exists a messenger ribonucleic acid (mRNA) containing the information, in the form of nucleotide triplets composed of adenine, uracil, guanine and cytosine, necessary for the proper placement of amino acids in a polypeptide. mRNA, of course, serves as the intermediary substance in translation and transcription of information contained in DNA. Consequently, the sequence and amino acid composition of any given polypeptide chain is a relatively direct expression of genetic information.

Protein structure and molecular configuration are a function of 1. the types and frequency of molecular interactions effecting polypeptide configuration and association, and 2. the size and number of polypeptides associated with a protein. The partial folding (secondary structure) of individual polypeptides and the final three dimensional shape of a polypeptide (tertiary structure) are dictated by the amino acid sequence (primary structure) of the polypeptide. Once formed, covalent bonds (i.e., sulfydryl bonds), hydrogen and hydrophobic bonds, coordination complexes with metal ions and dipole-dipole interactions cause the linear polypeptide to fold in both two dimensional and three dimensional space. Thus, not only size, but also shape of a polypeptide may be traced to a genetic origin. If a protein is composed of more than a single polypeptide, protein formation occurs by association of polypeptides to form a quaternary structure. A protein may be conformed of two, three, four or more polypeptides originating from a single gene, or two or more polypeptides originating from two or more genes. In summary, any given protein may represent in its simplest form, the information from a single gene, or in its most complex form, information originating from several genes. In addition, some proteins may also contain non-protein prosthetic groups and metal ions and proteins may form complexes with one another.

In general, proteins may be classified as 1. fibrous (i.e., collagens, keratins, etc.), 2. globular (i.e., albumins, globulins, histones, etc.), or as 3. enzymes (any catalytically active protein). Non-enzymatic proteins generally serve as structural or storage moieties. Most physiological and genetical research utilizes those proteins which are catalytically active, i.e., enzymes. Enzymes may be classified into several groups based upon the chemical reaction which they catalyze. The most general classification of enzymes is as follows:

1. Oxidoreductases, containing oxidases and dehydrogenases.

2. Transferases, including transaminases and other transferase enzymes.

3. Hydrolases, including esterases, phosphatases, glycosidases and peptidases.

4. Lyases, including decarboxylases, hydratases and lyases.

5. Isomerases, including various racemases, epimerases, intramolecular oxidoreductases and intramolecular transferases.

6. Ligases, including carboxylases.

Polypeptide Behavior in a Polarized Field. Since proteins are composed of amino acids, many chemical properties of proteins are similar to amino acids. Amino acids are zwitterions, hence may exist as dipolar ions. If placed in an acidic environment, amino acids become positively charged, in a basic environment, negatively charged. The degree of charge associated with a polypeptide at specific pH values is dependent upon the number and kinds of $-NH_2$ and -COOH functional groups on the peptide surface. Consequently, because proteins and enzymes are composed of one or more polypeptides, they may be charged by placement in an acidic or basic buffer solution.

If placed in a polarized field, most proteins will migrate toward the anode in solution of basic pH and toward the cathode in solutions of acidic pH. Relative rates of migration in solution will be proportional to the charge on the protein. Consequently, it is possible to separate mixtures of proteins by their relative rates of migration in solutions of specified pH values. This process is called "electrophoresis".

Application of Gel Media. Although proteins may be separated in solution by ionic charge, they may also be separated by size and shape if forced to migrate through a semi-porous medium saturated with an electrolytic buffer solution. Because additional resolution is achieved using various gel media, most electrophoretic protein separations are performed using semi-porous gels and fiber matrices with a variety of pore sizes, chosen to provide optimal separation for the materials being studied.

Historical Development

Techniques. The first effort to characterize proteins by electrophoretic mobility through a semi-porous medium was the separation of serum albumin and globulin in a water-cooled glass column packed with glass wool and filled with buffer (Coolidge, 1939). Durrum (1950) separated whole serum into five component proteins using paper strips.

Smithies (1955a, 1955b) introduced starch gel for zone electrophoresis and reviewed its theoretical advantages (Smithies, 1959). Starch gel provides resolution superior to the paper support media and additional new advantages. Starch gel absorbs protein molecules to a lesser extent than paper, and sample placement on the gel medium is easily controlled. Protein solutions to be analyzed are absorbed into filter paper and then placed in a slot in the gel, providing a narrower point of origin than previously obtainable on paper. The previous electrophoretic processes separated proteins by their ionic charge; starch gel added the capability of separating proteins by means of size and shape. The movement of proteins of large molecular weight is retarded by the small size of the gel pores, while proteins of small molecular weight are free to migrate according to their effective ionic charge. This "molecular sieving" effect was a primary advantage of the starch gel technique over earlier methods.

The use of acrylamide as a support medium for zone electrophoretic separation of proteins was introduced by Raymond and Weintraub (1959). This medium possesses the advantages of the "molecular sieving" effect of starch gels while providing ease of preparation and handling, gel clarity, and for some protein systems, improved resolution and increased reproducibility. The chemical structure of the gel is a three dimensional linear polymer of acrylamide cross-linked by methylene bridges (see Raymond and Wang, 1960) as indicated below:

$CH_2CH(CONH)CH_2CH(CONH_2)$
$\qquad\qquad\qquad |$
$\qquad\qquad\quad CH_2$
$\qquad\qquad\qquad |$
$CH_2CH(CONH)CH_2CH(CONH_2)$

Except for the terminal portions of the acrylamide polymer the gel media is free of hydroxyl and acetic groups, providing a more inert support media than starch. The range of pore sizes and the compatibility of a wide variety of buffers with polyacrylamide gels provides a versatility equal to that of starch support media.

Starch and acrylamide gels were first used in strip or slab form measuring 3 mm in thickness; electrophoresis was performed using continuous buffer systems (identical buffers in the electrode compartments and the support medium). Ornstein (1964) and Davis (1964) modified these techniques for the preparation of disc gels using discontinuous buffer systems.

Holmes (1967) adapted the techniques of Ornstein and Davis to discontinuous acrylamide gel slab electrophoresis. Other advances have led to use of acrylamide gel electrophoresis for the separation of lipo proteins (Raymond et al., 1966), ribonucleic acids (Richards et al., 1965), and for microelectrophoresis (Grossbach, 1965; Hyden et al., 1966; Felgenhauer, 1967). The application of histochemical staining procedures for enzymes directly in the gels has further expanded the use of the technique (Smith, 1968).

The precision with which gel electrophoresis is capable of separating similar protein and enzyme types has been amply demonstrated. The classic studies of Ingram (1957) and Hunt and Ingram (1958) relating genetic variation of haemoglobin types to their amino acid composition, and the use of electrophoresis by Henning and Yanofsky (1963) to demonstrate colinearity of desoxyribonucleic acid and tryptophan synthetase A amino acid sequence, adequately represent the analytical precision of the technique. Shaw (1965) stated that a single amino acid substitution in an average protein or enzyme will result in detectable differences in electrophoretic mobility in 27% of the cases. Hubby and Lewontin (1966) felt that the technique detected a "substantial fraction of the variation" in multiple forms of enzymes in a species.

Isoenzyme Terminology and Classification. The multiple forms of enzymes separable by electrophoretic procedures, occurring within the same organism and having similar or identical catalytic activities, are generally termed "isozymes or isoenzymes". The term "isozyme" was first introduced by Markert and Møller (1959), but the alternative form "isoenzyme" is also widely accepted.

Several proposals have been made concerning the restriction of the use of the terms isoenzyme and isozyme to enzymes originating from the same organ or tissue or acting catalytically on the same naturally occurring substrate (see for review: Scandalios, 1969; Shaw, 1969). A basic discrimination of isoenzyme types was proposed by Shaw (1969), who defined two classes, primary and secondary isoenzymes. While the first class includes those enzymes which are distinctly different molecules *in vivo*, the secondary isoenzymes result from artificial alterations in the structure of a single polypeptide species, perhaps during the extraction or separation procedures.

Of the various definitions and designations often used, only one has been established in the literature. The term "allozyme" designates the variant enzyme produced by different alleles of one and the same gene locus (Prakash et al., 1969). Hence we have two operational terms, isoenzyme, and after elucidation of the genetics of the isoenzyme system, allozyme.

After the development of the zymograms, it has become practice to distinguish the various isoenzyme bands by designating them with letters or numbers according to their electrophoretic mobilities. In most cases, the designation proceeds from the anode to the origin (cathodal end) of the gel, i.e., the different bands may be indexed A, B, C, or 1, 2, 3 in order of decreasing anodic mobility. If the isoenzyme patterns can be divided into several zones or regions on the basis of genetical or methodical relationships, zones may be designated with letters, and the individual bands within each zone with numbers. Formerly, two genetically defined isoenzyme variants and their corresponding genes have often been indexed 1^S or 1^F, indicating slow (S) and fast (F) migrating bands in the zymogram. This system, however, becomes extremely impractical if additional allozymes with faster or slower migration rates are found by other workers using identical tissues and electrophoretic methods. One excellent method for designating isoenzyme bands is to calculate a relative mobility (Rm) for each. Using a marker dye (or perhaps a borate ion front or sample) as a standard, mobility of individual isoenzymes may be calculated by dividing the distance between the origin and the band by the migration distance of the marker dye or standard.

Application in Research

Isoenzyme Variation - Historical Background. The first comprehensive investigation demonstrating multiple molecular enzyme forms was reported by Gomori et al. (1959) and Markert and Møller (1959). Tissue specific esterase and phosphatase enzymes were demonstrated with mouse tissues and comparison of embryonic and adult pig heart tissue demonstrated ontogenetic variation of lactic dehydrogenase. They proposed two genetic interpretations: 1. each isoenzyme could be attributed to a single gene, 2. each species could have a uniquely characteristic gene capable of synthesizing an array of isoenzymes. Present evidence indicates that the first proposal is the more correct. The literature concerning investigations of mammalian isoenzyme variation is voluminous and will not be reviewed here; however, for readers interested, the following references are suggested: Kaminshi and Gajos, 1964; Shaw, 1964; Ashton, 1965; Beckman and Nilson, 1965; and Johnson et al., 1967.

One of the earlier paper electrophoresis investigations of isoenzyme variation in higher plants by Jermyn and Thomas (1954) demonstrated different zymogram peroxidase patterns in horseradish (*Cochlearia armoracia* L.) and fig (*Ficus macrophylla* Roxb.). Other investigations have revealed three peroxidase isoenzymes in roots of turnip (*Brassica ropa*) (Hosoya, 1960) and six in maize (McCune, 1961). More recently a variety of isoenzyme systems have been reported in higher plants. Examples include malate dehydrogenase and amylase in barley (*Hordeum vulgare* L.) (Frydenberg and Nielson, 1965; Yue, 1966; Hamrick and Allard, 1972) catalase, esterase, phosphatase and dehydrogenase in corn (*Zea mays* L.) (Beckman et al., 1964; Schwartz and Endo, 1966; Brown and Allard, 1969a,b); tyrosinase in broad bean (*Vicia faba* L.) (Robb et al., 1965); and peroxidase and acid phosphatase in bean (*Phaseolus vulgaris* L.) (Staples et al., 1965; Racusen and Foote, 1966).

The presence of isoenzymes detected by the zymogram method quickly led researchers to extend the technique to investigation of interspecific enzyme variability. Jooste and Moreland (1963) compared esterase zymograms of cucumber (*Cucumis sativus* L.), soybean (*Glycine max* L. Merr.), corn (*Z. mays*), and wheat (*Triticum aestivum* L.). They elucidated interspecific differences and postulated that these differences were probably due to genetic diversity, variant growth conditions, variation in cellular differentiation, and tissue specificity. Schwartz et al. (1964) investigated the esterase variation in a variety of plant types including Cucurbitaceae, cabbage (*Brassica oleracea*), bean (*P. vulgaris*), carrot (*Daucus carota*), and potato (*Solanum* spp.). Zymogram analysis of these plant esterases demonstrated broad species differences as well as zymogram differences between strains of the same species and different parts of the same plant. In a similar study, Hall et al. (1969) made a comparative investigation of acid phosphatases, esterases, malate dehydrogenases, and peroxidases using eleven plant genera and various tissues within genera. They demonstrated that wide diversity between genera is a general rule for all isoenzymes studied and that tissue specific isoenzymes were common. The number of esterase isoenzymes resolved using polyacrylamide disc electrophoresis varied from ten in *P. vulgaris* stem tissue to two in *Beta vulgaris* roots and *Begonia argentoguttata* leaf tissue. Peroxidase isoenzymes also showed considerable variation. *B. vulgaris* leaves, *D. carota* root and *Cyrotmium falcatum* leaves all possessed only one while *Solanum microdontum* tuber had nine.

Studies of zymogram patterns have verified the taxonomic value of isoenzyme analysis. Hart and Bhatia (1967) demonstrated differences in enzyme composition between several species of Nicotiana. Chu and Oka (1967) distinguished two series of Oryza by their characteristic peroxidase zymograms. However, Desborough and Peloquin (1967) reported that esterase isoenzymes were not related to taxonomic groups or ploidy level of *Solanum* spp. Many interspecific differences were observed but because of the complex zymogram patterns in most species, taxonomic application was of limited value.

Many researchers have reported that electrophoretic protein patterns are modified by growth, developmental and environmental processes (e.g., Steward et al., 1965; Coleman et al., 1966; Morris, 1966; Gerloff et al., 1967; Hadacova and Sahulka, 1967). Macko et al. (1967) were among the first to demonstrate both protein and isoenzyme changes during plant growth and development. Extracts of wheat (*T. aestivum* var. Lee) from dormant seed, germinated seed, etiolated coleoptile and first green leaf were analyzed by polyacrylamide disc electrophoresis for changes in dehydrogenase, peroxidase, catalase, phosphatase, esterase, and amylase isoenzymes. 6-phosphoguconate dehydrogenase exhibited only quantitative differences in activity while for all other enzyme systems both qualitative differences in isoenzyme zymograms and quantitative differences in activity were noted.

McCown et al. (1969a) investigated leaf and stem isoenzymes in three Dianthus cultivars grown under summer and winter conditions. Seasonal differences were observed for peroxidase, acid phosphatase, and esterase while dehydrogenase and catalase isoenzyme profiles remained constant. In a latter study, McCown et al. (1969b) demonstrated peroxidase isoenzyme differences between winter acclimated and succulent plants of four plant genera including Sedum, Cornus, Lysimachia and Salix.

Isoenzyme changes were noted in *Lilium henryi* pollen during meiotic development (Linskins, 1966). Peroxidase and lactic dehydrogenase isoenzyme profiles were highly correlated with meiotic stages. In *T. aestivum*, Stutte and Todd (1969) reported that increasing water stress caused the appearance of certain new peroxidase isoenzymes and loss of others as well as loss of lactic dehydrogenase isoenzymes. Okerse et al. (1965) demonstrated that indol-3-acetic acid (IAA) had an effect on the peroxidase isoenzyme composition of mature Pisum stem tissue. One of the eight peroxidase isoenzymes could be suppressed by IAA application to excised stem sections. Stuber and Levings (1969) also demonstrated IAA control of peroxidase in oats (*Avena sativa* L.).

Although it had long been generally recognized that phytophatogens might modify the protein and amino acid composition of infected tissue, Staples and Stahmann (1963) working with bean (*P. vulgaris*) were the first to identify specific isoenzyme modification caused by pathogenic infection. They found that in addition to a *Uromyces phaseoli* Schw. malate dehydrogenase, a bean host enzyme was present during early stages of development but normally absent from mature tissue. In infected mature tissue, this host malate dehydrogenase was always present. In later reports (Staples and Stahmann, 1964; Rudolph and Stahmann, 1968) four additional isoenzyme changes in this host-pathogen system were discussed. With the exception of isocitrate dehydrogenase profiles, all other enzyme differences due to infection were related to the introduction of enzymes originating in *U. phaseoli* rather than to a stimulation of the host system. Johnson et al. (1968) arrived at similar conclusions after studying malate dehydrogenase in *T. aestivum* infected with *Puccinia recondita*. The relationship between peroxidase isoenzymes in *P. vulgaris* leaves and infection by southern bean mosaic virus (SBMV) has been intensively studied by Farkas and Stahmann (1966). Infection caused an increase in peroxidase activity and marked enzyme changes. These changes were associated with accelerated senescence caused by the viral infection. Andreev and Shaw (1965) observed two new peroxidases in flax (*Linum usitatissimum*) infected by *Melampsora lini* (pers.) lev. One new isoenzyme was associated with tissue senescence.

From this brief review it can be seen that enzyme changes in tissues infected with phytopathogens can be explained by one or all of the following: 1. the source of the new enzyme in the plant is the pathogen; 2. new isoenzymes arise in the host under influence of the pathogen and 3. isoenzyme changes arise as a result of host tissue degradation and senescence.

Genetic control of isoenzyme variation has been studied and reported for a wide variety of enzymes in several organisms. In higher plants genetic control of enzymes has been primarily investigated in the following genera: Zea, Phaseolus, Triticum, Avena and Hordeum. The first example of genetic control of a plant enzyme separated by electrophoresis was provided by Schwartz (1960) for an alkaline phosphatase in maize endosperm. The synthesis of the enzyme was found to be under the control of the Sh gene in chromosome 9 and was capable of being partially suppressed by a Ds gene adjacent to Sh. In triploid endosperm this alkaline phosphatase system was also shown to exhibit a dosage effect. In a later series of papers, the genetic control of esterases in maize was reported (Schwartz et al., 1965). The esterases were dimers (enzymes containing two polypeptide monomers), each monomer controlled by a polymorphic locus having a series of seven co-dominant alleles. Homozygous progeny contained only one electrophoretically distinguishable isoenzyme while heterozygotes contained a hybrid enzyme in addition to the two parental types.

Scandalios (1965) found that maize catalase isoenzymes were also under strong genetic control. His results strongly supported a hypothesis that maize catalase is a tetramer. When two homozygous individuals, each with a single but dissimilar catalase, were crossed, electrophoretic analysis of progeny indicated the presence of both parental enzymes and three new hybrid enzymes. Each monomer of the tetramer was controlled by a single allele and random association of monomers gave rise to five isoenzyme types.

The genetic control of two leucine aminopeptidase systems (Lap A and Lap D) was demonstrated in maize by Beckman et al. (1964). Both systems were shown to be controlled by single loci with co-dominant alleles. Heterozygotes were characterized by two isoenzymes (no hybrid enzyme was present) and homozygotes by a single isoenzyme. Brown and Allard (1969a,b) reported inheritance patterns for a total of nine maize isoenzyme systems in twelve inbred parents and two reciprocal recurrent selection populations. These systems included: alcohol dehydrogenase (Adh); two phosphatases (Phos 4, Phos 8); one peroxidase (Px 1) and five esterase systems (E_1, E_3, E_4, E_{12}, E_{16}). Adh, E_1 and E_3, were dimers controlled by a single locus with co-dominant alleles and demonstrated "hybrid" enzyme formation in heterozygotes. E_4, E_{12}, E_{16}, Phos 4 and Px 1 were also controlled by a single locus with co-dominant alleles but demonstrated no hybrid enzyme formation in heterozygotes. Phos 8 seemed to be controlled by co dominant alleles, but with an unexplained heterozygote phenotype. The papers by Brown and Allard (1969a,b) are particularly important in that this was the first report of a wide range of isoenzymes suitable for population analysis in a higher plant species. Thus, they were able to assign individuals a specific genotype and to follow genotypic changes in a population during the course of a breeding program. It is of interest to note that the inheritance patterns of some isoenzyme systems could only be elucidated using 4 to 6-day old corn seedlings and that Phos 8 could only be studied using extremely standardized growth conditions and sampling procedures. These data further emphasize the effects that growth and development may have on isoenzyme expression and subsequent analysis. Furthermore, this research group was able to analyze the genetics of similar isoenzyme systems in Avena and Hordeum (Marshall and Allard, 1969; Kahler and Allard, 1970; Clegg and Allard, 1973).

West and Garber (1967), using *P. vulgaris* and *Phaseolus coccineus*, compared esterase and leucine aminopeptidase zymograms of interspecific hybrids and backcross progenies. They found that two leucine aminopeptidase enzymes were controlled by a single locus with co-dominant alleles. Esterase isoenzyme patterns were also controlled by a single locus but the (Est V) allele was dominant over (Est C). In *T. aestivum* L., Barber et al. (1968) found that esterase isoenzymes were directly related to the homologous relationships of the chromosomes. In some nullisomics certain esterases were absent. Scandalios (1969) has reviewed much of the literature on genetic control of multiple molecular forms of enzymes in plant species.

Isoenzyme Variation in Forest Trees. Until 1970, no reports were available elucidating the genetic control of isoenzymes in forest-tree species. The technique of gel electrophoresis was utilized prior to 1970 to study seed proteins and enzymes in several forest-tree species. Lewis and Cech (1969) reported protein and enzyme electrophoretic variation in *Prunus serotina* Ehrh. seeds collected from a single stand in West Virginia and were able to resolve phosphatases, esterases, peroxidases and leucine aminopeptidases from eight single-tree, open-pollinated seed lots. Acid phosphatase and leucine aminopeptidase enzymes exhibited no variation; the former having one rapid and two slow-moving enzymes with minor inter-tree differences in activity, the latter exhibiting two slow-moving bands. Slight inter-tree differences in activity were demonstrated for peroxidases but no differences were observed between seed lots from a single tree. Esterase enzyme patterns exhibited variation between seed lots of single trees and between trees; three to four major bands were commonly observed. Twelve individual-tree bands were noted.

Hare and Switzer (1969) analyzed seed protein extracts of *Pinus echinata* Mill. and *Pinus taeda* L. Shortleaf pine seed-protein electropherograms were invariate over the entire range of the species but loblolly pine seed-proteins exhibited some variation. Three proteins were found unique to shortleaf, a single protein unique to loblolly, and fourteen proteins were common to both species. Using a point system for the presence of the four species specific proteins it was determined that loblolly pine from the western portion of its range possessed a greater protein similarity to shortleaf pine than did eastern sources. The authors concluded that introgression of shortleaf and loblolly pine in the west may account for the similarities.

Durzan (1966) compared disc electrophoretic patterns of embryo and female gametophyte tissue proteins from *Picea glauca* (Moench) Voss, *Pinus banksiana* Lamb, and *Pinus strobus* L. In all cases protein patterns differed between embryo and gametophyte tissue. In a later article the soluble protein content of *P. banksiana* seed was shown to be independent of the climate at the seed-source but protein differences were observed between seed-sources (Durzan and Chalupa, 1968). Pickering and Fairbrothers (1967) studied soluble seed proteins from four Magnolia taxa to determine their usefulness as taxonomic discriminators. A total of eleven proteins could be resolved on disc gels, but all taxa had identical protein patterns.

Pollen proteins have also been investigated in several species of Pinus. The electrophoretic mobilities of several incompatible and compatible species of pine pollen were studied by Hagman (1967). Many protein patterns were common to all species and interspecific differences were not correlated with incompatibility patterns. Pine pollen proteins were also investigated by Bingham et al. (1964), in six species and two interspecific hybrids. Differences in major protein patterns were not apparent although weakly staining proteins did exhibit interspecific variations. The authors questioned the significance of their results because of possible sample modifications after storage.

Several studies of enzyme activity were made prior to 1970 with forest-tree material to elucidate geographic patterns of genetic variation. Embryo and female gametophyte enzyme activities from 27 altitudinal provenances of *Picea abies* (L.) Karst were investigated by Bartels (1964). Enzyme activities measured included malate dehydrogenase, glutamate-pyruvate transaminase and aldolase. Although activities of multiple samples from single trees ranged as much as activities from different trees, clinal variation was evident. Sources from high altitudes possessed greater enzyme activity per gram fresh weight than sources from low altitudes. Both embryo and gametophyte tissue exhibited the same trends although embryo enzyme activities were three times as great as activities in the female gametophyte tissue. In a similar study of geographic variation, Williams (1967) noted that leaf homogenates of *Liquidambar styraciflua* L. from several seed-sources grown at a single location, demonstrated clinal variation in spring initiation of malate dehydrogenase synthesis.

Since 1970 several reports concerning forest-trees have appeared elucidating the genetic control of isoenzymes in natural and synthetic populations. Bartels (1971) and Bergmann (1971, 1973a) have isolated gene markers in *Picea abies*. By using analysis of haploid macrogametophyte tissue several gene loci were identified controlling esterase and leucine aminopeptidase isoenzyme phenotypes. While the genetic system controlling esterases was composed of co-dominant alleles, the leucine aminopeptidase system was controlled by both co-dominant and recessive alleles (the so-called null-alleles). Subsequent population studies demonstrated genetic variability both within and among different Norway spruce provenances (Bergmann, 1973b). These differences indicated that such gene markers could be used to genetically certify the origin of seed collections (Bergmann, 1972). Feret (1974) reported the isolation of a macrogametophyte esterase gene marker in small populations of *Pinus pungens*. Frequency distribution of the alleles controlling the esterase phenotypes indicated that the populations were in Hardy-Weinberg equilibrium.

Conkle (1971a) using embryo tissue obtained from seed of controlled-pollinated crosses, demonstrated the inheritance of alcohol dehydrogenase and leucine aminopeptidase in *Pinus attenuata*. He further demonstrated the effects of development on isoenzyme phenotypes (Conkle, 1971b). Feret and Stairs (1971a) and Feret (1972), also using diploid tissue derived from controlled pollinated crosses, demonstrated the inheritance patterns of peroxidase isoenzymes in *Ulmus pumila* and several *Ulmus* spp. interspecific hybrids. In addition, seed source differences in *U. pumila* indicated that some phenotypes were seed source specific and that probably as many as nine different alleles could be found in natural populations (Feret and Stairs, 1971b). Rudin and Rasmuson (1973) demonstrated the genetic control of esterase isoenzymes in *Pinus sylvestris* and showed isoenzyme frequency differences among provenances (Rasmuson and Rudin, 1971).

Other studies have demonstrated isoenzyme variation in natural and controlled populations. Esterase and peroxidase variability among selfed, full-sib and open-pollinated progeny of *P. glauca* were investigated by Feret (1971). Although no simple genetic hypotheses could be formulated to explain the isoenzyme phenotypes observed, wide variations among families illustrated that selfed progeny contained small amounts of variability while half-sib and open-pollinated progenies contained correspondingly greater variability. Sakai et al. (1971) demonstrated isoenzyme variability among subpopulations of *Thujopsis dolabrata*. In similar studies, Sakai and Park (1971) and Feret (1974) showed isoenzyme differences among subpopulations of *Cryptomeria japonica* and *P. pungens*.

A few reported studies have shown that isoenzyme frequency data may be effectively used for clonal and putative hybrid analysis. Miyazaki and Sakai (1969) demonstrated the usefulness of isoenzyme analysis for clonal identification of *C. japonica*. Hamaker and Snyder (1973) screened 53 enzymes in an effort to determine which would be useful for chemotaxonomic research in *Pinus palustris* and *P. taeda*. Among the enzymes studied, 9 were thought to be of potential taxonomic value.

Methods and Techniques

Methods of Tissue Preparation and Protein Extraction

Before techniques of tissue sampling and protein extraction can be fully understood, it is necessary to understand the interactions that can occur among proteins and other cellular constituents.

Any general review of the biochemical literature quickly brings the realization that many different procedures exist for the extraction and purification of enzymes. The choice of a single procedure is usually dictated by the chemical characteristics of the enzyme and the tissue from which it is to be isolated. Enzyme extraction from plant leaf tissue involves tissue maceration resulting in an admixture of substances that were rigidly compartmentalized *in vivo*. Among these substances are vacuolar acids, carbohydrates, proteases, "tannins" and cell wall fragments (Pirie, 1959; Stahmann, 1963). These materials are all capable of enzyme denaturation or precipitation and they could cause subsequent loss of enzyme activity in plant leaf enzyme preparations.

The presence of high concentrations of "tannins" (phenols and quinones that often precipitate proteins) in plant tissues (Harborne, 1964; Harborne and Simmonds, 1964) has been the principal cause of plant enzyme instability in leaf tissue homogenates (Loomis, 1969). The ability of "tannins" to react and denature proteins and enzymes has stimulated investigations of the mechanisms involved. These mechanisms and their relevance to plant enzyme extraction have been reviewed by Loomis and Battaile (1966). They suggest that phenols can combine with proteins by two mechanisms: 1. reversible hydrogen bonding between a phenolic hydroxyl group and the carboxyl group of the protein peptide linkage; and 2. oxidation of phenol compounds to sulfhydryl, free amino or imino groups of a protein. Protein and enzyme denaturation by these reactions in aqueous homogenates may be inhibited in several ways. Phenol oxidation and accumulation of quinones may be limited by inactivation of phenoloxidase enzymes or by reducing oxidized phenols as they are formed. Gell et al. (1960) used sodium hydrosulfite to inhibit oxidation while Macko et al. (1967) used ascorbic acid to reduce oxidized phenolics as they are formed. Pierpoint (1966) and Anderson (1968) have reviewed the use of a wide variety of additives effective for the prevention of phenol oxidation. Included among these are cysteine, 2-mercaptobenzothiazol, mercaptoethanol, polyvinylpyrrolidone (PVP) and diethyldithiocarbamate (DIECA). PVP has also been used for the removal of phenolics from aqueous plant extracts (McCown et al., 1968).

Many enzymes may be protected from the adverse effects of tanning by the addition of various protective reagents to the homogenizing medium. Goldstein and Swan (1965) found "Tween 80" (dipolyethylene glycol ether of sorbitan monooleate) effective in preventing protein-tanning complexes. Dithiothreitol (DTT), mercaptoethanol, and cysteine have been suggested as protective reagents for protein sulfhydryl bonds (Cleland, 1964). Sucrose and glycerol also inhibit denaturation of protein in aqueous solutions (Simpson and Kauzman, 1953). The broad differences in chemical structure between phenols and proteins has permitted separation of these compounds by chemical and physical methods. Butanol washings, acetone and ammonium sulfate precipitation, dialysis and column chromatography have been effectively used for the separation of protein from phenolics in plant extracts (Loomis, 1969).

A variety of the above procedures have been successfully utilized in the extraction of plant peroxidases and esterases for analysis by zone electrophoresis. Hall et al. (1969) macerated plant tissues in 0.1 M "Hepes" buffer (pH 7.3) containing 0.1 mM DTT and PVP. Following maceration, extracts were centrifuged to remove cell debris and dialized against 0.05 M Tris-HCl buffer for six hours, then concentrated. Peroxidase, esterase and other enzyme activities were then detected after electrophoretic separation. Macko et al. (1967) used 0.1 Tris-HCl buffer (pH 8.0) containing 0.5 M sucrose, 6 mM cysteine hydrochloride and 6 mM ascorbate for the extraction of peroxidases, esterases and other enzymes from *T. aestivum* L. Further purification was not required for enzyme separation by electrophoresis. Hart and Bhatia (1967) purified *Nicotiana* spp. peroxidases and esterases using the technique of Morton (1955). This method prescribed the washing of aqueous homogenates with acetone. The acetone precipitated protein was dispersed in butanol, centrifuged, and the resulting pellet rewashed several times in acetone, dried and resuspended in 0.05 M phosphate buffer (pH 7.8). The electrophoretic mobilities of Orchidaceae fruit peroxidases were studied by Alverez (1968). He used a 0.1 M Tris-glycine buffer (pH 8.2) for homogenization. After centrifugation of the homogenate, ammonium sulfate precipitations were made; precipitated pro tein was resuspended in buffer and dialized 15 h against the Tris-glycine. Extracts were then concentrated and analyzed by zone electrophoresis.

Other workers have found it unnecessary to use a buffered extraction media. Lanzani et al. (1967) extracted peroxidases from wheat germ (*T. aestivum* L.) in unbuffered 0.5 M sucrose. Homogenates were filtered, centrifuged, treated by reducing agents, dialized and purified by column chromatography prior to electrophoretic analysis. Schwartz et al. (1964) found buffer, additives and preparative procedures unnecessary for extraction of maize endosperm esterases. Similarly, Desborough and Peloquin (1967) successfully extracted *Solanum* spp. tuber esterases by presoaking tissue in sodium hydrosulfite followed by maceration without addition of buffer or protective reagents.

A complete analysis of plant enzyme extraction problems is provided by the review articles by Dixon and Webb (1964); Pierpoint (1966); Anderson (1968) and Loomis (1969). Probably no single isolation technique can be expected to be effective for all enzymes of a single species and no single procedure can be expected to be effective for the extraction of a particular enzyme from all plant species.

In addition to problems associated with the admixture of proteins and enzymes and other cellular chemical components, problems in sampling and homogenization may be caused by temperature denaturation. Because most proteins and enzymes are heat labile (and some may be denatured by freezing), storage of unextracted tissue must be evaluated for each enzyme. Some researchers place field collected material in liquid nitrogen immediately upon removal from the plant specimen. As a minimum, field collections should be stored on ice in a cooler until returned to the laboratory for freezing at -30°C or for protein extraction. Unless a priori knowledge of temperature stability is obtained, all extractions should be performed at 0-4°C to avoid temperature denaturation of proteins or enzymes of interest.

In general, most forest-tree research workers have utilized modified extraction procedures adapted for their material from previously published enzymatic research in the plant and animal sciences. References discussed above will provide the reader with the details of previously used procedures. Basically, two protocols are used. The first type, utilizes the preparation of dry acetone powders resuspended in buffer solution prior to electrophoresis (Hare and Switzer, 1969). The second and more commonly used technique involves the homogenization of tissue-buffer solutions containing enzymatic protective reagents.

Observation leads to the conclusion that most forest-tree genetics researchers do not utilize purification techniques such as column chromatography, dialysis, lypholization, "salting-in" and "salting-out" or preparative electrophoresis. Although these techniques are commonly used in most biochemical studies, the large number of samples usually required for genetic studies precludes their routine use.

Starch Gel Formulation and Techniques

Starch gel was first used in a so-called starch block electrophoresis by Kunkel and Slater (1952), but the relatively thick gel layer was well adapted only to preparative electrophoresis. In order to identify visually the different protein and enzyme bands in a suitable "gel sheet", Smithies (1955b) introduced the technique of starch gel zone-electrophoresis. While this technique can be used for both vertical and horizontal separation, in this chapter only the horizontal starch gel electrophoresis is described. However, it should be mentioned that most of the methodical details are pertinent to either separation type.

The most important components of the electrophoretic procedure are buffers or buffer systems and thus modification of starch gel electrophoresis in most cases refers to alteration of buffer systems. Based on the molecular structure of polypeptides, the behavior of protein molecules during gel electrophoretic separation depends primarily on the pH value, the ionic strength and the type of the buffers employed. Therefore, it has been found in various investigations that each individual protein and enzyme system requires a specific buffer system for optimal separation. In general, a buffer system consists of two parts: 1. the electrode or bridge buffer which is placed in the chamber vessels or electrode trays, and 2. the gel buffer used for the starch gel preparation. Usually the gel buffer has one-tenth the molarity of the electrode buffer. If both parts of the buffer system are composed of the same constituents, we have a so-called continuous system; if the gel buffer is different from the electrode buffer with respect to the salt compounds, it is a discontinuous buffer system (Poulik, 1957).

Prior to preparation of the starch gel, suitable gel forming plates are arranged according to the requirements of the particular system being used. Gel plates are usually simple plexiglass forms commercially available or may be custom-built for particular research needs. For microelectrophoresis on starch, glass slides (or plates) are placed in the gel trays and overlaid with starch gel. After gellation is complete, individual gels may be cut, using the slides as a gel support. In setting up the gel formation trays, care should be taken to insure that the trays are horizontal so uniformity of gel thickness is insured. Generally, starch gels used in horizontal systems are from 3-10 mm thick.

Depending on the size and thickness of the gel to be formulated, as well as on the concentration of the gel needed for separation, a defined quantity of partly hydrolyzed starch powder is suspended in the appropriate amount of gel buffer in a side-arm flask. This mixture is heated with vigorous stirring until it forms a translucent but homogeneous viscous liquid. Heating can then be stopped or continued for a short time (5-15 min). The hot solution is then degassed by means of a suction pump connected to the side-arm of the flask and poured into a gel tray and allowed to cool (2-3 h at room temperature). If forced cooling is desired, the gel tray may be placed in a refrigerator for about 20-30 min.

Once gels are cut or otherwise prepared for use in the electrophoresis chamber, a narrow slot is cut at the desired origin position. Previously prepared protein or enzyme extracts may be pipetted directly into the slots or soaked into filter paper strips, which are then placed into the slots. After the samples have been added, the whole gel tray or gel plate is placed in the electrophoresis chamber and the two ends of the gel are brought into contact with the electrode buffer by means of paper or synthetic wicks.

After closing the electrophoresis chamber by a fitting lid, electric current is supplied by a D.C. power supply. Depending on which type of electrophoresis is to be carried out, constant voltage or constant current is used for separation. Usually constant voltage is applied at a rate of 10 V/cm to 20 V/cm. Since the passing current causes an increase of gel temperature, it is advisable to cool the gel tray or even the whole chamber during electrophoresis to prevent enzyme denaturation. After a definite time (2 to 4 h) the electrophoretic separation is completed.

The gels are then removed from the trays or plates and prepared for the staining. Since the upper surface of the starch gel does not usually yield clear banding patterns, the gels are usually cut horizontally into two or more slices by means of a razor-blade or a commercially available gel slicer; only the lower slice or slices are stained. If gel thickness permits division of the gel into several slices, different isoenzyme systems can be identified from the same sample. Immediately after cutting, the gel slices are placed in staining solution or, just before staining, they are immersed in a buffer solution for pH adjustment. Depending on which enzyme system is to be analyzed, a specific staining solution is used for enzyme visualization. Zymogram development at room temperature or at 37°C in a dark incubator usually requires 1 to 3 h.

Incubator mixtures usually employed for staining consist of a specific buffer solution for maintaining the pH necessary for the enzyme reaction. The buffer contains the enzyme substrates, sometimes coenzymes, and all other cofactors essential for maximum enzyme activity. In general, there are two main enzymatic reaction types used for zymogram development. The first reaction type utilizes the oxidation-reduction mechanism of all NAD and NADP dependent dehydrogenase systems. The dye system linked to the reaction is based on the reduction of tetrazolium salt to formazan. With this method any dehydrogenase enzyme for which the substrate is available and some other enzyme systems linked to this dehydrogenase reaction can be identified with relatively high specificity. Comparatively unspecific methods represent the second reaction type, leading to the detection of esterases, phosphatases, etc. The corresponding enzyme splits an aromatic component (phenyl-, naphthyl-) from the substrate and this component is then coupled with a diazonium salt producing a typical "azo-dye". Further methods, explanations, and descriptions concerning buffer and staining systems are given by Shaw and Koen (1968), Shaw and Prasad (1970) and Brewer (1970).

The visualization of protein patterns in starch gels is relatively simple, but generally requires more time than most enzyme analyses. Usually the gels are treated with a 7-10% acetic acid solution containing 1% metachromatic compounds such as amido black or Comassie blue. After staining for a definite time (0.5 - 2 h depending on the protein concentration) at room temperature, the dark blue gels are destained by incubating them in repeated changes of 7% acetic acid for several hours or by washing in a destaining bath containing activated charcoal. Sometimes methanol is added to the acetic acid solution for faster fixation. For a quantitative study of protein bands in the gel, the following staining solution has been proposed: 1% fast green in 7% acetic acid.

Acrylamide Gel Formulation and Techniques

In contrast to the slab technique used with starch gels, acrylamide gels are usually used for the technique of disc gel electrophoresis. Gels are formulated in tubes (usually measuring 5 mm I.D. x 70 mm) and extract solution is placed at the top of the gel in the tube. After electrophoresis and staining, the proteins appear as "discs" in the acrylamide gel, hence the name "disc electrophoresis".

Acrylamide disc gel electrophoresis was originally described by Ornstein (1964) and Davis (1964) and their techniques have been only slightly modified. Acrylamide gels are composed of three parts, a small pore "running gel", a large pore "spacer gel" and a "sample gel". The small pore gel is composed of the following mixture (Smith, 1968):

1. Monomer (acrylamide, ($CH_2 = CH:CONH_2$) : 20-40% in water) Comonomer (methylenebisacrylamide (BIS), about 3% in monomer. solution).

2. Gel buffer (a variety of buffers are used) one example: trishydroxymethylaminomethane - HCl (Tris-HCl) pH 8.9, containing 0.23% initiator (Tetramethylelhylenediamine (TEMED)).

3. Water.

4. Catalyst (ammonium persulfate, 0.2 - 0.4% in water).

The above components are mixed in a 2:1:1:4 ratio, degassed to prevent formation of air bubbles in the running gel, and pipetted into the glass tubes placed vertically in serum stoppers mounted on a board. The acrylamide solution is then overlaid with distilled water to 1. remove the miniscus and 2. exclude oxygen which prevents gel formation. Gels form in 10-30 min depending upon the concentration of initiator and catalyst used, and buffer pH. Acidic gels require several hours for formation.

Following formation of the small pore running gel, a second gel, about 3-5 mm thick, is placed over the running gel after removal of the water layer. This spacer gel is formulated using a 3-5% monomer solution and in place of ammonium persulfate, riboflavin is usually used as the catalyst. Spacer gels are photopolymerized under fluorescent light and normally form in 5-10 min. The purpose of the spacer gel is to prevent direct contact of extract with ammonium persulfate residues in the running gel, and to provide a space in the tube where proteins can migrate and "stack" prior to separation in the running gel. Many researchers find that resolution of some proteins and enzymes is not enhanced by the spacer gel and consequently it is often times omitted from the procedures.

The sample gel (formulated on top of the spacer gel) is used to encapsulate the protein extract in the tube and is formulated in a fashion similar to the formulation of the spacer gel. The sample gel is rarely used in practice because it generally does not provide for better resolution, is time consuming to formulate and often times it is difficult to attain the gelling affect without causing denaturation of heat labile proteins or enzymes.

Once the disc gels are formulated they are cooled at 4°C. Protein extract (usually 0.05 ml) is then applied to the surface of the gel. If a sample gel is not used, it is important that 0.5 M sucrose be incorporated into the protein extract. This will permit layering of reservoir buffer over the extract in the tube to inhibit admixture of the extract with the reservoir buffer when tubes are placed in the electrophoretic chambers.

Following application of the extract to the disc tube, and overlayering with reservoir buffer, the tubes are placed in the chamber and electrophoressed. Separations are generally performed at 0.4 ma/cm with a usual voltage gradient of 100-200 VDC for 30-90 min at 4-6°C. The electrophoretic environment will vary depending upon buffers used, gel length and apparatus. A bromophenol blue marker dye (0.1%) is added to the upper reservoir buffer to ensure visual perception of the electrophoretic rate and to serve as a reference point when analyzing the gels.

Following electrophoresis, gels are removed from the tubes under distilled water by insertion of a dissecting needle between the tube interior surface and the gel. By rapid rotation of the needle about the gel circumference, the gel is easily removed. The gel is then cross-sliced at the front (location where the bromophenol blue marker dye band is located) and placed in a general protein stain.

For staining of proteins, Comassie Blue (Chrambach and Rodbard, 1971) may be used. For this procedure gels are placed in 8 x 100 mm test tubes and soaked in 0.5% Comassii Blue in 10% TCA for 0.5 - 2 h. They are then destained in 7% acetic acid and the zymogram analyzed. Zymograms are sketched and Rm values calculated. Repeatability of Rm calculations should be ±3% or less.

If enzymes are to be stained, it is usually necessary to presoak the gels before adding enzyme substrates, cofactors and diazonium salts. The presoaking period varies depending upon the pH changes required, the molarity of the reservoir buffer used and the gel density. Generally two hours is an optimal presoak period. Enzyme zymograms may be analyzed in the same fashion as general protein zymograms. In general, any enzyme that can be stained in starch gels can also be stained in polyacrylamide gels, although some technique modification may be necessitated.

Smith (1968) has provided a detailed outline of the disc electrophoretic technique, recipes for gel formulations and enzyme staining procedures. The reader should refer to Smith (1968) and Chrambach and Rodbard (1971) for a complete and detailed analysis of practical and theoretical aspects of executing disc electrophoretic techniques.

Discussion

Advantages and Disadvantages of Gel Media. Both starch and polyacrylamide gel behave as a "molecular sieve", as already mentioned, and thus electrophoretic protein separation on one of these two support media depends not only on the charge of the various compounds but also on their ability to penetrate the "sieve", i.e., on their sizes and shapes. Therefore, it is possible with these techniques to achieve excellent resolution of a wide variety of protein and enzyme systems relative to all other support media.

Starch gel as well as polyacrylamide gel possess some advantages and disadvantages when comparing them with one another. In most cases it becomes obvious that the appraisal of a certain gel method is subjectively influenced by the research worker's own preferences. Depending on the enzyme system to be analyzed as well as the plant material to be used for extraction, starch gel or polyacrylamide gel may be of greater resolving power. No published comparisons have been made with plant tissues. In general, the advantages of starch gel zone-electrophoresis over polyacrylamide disc electrophoresis concern the amount of extract to be separated and the exact comparison of the different zymogram patterns. Since the length and the diameter of the tubes employed for disc electrophoresis can reach only definite values, their capacity to contain plant homogenates or protein extracts is limited. In contrast to this, the width of the slot in the starch gel and the thickness of the starch gel layer may be increased according to the amount of the sample to be analyzed. If the starch gel is represented by a large slab, different samples can be separated in one gel migrating from the same origin towards the same front-line. This method thus lends itself better to side by side comparisons of the zymogram patterns than the disc electrophoresis which usually has small variations in zymogram configuration from one gel to another. Polyacrylamide gel slab techniques do, however, circumvent these problems.

Polyacrylamide gels provide a translucent gel and consequently, are, by far, more suitable for direct densitometric studies than starch gel. Furthermore, acrylamide gels are not as fragile as starch gels, making them easier to handle. Acrylamide gels do not need to be cut before staining and, compared to starch gel, can be stored without fixation for relatively long periods. Since the concentration of acrylamide may be varied from 5 to 30% for gel preparation, the total range of variation in pore size is wider than that of starch gel. This property has frequently been used for formulating gradient gels of several pore sizes in one and the same gel tube in order to achieve an effective fractionation of nearly identical protein molecules.

Analytical Problems. Since the first introduction of the zymogram technique, methods have been developed for the visual detection of more than 60 enzyme and protein systems. In recent years, the number of investigations has rapidly increased demonstrating the occurrence of multiple forms of proteins and enzymes in nearly all animal and plant tissues. However, it has been repeatedly found that different studies using the same plant material and analyzing the same isoenzyme system using different techniques result in different isoenzyme patterns. Bartels (1971) and Bergmann (1971, 1973a), for example, employed the same plant tissue (spruce endosperm) and identified the esterase enzyme system by means of the same staining procedure. The isoenzyme pattern resulting from polyacrylamide gel disc electrophoresis showed no exact accordance with the pattern from starch gel zone-electrophoresis although some bands appeared to be identical. These differences may be more or less dependent on the gel constitution, pore size and current behavior, varying from one electrophoretic technique to the other. However, the main factors responsible for the evident differences in isoenzyme patterns are the specific properties of the different electrophoretic and buffer systems employed.

It is well known that certain buffer types used for enzyme extraction or in electrophoretic separation are capable of modifying the native enzyme molecules. Some buffers may cause formation of artifacts in zymograms and such artifacts can complicate genetical analyses. For example, borate buffer ions interact reversibly with enzyme or protein molecules in the pH range 6-9 to give two or even more electrophoretic zones with the corresponding activity. In order to control the origin of the multiple enzymatic forms found in zymograms after electrophoretic separation, the major enzyme band can be eluted from the gel medium and is then reapplied and again electrophoretically separated. If in this case the multiple bands should again appear, their secondary formation is to be accepted in all further investigations. By using this method, Bartels (1971) found that an esterase pattern consisting of three active bands originated from one native enzyme only. The complicating factors arising from protein-phenolic interactions have already been discussed.

Data Evaluation

Zymogram Survey and Characterization. After the development of the zymogram the investigator may find varying numbers of bands arising from the enzymatic action on the same synthetic substrate. Assuming that all bands represent primary isoenzymes, a preliminary characterization of the banding pattern is possible by analysis of substrate specificity. This is particularly advantageous if the enzyme systems being investigated include the unspecific esterases, phosphatases or oxidases. By using more specific substrates, a subset of the original bands becomes visible, and hence, isoenzymes may be classified according to their substrate specificity. For example, the substrate β-naphthyl acetate is more reactive with cholinesterases, while α-naphthyl acetate is most reactive with aliesterases. Sometimes the utilization of enzymatic inhibitors may also help to achieve a correct classification of the various isoenzymes found. With these methods, it is often possible to divide the whole zymogram into several parts or zones of identically reacting enzyme bands, thus providing a basis for a later genetic analysis.

When zymogram partitioning is not possible, and a large number of isoenzymes are isolated, a genetic analysis becomes complicated. In this case, it is advisable to mark the individual bands by their migration rates relative to some standard isoenzyme in all samples, or relative to a marker dye such as bromophenol blue which migrates at the buffer front. Using this method, isoenzymes may be individually labeled as they appear while alternative labeling systems using letters and numbers require designation changes when new isoenzymes appear.

Analyses of the Genetic Control of Isoenzyme Variations. A prerequisite of any genetic analysis of isoenzymes is the occurrence of variation among the isoenzyme patterns found in the material under study, i.e., the banding pattern of one individual must differ from the pattern of another in the appearance of at least one isoenzyme band. In general, most of the enzyme and protein systems so far investigated possess a high variability in both animal and plant species. If it is ascertained that a variation in banding patterns found among different individuals of a species is not caused by environmental or developmental factors, it must be correlated with allelic variability. Thus, in the simplest case, each isoenzyme band in the zymogram represents the phenotypic expression of a gene according to the "one-gene-one-polypeptide" hypothesis. By means of classical genetic tests, it must be then determined which bands are coded by alleles and which are specified by genes at different loci. For this purpose forest geneticists require full-sib families. In many cases, the bands segregate as single Mendelian factors, where the simplest type of diploid variation is represented by the presence of two co-dominant alleles, one specifying a fast-moving band, the other a slow-moving band. Consequently, an individual homozygous for either allele will have a single band on the gel whereas the heterozygote will have both bands.

More complex patterns are shown by enzymes composed of more than one polypeptide (with polymeric structure) and by those coded by two different gene loci. For example, dimeric enzymes consisting of two subunits will exhibit a three-band pattern in the heterozygous condition, if the two types of subunits coded by the two alleles randomly combine. Two bands of this pattern consist of homologous subunits also found in the homozygotes, while the third isoenzyme band (hybrid band) showing an intermediate migration is composed of different subunits each derived from a different allele (AA, AB, BB). The subunit composition of tetramers (enzymes with four polypeptides) leads to a five-band pattern in zymograms of heterozygote individuals (AAAA, AAAB, ABBB, BBBB). In order to determine rapidly the number of isoenzyme bands to be observed with varying polymeric stages and varying numbers of subunit types, Shaw (1964) has given the following formula:

$$i = \frac{(s + p - 1)!}{p!\ (s - 1)!}$$

where: "i" equals the number of isoenzymes, "p" the number of polymers and "s" the number of different subunits.

If coniferous species are to be investigated, the elucidation of the genetics of isoenzyme polymorphism is faster and simpler by using the haploid gametophyte tissue of dry seeds as test material rather than full-sib families. Since seed storage tissue develops from the female gametophyte after fertilization, it may be regarded as a common gamete, provided that the four macrospores are randomly distributed after meiosis. Thus, it is possible to correlate isoenzyme variation found in the seed sample of an open-pollinated tree with a simple Mendelian segregation pattern. For example, two alternatively appearing isoenzyme variants found in a 1:1 ratio in one seed sample must be coded by two alleles of one gene locus. This method, though only recently introduced, has already led to manifold isoenzyme gene identifications (Bartels, 1971; Bergmann, 1971, 1973a, 1974; Tigerstedt, 1973; Feret, 1974; Rudin et al., 1974).

Statistical Evaluation of Population Differences. Evaluation of isoenzymes located in zymograms of either starch or polyacrylamide gels may be difficult if reproducibility of relative mobility values is low, if a large number of isoenzymes are present in individual gels, or if enzyme activity levels are variable. However, if a reproducible system is considered, a range of statistical tests are available for testing differences in isoenzyme frequencies among experimental treatments (provenances, seed sources, half-sib or full-sib families, etc.). When segregating phenotypes are being analyzed in full-sib families or in populations of macrogametophytes, traditional chi square analyses are appropriate. Because of the difficulties in generating full-sib progeny in many forest-tree species, or if the use of isoenzyme analysis is to serve primarily as a biochemical indicator or genetic variation, nonparametric statistical procedures appropriate for analysis of binomial data may be used for testing interpopulation differences in isoenzyme frequencies. Such procedures do not require the a priori assumption that each isoenzyme is the direct expression of an allele or series of alleles. In this sense (i.e., where segregation ratios are not considered) analysis of isoenzymes is no more complex than analysis of phenolic or terpene variation.

Three statistical procedures are available for testing significant isoenzyme frequency differences between treatment groups. Chi square tests for homogeneity may be used for testing independence in 2 x 2 contingency tables. The formula used takes the form:

$$x^2 = \frac{(n_{11}n_{22} - n_{12}n_{21})^2 \, n_{..}}{n_{1.}n_{2.}n_{.1}n_{.2}} \quad \text{where:}$$

n_{11} = # individuals in population A with isoenzyme

n_{12} = # individuals in population A lacking isoenzyme

n_{21} = # individuals in population B with isoenzyme

n_{22} = # individuals in population B lacking isoenzyme

$n_{..} = (n_{11} + n_{12} + n_{21} + n_{22})$

$n_{.1} = n_{11} + n_{21}$

$n_{.2} = n_{12} + n_{22}$

$n_{1.} = n_{11} + n_{12}$

$n_{2.} = n_{21} + n_{22}$

When, for instance, many provenances are being analyzed, chi square analysis of r x c contingency tables may be used (Steele and Torrie, 1960, or any other statistical handbook). Isoenzyme analyses in forest tree research generally involve the recording of more than one isoenzyme from an individual genotype. Most chi square tables assume independence among classification variables but such may not be the case when multiple "responses" are elicited from one genotype (i.e., because of linkage, etc.). Under these circumstances, the more rigorous chi square tables of Jensen et al. (1968) may be more appropriately used.

Several authors have used Berry's D^2 (Berry, 1961) to test the differences among treatment isoenzyme frequencies. Using this technique, frequency (P) of isoenzyme occurrence is transformed to angular values using the following formula:

$$\Theta = \sin^{-1}(1-2p)$$

Berry's measure of distinctiveness between two populations is estimated by:

$$D^2 = (\Theta_1 - \Theta_2)^2 - V \quad \text{where:}$$

V = variance of $\Theta_1 - \Theta_2$ and is estimated by

$$\frac{1}{N_1} + \frac{1}{N_2} \quad \text{where:}$$

N_1 = size of first population

N_2 = size of second population.

D^2 is significant (i.e., populations are significantly different) when > 2.84V (5% level) and when > 5.6V (1% level). Muhs (1974) has expanded use of D^2 to account for more than one character at a time.

The properties of the Z distribution have been utilized by Sakai and Park (1971) to test for significant differences between populations.

Methods of Expressing Isoenzyme Variability. When the genetic control of isoenzyme variability does not conform to Mendelian hypotheses, variation may be quantitated by using the Polymorphism Index (PI) of Singh and Jain (1971). The formula for calculating the PI for any one population is:

$$\text{Polymorphism Index}_j = \sum_{i=1}^{I} Pij\ (1-Pij)(Tj) \quad \text{where:}$$

I = total numbers of isoenzyme bands in the population

Pij = frequency of individuals in the jth population with the ith isoenzyme

Tj = number of individuals analyzed in the population j.

A second method for quantitating isoenzyme variation may be called an Average Isoenzyme Deviation Unit (AIDU). Reasoning that an isoenzyme with a frequency approaching 1.0 or 0.0 contributes little to variation while those with intermediate frequencies in a population contribute most of the variation, the following formula may be applied:

$$AIDU_j - I \sum_{i=i}^{I} D_{ij}/N_j \quad \text{where:}$$

I = number isoenzymes analyzed

N_j = number of individuals analyzed

D_i = 0 if X_{ij} = 0 or X_{ij} = N

D_i = $N_j - X_{ij}$ if $X_{ij} = N_j/2$

D_i = X_{ij} if $X_{ij} = N_j/2$ where:

X_{ij} = the number of individuals in population j exhibiting isoenzyme i.

Populations with a great deal of variation will possess an AIDU approaching 1.0 while those with low amounts of variation will have an AIDU approaching 0.0. Genetic factors affecting both the PI and AIDU would include heterozygosity, multiple allelism, and allelic interactions. Both environmental and developmental factors would, of course, also affect PI and AIDU.

Applications of Enzyme Electrophoresis in Forest Genetics and Tree Breeding

Ontogenetic Studies

It has been demonstrated in several plant species that isoenzyme profiles of a single individual may vary both by organ or tissue and by time (stage of development). Generally, electrophoretically identical isoenzymes may be present in different tissues at varying concentrations or a given isoenzyme may be present at one developmental stage, but absent at another. Also, some isoenzymes may only be found in specific tissues and absent in others (Scandalios, 1969). Although variability in isoenzyme expression over time and tissue may be obscure variability due to genetic factors, such variability may be used to monitor developmental progress of an individual. More specifically, if the genetic control of a set of isoenzymes is known, the modification of isoenzyme expression by ontogenetic factors may be used to study the control of organ and tissue differentiation and the metabolic changes accompanying differentiation.

The literature describing developmental aspects of isoenzyme expression has been reviewed by Scandalios (1974) and in forest trees is limited. Conkle (1971a) demonstrated differences in isoenzymes at different stages of seed germination and early embryo growth of *P. attenuata*. Seasonal variability in isoenzyme expression has been demonstrated in *Acer rubrum* by Perry (1971), and the effects of environmental factors and genotype on peroxidase isoenzyme expression in poplar clones have been investigated by Wray (1974).

The potential for gaining an understanding of the developmental genetics of forest trees appears to be great using techniques of isoenzyme electrophoresis. Analyses of isoenzymes with known genetic control over varying developmental stages in varying environments should significantly aid in the development of genetic models of tree ontogeny.

Analysis of Genetic Variability

A fundamental problem confronting the forest tree geneticist continues to be the determination of the distribution of allelic variation in natural populations. McNaughton and Wolf (1970) discussed this question relative to pioneer and climax species and concluded that in the former allelic variation is distributed between individuals while in the latter allelic variability is distributed within individuals by means of a high percentage of heterozygous loci. The technique of enzyme electrophoresis permits analysis of allelic variability. Consequently, the traditional investigations based on techniques of quantitative genetic analysis of metric traits now can be complemented by studies of allelic variation in forest trees.

Polymorphism. It has long been known that most natural populations contain genetic polymorphism. Ecological genetics theory hólds that such polymorphisms are a part of the adaptive strategies of populations in heterogeneous environments. However, few polymorphisms have been genetically defined in terms of the number of polymorphic loci in a population, or in terms of the number of heterozygous loci in individuals. The relatively recent finding that isoenzyme variability can be directly equated with allelic variation makes it possible to define genetic polymorphism by analysis of isoenzyme polymorphism.

The presence of polymorphic enzyme and protein loci within natural populations of Drosophila was first demonstrated by Lewontin and Hubby (1966) and Hubby and Lewontin (1966). Later plant scientists discovered similar variability in populations of several plant species (Brown and Allard, 1969a,b; Clegg and Allard, 1972, 1973; Hamrick and Allard, 1972). Similar studies have revealed the occurrence of multiple alleles at single isoenzyme loci in several forest tree species. In *U. pumila*, Feret and Stairs (1971b) demonstrated the presence of nine alleles for a peroxidase locus. Alleles were distributed differently among the seed sources analyzed. Bergmann (1973b, 1974) showed that polymorphism occurs in *P. abies* populations for both esterase and leucine aminopeptidase loci. Rudin et al. (1974) found highly polymorphic loci for a number of systems in *P. sylvestris* populations. Because only a limited number of investigations in forest trees have revealed several instances of polymorphism, it appears that directed study of polymorphism in forest trees will be highly successful. Eventually quantification of the proportion of polymorphic loci in trees of several species and the number of alleles at each locus would greatly assist in the understanding of the ecological genetics of forest trees. The acquisition of such knowledge would be of immense help in understanding and developing breeding methods to improve forest trees.

Gene Flow and Isolation. Understanding polymorphism in trees through isolation of isoenzyme systems of known genetic control should permit forest tree geneticists to study gene frequencies in natural and selected forest tree populations. Study of variation in gene frequencies in forest trees from differing physiographic regions, provenances, populations and subpopulations should greatly expand knowledge of the genetic relationships among population types. The direction, rate and distance of gene migration in continuous, discontinuous and isolated populations can be elucidated by isoenzyme gene frequency distributions for numerous tree species. Thus, many problems in forest tree genetics and breeding may be solved. The use of isoenzyme analysis in studies of genetic differentiation, hybridization and introgression, inbreeding, and effective breeding sizes of populations will increase the efficiency of breeding programs. Additionally, after a large number of enzyme systems can be genetically defined, relative amounts of homozygosity and heterozygosity may be calculated. Knowledge of provenances containing a great deal of heterozygosity would be useful for defining areas to be set aside for genetic conservation, and could serve to define areas for selection of trees to be used for increasing the genetic base in previously selected breeding populations.

Geographical and Clinal Variation. Analysis of frequency distributions of several gene markers throughout the range of a forest tree species will add to and either support or refute current information on geographical patterns of genetic variability. Identification of continuous and discontinuous patterns of allelic variation could aid in the definition of races or ecotypes. Analysis of genotype-environment interactions with respect to altitudinal, latitudinal and longitudinal gradients may lead to development of more sophisticated models of forest tree genecology. Of interest to many biologists is the question of the distribution of heterozygosity in populations in the center of a species range relative to the distribution in populations in the marginal zones. A broad understanding of the genetic structure of forest tree populations will aid in the planning of optimal breeding strategies for tree improvement and domestication.

A few investigations have already shown that isoenzyme techniques are suitable for solving genecological problems. Among different *P. abies* provenances from Sweden, clinal variation has been demonstrated for two esterase loci (Bergmann, 1973b). Tigerstedt (1973) used three isoenzyme systems for investigations of the genetic structure in a marginal and a central population of *P. abies*.

Characterization of Individuals and Populations

The use of isoenzyme analysis may be extended to uses beyond genetical analysis of individuals and populations. Because, at least theoretically, each individual of a population possesses a unique genotype, so also should it possess a unique set of isoenzymes. Although identification of an individual tree may require a composite profile of several isoenzyme zymograms such an identification could serve well for clonal registration. The identification of clones, using isoenzyme profiles, could be useful for verifying the identity of suspected mislabeled or unlabeled clones and may in some areas serve as a means of enforcing patent or contractual laws regarding shipment and selling of registered cultivars. Isoenzymes so used would, of course, have to be environmentally and physiologically stable.

The use of isoenzyme analysis for seed source certification has been discussed by Bergmann (1972). Any method used for seed certification must be directly dependent upon the genetic (allelic) composition of a population. In addition any analysis of seed of unknown origin must be rapid and should not require large outplantings demanding years for evaluation. Bartels (1971) and Bergmann (1971, 1973a) have isolated allelic variants of isoenzyme systems in haploid macrogametophyte tissue. Using techniques of macrogametophyte analysis of a number of isoenzyme systems, allelic frequency data may be generated. In many instances provenances will differ in allelic frequencies, and perhaps in the presence or absence of specific alleles. Under these circumstances, if an inventory of seed source alleles and allelic frequencies is established, seed of unknown origin can easily be analyzed and its origin identified.

Conclusions

Although the techniques of gel electrophoresis of proteins and enzymes are not new, their use in the study of forest tree genetics is recent. The potential uses of the techniques are considerable, yet by no means are they currently being exploited by forest tree geneticists. Although rudimentary knowledge of genetically caused isoenzyme variability is available for a few Pinus species, *P. abies*, *Ulmus* spp., most tree genera have not been investigated. Clearly the technique can provide solutions to a number of basic and applied problems in forest tree genetics and tree improvement. In the future, serious application of gel electrophoresis should greatly expand the body of knowledge in forest genetics.

References

Alverez, M.R.: Temporal and spatial changes in peroxidase activity during fruit development in *Encyclia tampensis* (Orchidaceae). Am. J. Botany 55, 619-625 (1968).

Anderson, J.W.: Extraction of enzymes and subcellular organelles from plant tissues. Phytochemistry 7, 1973-1988 (1968).

Andreev, L.N., Shaw, M.: A note on the effect of rust infection on peroxidase isoenzymes in flax. Can. J. Botany 43, 1470-1483 (1965).

Ashton, G.C.: Serum amalyase (thread protein) polymorphism in cattle. Genetics 51, 431-437 (1965).

Barber, H.J., Driscoll, C.J., Long, P.M., Vickery, R.S.: Protein genetics of wheat and homologous relationships of chromosomes. Nature 218, 450-452 (1968).

Bartels, H.: Über die Charakterisierung von Fichtensaatgut aus verschiedenen Höhenlagen durch Enzymaktivitäten. In: Forstsamengewinnung und Pflanzenanzucht für das Hochgebirge (ed. H. Schmidt-Vogt), pp. 83-92. Munich: BLV, Bayer. Landwirtschaftsverlag 1964.

Bartels, H.: Genetic control of multiple esterases from needles and macrogametophytes of *Picea abies*. Planta (Berl.) 99, 283-289 (1971).

Beckman, L., Nilson, L.R.: Variation in serum enzymes in bird species and hybrids. Hereditas 53, 221-230 (1965).

Beckman, L., Scandalios, J.G., Brewbaker, J.L.: Genetics of leucine aminopeptidase isozymes in maize. Genetics 50, 899-904 (1964).

Bergmann, F.: Genetische Untersuchungen bei *Picea abies* mit Hilfe der Isoenzym-Identifizierung. I. Möglichkeiten für genetische Zertifizierung von Forstsaatgut. Allg. Forst u. J. Ztg. 142, 278-280 (1971).

Bergmann, F.: Experiments on possibilities for genetic certification of forest seed. Proc. IUFRO and SABRO joint meeting. Tokyo, 1972.

Bergmann, F.: Genetische Untersuchungen bei *Picea abies* mit Hilfe der Isoenzym-Identifizierung. II. Genetische Kontrolle von Esterase- und Leucinaminopeptidase-Isoenzymen im haploiden Endosperm ruhender Samen. Theoret. Appl. Genet. 43, 222-225 (1973a).

Bergmann, F.: Genetische Untersuchungen bei *Picea abies* mit Hilfe der Isoenzym-Identifizierung. III. Geographische Variationen an 2 Esterase- und 2 Leucinaminopeptidase-Loci in der schwedischen Fichtenpopulation. Silvae Genetica 22, 63-66 (1973b).

Bergmann, F.: Genetischer Abstand zwischen Populationen. II. Die Bestimmung des genetischen Abstands zwischen europäischen Fichtenpopulationen (*Picea abies*) auf der Basis von Isoenzym-Genhäufigkeiten. Silvae Genetica 23, 28-32 (1974).

Berry, R.J.: Epigenetic polymorphogenesis in wild populations of *Mus musculus*. Genet. Res. 7, 193-200 (1961).

Bingham, L., Krugman, S.L., Esterman, E.F.: Acrylamide electrophoresis of pine pollen proteins. Nature 202, 923-924 (1964).

Brewer, G.J.: An introduction to isozyme techniques. New York, London: Academic Press 1970.

Brown, A.H.D., Allard, R.W.: Inheritance of isozyme differences among inbred parents of a reciprocal recurrent selection population of maize. Crop Sci. 9, 72-75 (1969a).

Brown, A.H.D., Allard, R.W.: Further isozyme differences among inbred parents of a reciprocal recurrent selection population of maize. Crop Sci. 9, 643-644 (1969b).

Chrambach, A., Reisfield, R.A., Wycoff, M., Zaccari, J.: A procedure for rapid and sensitive staining of protein fractionated by polyacrylamide gel electrophoresis. Anal. Biochem. 20, 150-154 (1967).

Chrambach, A., Rodbard, D.: Polyacrylamide gel electrophoresis. Science 172, 440-451 (1971).

Chu, Y.E., Oka, H.I.: Comparison of variations in peroxidase isozymes between *perennis-sative* and *breviligulata-glaberrima* series of Oryza. Botan. Bull. Acad. Sinica (Taiwan) 8, 261-270 (1967).

Clegg, M.T., Allard, R.W.: Patterns of genetic differentiation in the slender wild oak species *Avena barbata*. Proc. Nat'l. Acad. Sci. (USA) 69, 1820-1824 (1972).

Clegg, M.T., Allard, R.W.: The genetics of electrophoretic variants in Avena. II. The esterase E_1, E_2, E_4, E_5, E_6 and anodal peroxidase APX_4 loci in *A. fatua*. J. Heredity 64, 1-6 (1973).
Cleland, W.W.: Dithiothreitol, a new protective reagent for Sh groups. Biochemistry 3, 480-482 (1964).
Coleman, E.A., Bula, R.J., Davis, R.L.: Electrophoretic and immunological comparisons of soluble root proteins of *Medicago sativa* L. genotypes in cold hardened and non-hardened condition. Plant Physiol. 41, 1681-1685 (1966).
Conkle, M.T.: Inheritance of alcohol dehydrogenase and leucine aminopeptidase isoenzymes in knobcone pine. Forestry Sci. 17, 190-195 (1971a).
Conkle, M.T.: Isoenzyme specificity during germination and early growth of knobcone pine. Forestry Sci. 17, 494-498 (1971b).
Coolidge, T.B.: A simple cataphoresis apparatus. J. Biol. Chem. 127, 551-553 (1939).
Davis, B.J.: Disc electrophoresis. II. Methods and application to human serum proteins. Ann. New York Acad. Sci. 121, 404-427 (1964).
Desborough, S., Peloquin, S.J.: Esterase isozymes from Solanum tubers. Phytochemistry 6, 989-994 (1967).
Dixon, M., Webb, E.C.: Enzymes (2nd Edition), p. 782. New York: Academic Press 1964.
Durrum, E.L.: A microelectrophoretic and microionophoretic technique. J. Am. Chem. Soc. 72, 2943-2948 (1950).
Durzan, D.J.: Disc electrophoresis of soluble protein in the female gametophyte and embryo of conifer seed. Can. J. Botany 44, 359-360 (1966).
Durzan, D.J., Chalupa, V.: Free sugars, amino acids, and soluble proteins in the embryo and female gametophyte of jack pine as related to climate at the seed source. Can. J. Botany 46, 417-428 (1968).
Farkas, G.L., Stahmann, M.A.: On the nature of changes in peroxidase isozymes in bean leaves infected by southern bean mosaic virus. Phytopathology 56, 669-677 (1966).
Felgenhauer, A.: Microelectrophoresis on polyacrylamide gel. Biochem. Biophys. Acta 133, 165-167 (1967).
Feret, P.P.: Isoenzyme variation in *Picea glauca* (Moench.) Voss seedlings. Silvae Genet. 20, 46-50 (1971).
Feret, P.P.: Peroxidase isoenzyme variation in interspecific elm hybrids. Can. J. Forestry Res. 2, 264-270 (1972).
Feret, P.P.: Genetic differences among three small stands of *Pinus pungens*. Theoret. Appl. Genet. 44, 173-177 (1974).
Feret, P.P., Stairs, G.R.: Peroxidase inheritance in *Ulmus pumila*. Forestry Sci. 17, 472-475 (1971a).
Feret, P.P., Stairs, G.R.: Enzyme electrophoresis - Application of molecular biology to forest-genetics research. Proc. 18th N.E. Forest Tree Imp. Conf., pp. 72-80 (1971b).
Frydenberg, D., Nielson, G.: Amylase isoenzymes in germinating barley seeds. Hereditas 54, 123-139 (1965).
Gell, P.G.H., Hawks, J.G., Wright, S.T.C.: The application of immunological methods to the taxonomy of species within the genus Solanum. Proc. Roy. Soc. Botany 151, 364-383 (1960).
Gerloff, E.D., Stahmann, M.A., Smith, D.: Soluble proteins in alfalfa roots as related to cold hardiness. Plant Physiol. 42, 895-899 (1967).
Goldstein, J.L., Swan, T.: The inhibition of enzymes by tannins. Phytochemistry 4, 185-192 (1965).
Gomori, C., Markert, C.L., Hunter, R.L.: The distribution of esterases in mouse tissues. J. Histochem. Cytochem. 7, 42-49 (1959).
Grossbach, U.: Acrylamide gel electrophoresis in capillary columns. Biochem. Biophys. Acta 107, 180-182 (1965).
Hadacova, W., Sahulka, S.J.: Electrophoretic investigation of proteins in different root zones in *Vicia faba* L. Biol. Plant. 9, 396-402 (1967).

Hagman, M.: Serological studies of pollen, and the incompatibility in forest trees. In: Proc. 14th IUFRO-Congress, Section 22-AG 22/24, p. 60-71 (1967).
Hall, T.C., McCown, B.H., Desborough, S., McLeester, R.C., Beck, G.E.: A comparative investigation of isozyme fractions separated from plant tissues. Phytochemistry 8, 385-391 (1969).
Hamaker, J.M., Snyder, E.B.: Electrophoresis patterns of needle enzymes in longleaf and Sonderegger pines. U.S. Forest Service Res. Notes SO-151, 8 pp. (1973).
Hamrick, J.L., Allard, R.W.: Microgeographical variation in allozyme frequencies in *Avena barbata*. Proc. Nat'l. Acad. Sci. (USA) 69, 2100-2104 (1972).
Harborne, J.B.: Phenolic glycosides and their natural distribution. In: Biochemistry of Phenolic Compounds (ed. J.B. Harborne), pp. 129-169. London: Academic Press 1964.
Harborne, J.B., Simmonds, N.W.: The natural distribution of the phenolic glycones. In: Biochemistry of Phenolic Compounds (ed. J.B. Harborne), pp. 77-127. London: Academic Press 1964.
Hare, R.C., Switzer, G.L.: Introgression with shortleaf pine may explain rust resistance in western loblolly pine. U.S. Forest Service Res. Notes SO-88, 2 pp. (1969).
Hart, G.E., Bhatia, C.R.: Acrylamide gel electrophoresis of soluble leaf proteins and enzymes from Nicotiana species. Can. J. Genet. Cytol. 9, 367-374 (1967).
Henning, U., Yanofsky, C.: An electrophoretic study of mutationally altered A proteins of tryptophan synthetase of *E. coli*. J. Molec. Biol. 6, 16-21 (1963).
Holmes, R.: Discontinuous acrylamide-gel plate electrophoresis. Biochem. Biophys. Acta 133, 174-177 (1967).
Hosoya, T.: Turnip peroxidase. I. Purification and physiochemical properties of multiple components in turnip peroxidase. J. Biochem. 47, 369-381 (1960).
Hubby, J.L., Lewontin, R.C.: A molecular approach to the study of genic heterozygosity in natural populations. I. The number of alleles at different loci in *Drosophila pseudoobscura*. Genetics 54, 577-594 (1966).
Hunt, J.S., Ingram, U.M.: Allelomorphism and the chemical differences of the human haemoglobins A and C. Nature 181, 1062-1063 (1958).
Hyden, H., Bjurstam, K., McEwen, B.: Protein separation at the cellular level by micro disc electrophoresis. Anal. Biochem. 17, 1-15 (1966).
Ingram, U.M.: Gene mutations in human haemoglobin: The chemical differences between normal and sickle cell haemoglobin. Nature 180, 326-328 (1957).
Jensen, D.R., Beus, G.B., Storm, G.: Simultaneous statistical tests on categorical data. J. Exp. Educ. 36, 46-56 (1968).
Jermyn, M.S., Thomas, R.: Multiple components in horseradish peroxidase. Biochem. J. 56, 631-639 (1954).
Johnson, F.J., Kanapi, C.G., Richardson, R.H., Sakai, R.K.: Isoenzyme variability in species of the genus Drosophila. I. A multiple allelic isozyme system in *Drosophila busckii*: inheritance and general considerations. Biochem. Genet. 1, 35-40 (1967).
Johnson, L.B., Brannaman, B.L., Zsheile, Jr., F.P.: Protein and enzyme changes in wheat following infection with *Puccinia recondita*. Phytopathology 58, 578-583 (1968).
Jooste, J.W., Moreland, D.E.: Preliminary characterization of some plant carboxylic ester hydrolases. Phytochemistry 2, 263-271 (1963).
Kahler, A.L., Allard, R.W.: Genetics of isozyme variants in barley. I. Esterases. Crop Sci. 10, 444-448 (1970).
Kaminshi, M., Gajos, E.: Comparative examination of carboxylic esterases in sera of horse, donkey and their hybrids. Nature 201, 716-718 (1964).

Kunkel, H.G., Slater, R.J.: Zone electrophoresis in a starch gel supporting medium. Proc. Soc. Exp. Biol. Med. 80, 42-44 (1952).

Lanzani, G.A., Marchesini, A., Galante, E., Manzocchi, L.A., Sequi, P.: Peroxidase isoenzymes in wheat germs. Enzymologia 33, 361-372 (1967).

Lewis, R.A., Cech, F.C.: Electrophoresis separation of general proteins and isoenzymes of black cherry seed (*Prunus serotina* Ehrh.). In: Proc. 10th Southern Conf. Forest Tree Improve., p. 140-148 (1969).

Lewontin, R.C., Hubby, J.L.: A molecular approach to the study of genic heterozygosity in natural populations. II. Amount of variation and degree of heterozygosity in natural populations of *Drosophila pseudoobscura*. Genetics 54, 595-609 (1966).

Linskins, H.F.: Die Änderung des Protein- und Enzym-Musters während der Pollenmeiose und Pollenentwicklung. Planta (Berl.) 69, 79-91 (1966).

Loomis, W.D.: Removal of phenolic compounds during the isolation of plant enzymes. In: Methods of Enzymology (ed. J.M. Lowenstein), Vol. 13, pp. 555-563. New York: Academic Press 1969.

Loomis, W.D., Battaile, J.: Plant phenolic compounds and the isolation of plant enzymes. Phytochemistry 5, 423-438 (1966).

Macko, V., Honald, G.R., Stahmann, M.A.: Soluble proteins and multiple enzyme forms in early growth of wheat. Phytochemistry 6, 465-471 (1967).

Markert, C.L., Møller, F.: Multiple forms of enzymes: tissue, ontogenetic, and species specific patterns. Proc. Nat'l. Acad. Sci. (USA) 45, 753-763 (1959).

Marshall, D.R., Allard, R.W.: The genetics of electrophoretic variants in Avena. I. The esterase E_4, E_9, E_{10}, phosphatase P_5 and anodal peroxidase APX_5 loci in *Avena barbata*. J. Heredity 60, 17-19 (1969).

McCown, B.H., Beck, G.E., Hall, T.C.: Plant leaf and stem proteins. I. Extraction and electrophoretic separation of the basic, water-soluble fraction. Plant Physiol. 43, 578-582 (1968).

McCown, B.H., Hall, T.C., Beck, G.E.: Plant leaf and stem proteins. II. Isozymes and environmental change. Plant Physiol. 44, 210-216 (1969a).

McCown, B.H., McLeester, R.C., Beck, G.E., Hall, T.C.: Environment-induced changes in peroxidase zymograms in the stems of deciduous and evergreen plants. Cryobiology 5, 410-412 (1969b).

McCune, D.C.: Multiple peroxidases in corn. Ann. New York Acad. Sci. 94, 723-730 (1961).

McNaughton, S.J., Wolf, L.L.: Dominance and the niche in ecological systems. Science 167, 131-139 (1970).

Miyazaki, Y., Sakai, K.I.: Use of zymography for identification of a clone in *Cryptomeria japonica* D. Don. J. Jap. Forestry Soc. 51, 235-239 (1969).

Morris, R.O.: Changes in the pea root proteins associated with tissue differentiation: Effect of 2-4 dichlorophenozyacetic acid. Biochem. Biophys. Acta 127, 271-273 (1966).

Morton, R.K.: Methods of extraction of enzymes from animal tissue. In: Methods in Enzymology (eds. S.P. Colowick, N.O. Kaplan), Vol. 1, p. 25-51. New York: Academic Press 1955.

Muhs, H.J.: Distinction of Douglas-fir provenances using peroxidase isoenzyme patterns of needles. Silvae Genet. 23, 71-76 (1974).

Okerse, R., Siegel, B.Z., Galston, A.W.: Hormone induced repression of a peroxidase isozyme. Science 151, 452-453 (1965).

Ornstein, B.J.: Disc electrophoresis. I. Background and theory. Ann. New York Acad. Sci. 121, 321-349 (1964).

Perry, T.O.: Seasonal and genetic differences in fats, phenols, isoenzymes, and pigments of red maple. Forestry Sci. 17, 209-212 (1971).

Pickering, J.L., Fairbrothers, D.E.: A serological and disc electrophoretic investigation of Magnolia taxa. Bull. Torrey Botan. Club 94, 468-479 (1967).

Pierpoint, W.S.: The enzymic oxidation of chlorogenic acid and some reactions of the quinone produced. Biochem. J. 98, 567-580 (1966).

Pirie, N.W.: Leaf proteins. Ann. Rev. Plant Physiol. 10, 33-53 (1959).

Poulik, M.D.: Starch gel electrophoresis in a discontinuous system of buffers. Nature 80, 1477 (1957).

Prakash, S., Lewontin, R.C., Hubby, J.L.: A molecular approach to the study of genic heterozygosity in natural populations of *Drosophila pseudoobscura*. Genetics 61, 841-858 (1969).

Racusen, D., Foote, M.: Peroxidase isozymes in bean leaves by preparative disc electrophoresis. Can. J. Botany 44, 1633-1638 (1966).

Rasmuson, B., Rudin, D.: Variations in esterase zymogram patterns in in needles of *Pinus sylvestris* from provenances in northern Sweden. Silvae Genet. 20, 39-41 (1971).

Raymond, S., Weintraub, L.: Acrylamide gel as a support medium for zone electrophoresis. Science 130, 711 (1959).

Raymond, S., Wang, Y.: Preparation and properties of acrylamide gel for use in electrophoresis. Anal. Biochem. 1, 391-396 (1960).

Raymond, S., Miles, J.L., Lee, J.C.J.: Lipoprotein patterns in acrylamide gel electrophoresis. Science 151, 346-347 (1966).

Richards, E.G., Coll, J.A., Gratzer, W.B.: Disc electrophoresis of ribonucleic acid in polyacrylamide gels. Anal. Biochem. 12, 452-471 (1965).

Robb, D.A., Mapson, L.W., Swain, T.: On the heterogenity of the tyrosinase of broad bean *Vicia faba* L. Phytochemistry 4, 731-740 (1965).

Rudin, P., Rasmuson, B.: Genetic variation in esterases from needles of *Pinus sylvestris*. Hereditas 73, 89-98 (1973).

Rudin, D., Eriksson, G., Ekberg, I., Rasmuson, M.: Studies of allele frequencies and inbreeding in Scots pine populations by the aid of the isozyme technique. Silvae Genet. 23, 10-13 (1974).

Rudolph, K., Stahmann, M.A.: Changes in activity and isozyme pattern of soluble dehydrogenase in bean leaves after infection with *Pseudomonas phaseolicola* and *Uromyces phaseoli*. (Abstr.) Phytopathology 58, 1065 (1968).

Sakai, K.I., Miyazaki, Y., Matsuura, T.: Genetic studies in natural populations of forest trees. I. Genetic variability on the enzymatic level in natural forests of *Thujopsis dolabrata*. Silvae Genet. 20, 168-173 (1971).

Sakai, K.I., Park, Y.G.: Genetic studies in natural populations of forest trees. III. Genetic differentiation within a forest of *Cryptomeria japonica*. Theoret. Appl. Genet. 41, 13-17 (1971).

Scandalios, J.G.: Subunit dissociation and recombination of catalase isozymes. Proc. Nat'l. Acad. Sci. (USA) 53, 1035-1040 (1965).

Scandalios, J.G.: Genetic control of multiple molecular forms of enzymes in plants: A review. Biochem. Genet. 3, 37-79 (1969).

Scandalios, J.G.: Isoenzymes in development and differentiation. Am. Rev. Plant Physiol. 25, 225-258 (1974).

Schwartz, D.: Electrophoretic and immunochemical studies with endosperm proteins of maize mutants. Genetics 45, 1419-1427 (1960).

Schwartz, D., Fuchsmand, L., McGrath, K.H.: Allelic isozymes of the pH 7.5 esterase in maize. Genetics 52, 1265-1268 (1965).

Schwartz, D., Endo, T.: Alcohol dehydrogenase polymorphism in maize - simple and compound loci. Genetics 53, 709-715 (1966).

Schwartz, H.M., Biedron, S.I., von Holdt, M.N., Rehm, S.: A study of some plant esterases. Phytochemistry 3, 189-200 (1964).

Shaw, C.R.: The use of genetic variation in the analysis of isozyme structure. Brookhaven Symposia Biol. 17, 117-130 (1964).

Shaw, C.R.: Electrophoretic variation in enzymes. Science 149, 936-943 (1965).

Shaw, C.R.: Isozymes: Classification, frequency and significance. Intern. Rev. Cytol. 25, 297-332 (1969).

Shaw, C.R., Koen, A.L.: Starch gel zone electrophoresis of enzymes. In: Chromatographic and Electrophoretic Techniques (ed. J. Smith), Vol. II, pp. 325-359. Great Britain: Pitman Press 1968.

Shaw, C.R., Prasad, R.: Starch gel electrophoresis of enzymes - A compilation of recipes. Biochem. Genetics 4, 297-320 (1970).

Simpson, R.B., Kauzman, W.: The kenetics of protein denaturation. I. The behavior of the optical rotation of ovalbumin in urea solutions. J. Am. Chem. Soc. 75, 5139-5152 (1953).

Singh, R.S., Jain, S.K.: Population biology of *Avena*. II. Isoenzyme polymorphisms in populations of the Mediterranean region and central California. Theoret. Appl. Genet. 41, 79-84 (1971).

Smith, I.: Chromatographic and electrophoretic techniques. Zone electrophoresis. Vol. II. London: Pitman Press 1968.

Smithies, O.: Grouped variations in the occurrence of new protein components in normal human serum. Nature 175, 307-308 (1955a).

Smithies, O.: Zone electrophoresis in starch gels: Group variation in the serum proteins of normal human adults. Biochem. J. 61, 629-641 (1955b).

Smithies, O.: Zone electrophoresis in starch gels and its application to studies of serum proteins. Adv. Prot. Chem. 14, 65-113 (1959).

Stahmann, M.A.: Plant Proteins. Ann. Rev. Plant Physiol. 13, 137-158 (1963).

Staples, R.C., Stahmann, M.A.: Malate dehydrogenase in the rusted bean leaf. Science 140, 1320-1321 (1963).

Staples, R.C., Stahmann, M.A.: Changes in proteins and several enzymes in susceptible bean leaves after infection by bean rust fungus. Phytopathology 54, 760-764 (1964).

Staples, R.C., McCarthy, W.S., Stahmann, M.A.: Heat stabilities of acid phosphatases from pinto bean leaves. Science 149, 1248-1249 (1965).

Steele, G.D., Torrie, J.H.: Principles and procedures of statistics, 481 p. New York: McGraw-Hill 1960.

Steward, F.C., Lyndon, R.F., Barker, J.T.: Acrylamide gel electrophoresis of soluble plant proteins: A study on pea seedlings in relation to development. Am. J. Botany 52, 155-164 (1965).

Stuber, C.W., Levings, III, C.S.: Auxin induction and repression of peroxidase isozymes in oats (*Avena sativa* L.). Crop. Sci. 9, 415-416 (1969).

Stutte, C.A., Todd, G.W.: Some enzyme and protein changes associated with water stress in wheat leaves. Crop Sci. 9, 510-512 (1969).

Tigerstedt, P.M.A.: Studies on isozyme variation in marginal and central populations of *Picea abies*. Hereditas 75, 47-60 (1973).

West, N.B., Garber, D.E.: Genetic studies of variant enzymes. II. The inheritance of esterases and leucine aminopeptidases in *Phaseolus vulgaris* and *P. coccineus*. Can. J. Genet. Cytol. 9, 646-655 (1967).

Williams, G.J., III: Respiration and respiratory enzyme activity in latitudinal ecotypes of *Liquidambar styracifula* L. (Abstr.) Bull. Ecol. Soc. Am. 48, 74 (1967).

Wray, P.H.: Peroxidase and growth in hybrid poplar. Ph.D. Thesis, 154 pp. Ames, Iowa: Iowa State University 1974.

Yue, S.B.: Isoenzymes of malate dehydrogenase from barley seedlings. Phytochemistry 5, 1147-1152 (1966).

CHAPTER 4

Extraction and Analysis of Free and Protein-Bound Amino Acids from Norway Spruce Foliage

J. LUNDERSTÄDT

Introduction

Amino acids are basic constituents of all living things; they are the building blocks of proteins and are intermediates in metabolism. A forest geneticist will look at amino acid quality (types present) and quantity (amount) for a variety of reasons. Variation in the quality of amino acids may serve as gene markers, for clone identification, and as a means of measuring population dynamics. In addition, amino acids are one of the nutrients present in trees that can be used by tree pests and thus they have implications in breeding for insect and disease resistant varieties of trees.

Another field of potential great interest may be the use for feeding purposes of protein originating from tree leaves of both conifers and hardwoods, thus leading to challenging tasks for tree breeders. In a pilot experiment the contents of free and bound amino acids have been determined in the leaves of various forest tree species grown under semicontrolled conditions as compared to two herbs (Siren et al., 1970). These results reveal not only rather high levels of total protein but also very favorable contents of essential amino acids.

Although other branches of the biological sciences have long used measures of amino acids as a screening tool, very little information on the subject has appeared in forestry literature, with the possible exception of information about the specialized proteins known as enzymes. Possible reasons for the lack of information related to forestry are: 1. substances such as phenols are often present in high concentrations and may interfere with analytical techniques, and 2. analysis of amino acids requires specialized separation techniques.

For these reasons, our knowledge of the role of amino compounds in the nutrition of insects feeding on forest trees is limited, which hampers research on breeding for insect resistant tree varieties. Review articles on insect nutrition (Lipke and Fraenkel, 1956; House, 1961; Dadd, 1973) have shown that amino acids and proteins from plant hosts are important to the insects, and have also detailed objectives to be followed in breeding for insect resistant plants (Thorsteinson, 1960; Beck, 1965). However, in a listing of studies on resistance to damaging insects that included 20 plant species (Painter, 1958), no tree species were mentioned.

Recently, the amino acid composition of Norway spruce provenances has been related to the occurrence of spruce spider mites on these provenances (Löyttyniemi and Tulisalo, 1972), a work which has clearly shown the relevance of such research.

Therefore, I propose to set forth some of the methods that we, as forest biologists, may use to quantify the amino acid composition found in various plant extracts.

An organic substance must satisfy several criteria to be used as a marker for a specific genotype, or as a source of resistance, or a marker for resistance. Among these are 1. the quantity of the substance must not be affected greatly by environmental factors, and 2. the compound must have a highly specific effect on the insects that feed on the plants.

As a general rule, the biological specificity of an organic molecule increases with increasing molecular weight, since the diversity, configuration, and conformation of functional groups also increases (Hess and Rupley, 1971). The proteins are examples. Low molecular weight molecules like amino acids appear to be suitable for use in genetic characterization when criteria such as occurrence (qualitative) within species and absolute amount of a specific amino acid (quantitative) are measured (Adams, 1962; Fowden, 1964; Thompson et al., 1969).

Amino acids are primary catabolic products of plants. Therefore, they have shorter half-lives and can be expected to be more affected by environmental changes than proteins, to which the amino acids are connected by a series of steady state equilibria. Qualitative and quantitative changes of amino acids and proteins in forest plant material are influenced by different physiological conditions such as age of needles (Kaludin and Kaludin, 1968; Lunderstädt, 1973), age of trees (Kaludin and Kaludin, 1968), portion of tree sampled (Materna and Ryskova, 1965), time of day (Durzan, 1968b,c), annual rhythm (Wehrmann, 1959a; Durzan, 1968a; Feeny, 1970; Lunderstädt, 1973), stand composition (Wehrmann, 1959a), fertilization (Wehrmann, 1959b; Durzan and Steward, 1967), and poisonous agents (Lück and Pavlik, 1966). These clearly indicate the need for standardized growth conditions and analytical procedures so that results from various laboratories may be evaluated and compared. Therefore as many factors as possible should be kept constant, a condition not easily satisfied when large numbers of samples are needed. The effect of some of the conditions such as time of day, portion of tree sampled, and age of needles may be evaluated by control experiments with material from a standard tree (Durzan, 1968b,c).

Physiological processes should be stopped immediately after collection by immersing the material in liquid nitrogen and subsequently storing the material at -20°C or lower to avoid major changes in the material. Samples can then be ground in liquid nitrogen (N_2) or solid CO_2 (dry ice) followed by freeze-drying. The procedure has several distinct advantages:

1. Large numbers of samples can be processed and stored safely, independent of harvesting time.

2. All samples will have nearly comparable particle size, independent of size and thickness of the original material. Thus, comparable extractions may be made with different batches of material.

3. Small samples may be handled quantitatively because almost no material will be lost in the grinding process.

Methods

Quantitative Determination of Free and Protein-Bound Amino Acids from Spruce Needles

Tissue Preparation. About 2 g of spruce needles are removed from twigs with forceps while wearing rubber gloves to avoid contamination of the sample with amino acids from perspiration. As they are collected, the needles are dropped into small Dewar flasks that contain liquid N_2. The sample is transferred to a precooled porcelain mortar that contains a small amount of dry ice and is ground. Evaporating dry ice is replaced as needed. Large quantities of dry ice should be avoided because it causes poor grinding. Liquid N_2 may be used in the mortar instead of dry ice. The ground sample is transferred to a suitable container and freeze-dried for 24 h with a cold trap temperature of -50°C and a final vacuum of 10^{-3} Torr. The dry powder is stored at room temperature in a desiccator.

Extraction of Soluble Amino Acids. A 250 mg subsample of freeze-dried material is extracted with 100 ml of boiling 80% aqueous ethanol for 20 min with reflux. After the supernatant is removed, the insolubles are re-extracted with additional boiling 80% ethanol for 20 min with reflux. Then the insoluble material is recovered by filtration and retained for hydrolysis of protein-bound amino acids. The ethanol extract, which contains soluble amino acids, is also retained.

Concentration and Purification of Soluble Amino Acids. The combined 80% ethanol extract is absorbed to 1 g Celite *in vacuo* at 50°C (El-Basyouni et al., 1964). The celite cake is then suspended in 100 ml of water and re-extracted by boiling for 30 min under reflux. The cooled slurry is vacuum-filtered, the effluent concentrated *in vacuo* at 80°C to near dryness, and the residue re-dissolved in 3 x 1 ml distilled water. No losses were found with this technique when used for plant extracts as shown by radioactive labeling (Rohringer et al., 1967).

The resulting solution/suspension is applied to a Sephadex G-15 column (150 x 9 mm, maintained at 4°C) by allowing each ml to be taken up before the next one is added. Thirty fractions at 20 drops/fraction are collected. A large portion of the water-soluble phenolics will be removed if the temperature is kept at 4°C. An 100 µl aliquot of each fraction is tested for the presence of amino acids (Stegemann, 1960). The ninhydrin-positive fractions are pooled, concentrated to dryness, and re-dissolved in 3 x 1 ml distilled water for final isolation of amino acids.

Acidic Hydrolysis of Protein-Bound Amino Acids. The material insoluble in boiling 80% ethanol is dried at 105°C for 1 h, cooled to room temperature in a desiccator, and weighed. Four 20 mg subsamples of the dried insoluble material are placed in separate thick-walled test tubes (130 x 20 mm) that contain 3 ml of 6 N HCl, sealed with a glass stopper and clamp, and hydrolyzed at 105°C for 24 h (Cramer, 1958).

The dark brown hydrolysate is removed from the test tube with 5 x 1 ml methanol, added to and adsorbed on 0.5 g Celite *in vacuo* at 50°C. Amino acids are removed from the Celite with 7 x 2 ml hot water, vacuum filtered, concentrated to dryness, and re-dissolved in 3 x 1 ml distilled water.

Isolation of Amino Acids. Both soluble and hydrolyzed amino acid-containing fractions are separately applied to Dowex 50 x 8 columns (175 x 9 mm, 50-100 mesh, H^+ form). The columns are washed with 100 ml water (1 drop/5 sec) to remove sugars, carboxylic acids, and neutral compounds, then eluted with 60 ml of 0.4 N NH_4OH followed by 40 ml of 4.0 N NH_4OH. These eluates contain the amino acids. The combined NH_4OH eluate is concentrated to dryness *in vacuo* and re-dissolved in 2 x 0.7 ml solvent (soluble amino acids) or 3 x 1 ml solvent (amino acids from hydrolysate). The solvent is hot water if the sample is to be analyzed by paper chromatography, or elution buffer (sodium citrate buffer pH 2.2 (0.2 N in respect to Na)) if column chromatography on an amino acid analyzer is to follow.

Identification and Quantitative Determination of Amino Acids

Paper Chromatography

Two-Dimensional Paper Chromatography. 100 µl of the soluble amino acid fraction (approx. 0.5-2.0 µg of amino acids), 30 µl of the hydrolysate amino acid fraction (approx. 10 µg amino acids), and 5 µl of a solution of amino acid standards (2 mg/ml) are applied to a corner of separate 20 x 20 cm sheets of Schleicher and Schüll chromatography paper No. 2043B (or its equivalent). The chromatograms are equilibrated over solvent overnight, then developed in the ascending mode at room temperature with the organic phase of n-butanol/acetic acid/water (4/1/5,v/v) for 6 h, air dried, rotated 90° so the separated amino acids are now along the bottom edge, and redeveloped in phenol (re-distilled)/water/formic acid (75/26/1, v/v) for 7 h.

Round Filter Chromatography. A circle 6 cm diam. is drawn in the center of a 40 x 40 cm sheet of chromatography paper and used as the starting line. Up to 20 samples may be applied to one chromatogram. After overnight equilibrium over the butanol solvent, the chromatogram is developed for 7 h with the organic phase of n-butanol/acetic acid/water (4/1/5, v/v). Solvent is applied to the center of the chromatogram and amino acids are developed toward the periphery. Rerunning the chromatogram after air-drying may improve the separation of amino acids considerably.

In order to achieve comparable separation over a number of runs, fresh butanol solvent must be prepared for every run, since esterification occurs as the solvent ages (Cramer, 1958), and may lead to variable Rf values for individual amino acids.

Amino acids are detected by spraying either chromatogram with a ninhydrin solution (Cramer, 1958), heating the chromatogram to 105°C for 10 min, and then spraying the chromatogram with a solution of $Cu(NO_3)_2$ to stabilize the spots (Kawerau and Wieland, 1951) for densitometric determination.

Since many amino acids, e.g., β-alanine or proline, react with ninhydrin only weakly or not at all, their evidence has to be established by additional or special spray reagents as can be found in laboratory manuals (Block et al., 1958; Block, 1960; Bailey, 1967). Also the different sensitivity of various reagents with individual acids is most useful, for example, to detect acids whose concentrations as compared with the whole fraction are very low. The presence of low molecular peptides can be checked by making them visible according to the method of Rydon and Smith (1952) or by hydrolysis of a part of the fraction of soluble compounds and comparative chromatography of both to establish qualitative and quantitative differences in the resulting pattern.

Quantifying the Amino Acids. The optical density of amino acids detected by ninhydrin is either estimated visually or done at 515 nm with a densitometer equipped with an integrator. A blank of equally developed, sprayed, and heated paper is used for comparison. In both cases known amounts of the respective acids are run on the same sheet of paper. For round-filter chromatograms, the optical density is measured along a line through the middle of every compound of the respective sample. At this point of highest optical density of symmetric spots, the optical density is proportional to the optical density of the entire spot (McFarren et al., 1951).

Rubber gloves should be worn whenever handling the chromatograms to avoid contamination. In some cases it might be necessary to check for ninhydrin-positive contaminations by spraying a representative part of the same sheet as used for chromatographic separation.

Amino Acid Analyzer

Based on the early work of Moore and Stein in 1951 and Spackman et al. in 1958, various types of such analyzing systems are produced. They all combine the chromatographic separation of mixtures of amino acids on synthetic ion exchange resins (Hamilton, 1957; Schroeder, 1957; Dorfner, 1964; Walton, 1967a,b) with their immediate photometric detection, quantitative evaluation, and electronic registration. They are of greatest use whenever highly reproducible qualitative separations and very exact quantitative determinations of amino acids are to be carried out, as may be the case, i.e., in breeding projects. Further advantages are: 1. the analyses can be carried out in moderately equipped laboratories, 2. large numbers of samples can be analyzed since by careful planning several batches can be analyzed and prepared for analysis simultaneously (see also Table 1), 3. the original data can be easily processed for further evaluation in a computer.

Five hundred microliters of either soluble or hydrolyzed amino acids (2-150 μg of amino acids) are applied to the column of an automatic amino acid analyzer. Separation requires 4 to 8 h, depending upon the make of the analyzer, the number of amino acids to be separated, and the quality of separation required. The concentration of individual amino acids can be determined by a comparison of the peak area of each acid with that of areas derived from runs with known quantities of standard amino acids.

Calculation of amino acid concentration on a tissue weight basis:

a) Calibration of amino acid analyzer results with standards

$$\left[a_s\right] = \frac{P_s}{Z_s} \quad \left[cm^2/\mu M\right] \qquad \text{where:}$$

P_s = area under the peak of the standard amino acid in cm^2

Z_s = quantity of the standard amino acid in μM

Table 1. Average time request in hours for the synchronous analysis of a batch of four samples for free and protein-bound amino acids by one person

Aim	Procedure	Day 1	2	3	4	5	6	7	8	9	10	11
Extraction of the samples	Isolation of soluble and insoluble matter	4										
Separation of hydrophilic and hydrophobic compounds of the soluble matter	Celite filtration		4									
	Concentration of the hydrophilic compounds		5									
Crude separation of soluble phenolics and further water soluble compounds	Sephadexfiltration			1/2								
	Location of the fractions containing amino acids			1/2								
	Concentration			3								
Isolation of the soluble amino acids	Dowexchromatography				1/2							
	Concentration				2							
Hydrolysis of the insoluble matter	Determination of weight			1								
	Celitefiltration					2						
	Concentration						3					
Isolation of the hydrolyzed amino acids	Dowexchromatography							1				
	Concentration							4				
Chromatography of the amino acids by ion exchange chromatography	Chromatography of soluble amino acids								24	8		
	Insoluble amino acids									16	16	
Calculation of the concentrations per total sample	Soluble amino acids (determination of the peak areas manually)											5
	Hydrolyzed amino acids (determination of the peak areas electronically)											4
	TOTAL	4	9	5	2.5	2	3	5	24	24	16	9

b) Concentration of an amino acid in an aliquot of the sample

$$[a_x] = \frac{P_x}{P_s} \cdot Z_s \quad [(cm^2/cm^2)\ \mu M] \qquad \text{where:}$$

P_x = area under peak of amino acid from sample

c) Concentration of the amino acid per unit weight of total sample

$$[A_x] = [a_x]\ F$$

$$F = \frac{A \times C}{B \times D \times E}$$

$$[A_x] = \left[\frac{cm^2 \cdot \mu M \cdot \mu l \cdot mg}{cm^2 \cdot mg \cdot \mu l \cdot mg}\right] = [\mu M/mg] \qquad \text{where:}$$

A = total volume of amino acid solution from "D" mg tissue

B = volume of aliquot used for analysis

C = weight of insoluble matter per "E" mg tissue

D = weight of insoluble matter subjected to hydrolysis

E = weight of freeze-dried plant material subsample

Conclusions

Other methods of amino acid analysis are available; examples of some are paper electrophoresis (Clotten and Clotten, 1962; Blackburn, 1966; Bailey, 1967), enzymatic determination (Bergmeyer, 1970), microbiological determination (Block, 1960), gas chromatography (Burchfield and Storrs, 1962; Pierce, 1968), mass spectrometry (Vetter, 1972), and high pressure liquid chromatography. I have presented methods for paper chromatography and amino acid analyzers because these techniques are most likely to be available to forest biologists and because they have been widely used in the past.

At various points in the sample preparation and extraction, alternate techniques may be used; choice of technique depends largely upon the type of amino acids to be detected or the degree of accuracy needed. For example, in many cases simple chemical glassware can be substituted for reflux equipment at least as long as no further components of the leaf material are to be analyzed at the same time. Under some conditions, ion-exchange chromatography with Dowex resin may be avoided (Materna and Ryskova, 1965). To separate various classes of compounds, one may have to choose between counter-current extraction (Lück and Pavlik, 1966), filtration with lead acetate (Feeny, 1970), or the Celite filtration techniques described here. In every case control experiments should show clearly the possible losses with the different techniques.

The removal of water-soluble phenolics may not be necessary if an amino acid analyzer is used but is required before paper chromatography. Well over 100 solvent systems for paper chromatographic separation of amino acids have been described (Block et al., 1958; Smith, 1969), and the choice of solvent system is dictated by which amino acids require best separation. Two dimensional paper chromatography should be applied when complete separation of amino acids is desired, but round-filter chromatography may be used as an easy semi-quantitative method to determine amino acids without eluting spots from the chromatogram. Elution of amino acids from the chromatogram may be necessary as a preliminary step to determine the amount of extract to apply to an amino acid analyzer.

Great care is to be taken when leaf proteins are hydrolyzed. By comparing the N-contents of the sample before and after hydrolysis according to the Kjeldahl method, the degree of separation of proteins from their matrix can be adjusted. By following the time course of liberation of ninhydrin-positive compounds until its maximum, the completeness of hydrolysis is found. As with other proteins (Rees, 1946) acidic hydrolysis of plant material (spurce needles) also causes loss of several acids especially serine and threonine and a complete destruction of tryptophane. By replacing acidic hydrolysis as described above by hydrolysis using $Ba(OH)_2$ (Cramer, 1958) it was found that not only tryptophane but also large amounts of methionine were destroyed (unpublished data).

Asparaginyl and glutaminyl groups in the side chains of proteins may be determined either by deamination in strong acid at moderate temperatures and long incubation periods (Hamilton, 1960) or by conversion of the amides into γ-hydroxy-β-aminobutyric acid and δ-hydroxy-γ-amino valeric acid, respectively, followed by separation of these acids, aspartic acid and glutamic acid by ion exchange chromatography (Chibnall et al., 1958; Bailey, 1967).

Depending on the purpose of research, the presentation of the data will vary. Usually the results are expressed in μM/mg dry weight; the fresh weight may serve as a basis of reference with more physiological data. For genetic purposes an evaluation of the amino acid quantities in μM facilitates a survey of their distribution, e.g., in different species, while the expression in μg may be more useful for physiologists and breeders interested in the amounts of constituents to be produced by a given plant. Frequently considerable difficulties arise when unidentified ninhydrin-positive compounds have to be expressed as % of the total fraction. Depending on the quantities of the fractions under consideration, they can be 1. calculated on a weight basis, 2. their N-content as % of the N-content of the whole fraction can be determined, or 3. the amounts of liberated ammonia resulting from oxidative deamination of amino acids with ninhydrin (Sobel et al., 1945) in the unknown subfraction and the mother fraction can be compared. Finally, corrections for losses during hydrolysis are made on the grounds of the results gained for fractions of the same sample which have been exposed for different incubation periods.

References

Adams, E.: Amino acid metabolism. Ann. Rev. Biochem. 31, 173-212 (1962).

Bailey, J.L.: Techniques in Protein Chemistry, 2nd edition. Amsterdam, London, New York: Elsevier Publishing Company 1967.

Beck, S.D.: Resistance of plants to insects. Ann. Rev. Entomol. 10, 207-232 (1965).

Bergmeyer, H.U. (ed.): Methoden der enzymatischen Analyse, II. Auflage, pp. 1610-1682. Weinheim/Bergstr.: Verlag Chemie 1970.

Blackburn, S.: Micro techniques for amino acid analysis and peptide separation based on high voltage electrophoresis. In: A Laboratory Manual of Analytical Methods of Protein Chemistry Including Polypeptides (eds. P. Alexander, H.P. Lundgren), Vol. IV, pp. 78-102. Oxford, London, Edinburgh, New York, Toronto, Paris, Frankfurt: Pergamon Press 1966.

Block, R.J.: Amino acid analysis of protein hydrolysates. In: A Laboratory Manual of Analytical Methods of Protein Chemistry Including Polypeptides (eds. P. Alexander, R.J. Block), Vol. II, pp. 44-55. Oxford, London, New York, Paris: Pergamon Press 1960.

Block, R.J., Durrum, E.L., Zweig, G.: A Manual of Paper Chromatography and Paper Electrophoresis, 2nd ed. New York, Academic Press 1958.

Burchfield, H.P., Storrs, E.E.: Biochemical Applications of Gas Chromatography, pp. 573-588. New York, London: Academic Press 1962.

Chibnall, A.C., Haselbach, C., Mangan, J.L., Rees, M.W.: Studies of the amide and C-terminal residues in proteins. 5. Estimation of asparagine and glutamine residues. Biochem. J. 68, 122-128 (1958).

Clotten, R., Clotten, A.: Hochspannungselektrophorese. Stuttgart: Georg-Thieme-Verlag 1962.

Cramer, F.: Papierchromatographie, 4. Aufl. Weinheim/Bergstr.: Verlag Chemie 1958.

Dadd, R.H.: Insect nutrition: Current developments and metabolic implications. Ann. Rev. Entomol. 18, 381-420 (1973).

Dorfner, K.: Ionenaustauscher, Eigenschaften und Anwendungen, 2. Aufl. Berlin: Walter de Gruyter 1964.

Durzan, D.J.: Nitrogen metabolism of *Picea glauca*. I. Seasonal changes of free amino acids in buds, shoot apices, and leaves and the metabolism of uniformly labelled ^{14}C-L-arginine by buds during the onset of dormancy. Can. J. Botany 46, 909-919 (1968a).

Durzan, D.J.: Nitrogen metabolism of *Picea glauca*. II. Diurnal changes of free amino acids, amides, and guanidino compounds in roots, buds, and leaves during the onset of dormancy of white spruce saplings. Can. J. Botany 46, 921-928 (1968b).

Durzan, D.J.: Nitrogen metabolism of *Picea glauca*. III. Diurnal changes of amino acids, amides, protein, and chlorophyll in leaves of expanding buds. Can. J. Botany 46, 929-937 (1968c).

Durzan, D.J., Steward, F.C.: The nitrogen metabolism of *Picea glauca* (Moench) Voss and *Pinus banksiana* (Lamb.) as influenced by mineral nutrition. Can. J. Botany 45, 695-710 (1967).

El-Basyouni, S.Z., Neish, A.C., Towers, G.H.N.: The phenolic acids in wheat - III. Insoluble derivatives of phenolic cinnamic acids as natural intermediates in lignin biosynthesis. Phytochemistry 3, 627-639 (1964).

Feeny, P.: Seasonal changes in oak leaf tannins and nutrients as a cause of spring feeding by winter moth caterpillars. Ecology 51, 565-581 (1970).

Fowden, L.: The chemistry and metabolism of recently isolated amino acids. Ann. Rev. Biochem. 33, 173-204 (1964).

Hamilton, L.D.G.: The estimation of side chain groups in the intact protein. In: A Laboratory Manual of Analytical Methods in Protein Chemistry Including Polypeptides (eds. P. Alexander, R.J. Block), Vol. II. Oxford, London, New York, Paris: Pergamon Press 1960.

Hamilton, P.B.: Separation of amino acids. In: Ion Exchangers in Organic and Biochemistry (eds. C. Calmon, T.R.E. Kressman), pp. 255-289. New York: Interscience Publ. Inc. 1957.

Hess, G.P., Rupley, J.A.: Structure and function of proteins. Ann. Rev. Biochem. 40, 1013-1044 (1971).

House, H.L.: Insect nutrition. Ann. Rev. Entomol. 6, 13-26 (1961).

Kaludin, K., Kaludin, I.: Untersuchungen über den Stickstoff- und Aminosäuregehalt in den Nadeln von *Pinus silvestris*. Arch. Forstw. 17, 839-844 (1968).

Kawerau, E., Wieland, T.: Conservation of amino-acid chromatograms. Nature (London) 168, 77-78 (1951).

Lipke, H., Fraenkel, G.: Insect nutrition. Ann. Rev. Entomol. 1, 17-44 (1956).

Löyttyniemi, K., Tulisalo, U.: Amino acid composition of the needles of four Norway spruce provenances, and their effect on the occurrence of *Oligonyohus ununguis* (Jacobi) (Acarina, Tetranychidae). Ann. Ent. Fenn. 38, 122-126 (1972).

Lück, H., Pavlik, A.: Die Aminosäuren-Zusammensetzung des Eiweißes rauchgeschädigter Fichtennadeln. Zeitschr. Pflanz. Ernähr. Düng. Bodenk. 114, 113-116 (1966).

Lunderstädt, J.: Zur Nahrungsqualität von Fichtennadeln für forstliche Schadinsekten. 3. Über physiologisch bedingte Veränderungen in Fichtennadeln. Z. angew. Entomol. 73, 340-351 (1973).

Lunderstädt, J., Claus, G.: Zur Nahrungsqualität von Fichtennadeln für forstliche Schadinsekten. 2. Vergleichende biochemische Analysen von Fichtennadeln und Kot der Larven von *Gilpinia hercyniae* Htg. Z. angew. Entomol. 70, 386-403 (1972).

Materna, J., Ryskova, L.: Amino acids in Norway spruce. Comm. Inst. For. Cechosloveniae 4, 76-81 (1965).

McFarren, E.F., Brand, K., Rutkowski, H.R.: Quantitative determination of sugars on filter paper chromatograms by direct photometry. Anal. Chem. 23, 1146-1149 (1951).

Moore, S., Stein, W.H.: Chromatography of amino acids on sulfonated polystyrene resins. J. Biol. Chem. 192, 663-681 (1951).

Painter, R.H.: Resistance of plants to insects. Ann. Rev. Entomol. 3, 267-290 (1958).

Pierce, A.E.: Silylation of organic compounds, pp. 218-243. Rockford, Illinois: Pierce Chemical Company 1968.

Rees, M.W.: The estimation of threonine and serine in proteins. Biochem. J. 40, 632-639 (1946).

Rohringer, R., Fuchs, A., Lunderstädt, J., Samborski, D.J.: Metabolism of aromatic compounds in healthy and rust-infected primary leaves of wheat. I. Studies with $^{14}CO_2$, Quinate-U-^{14}C, and Shikimate-U-^{14}C as precursors. Can. J. Botany 45, 863-889 (1967).

Rydon, H.N., Smith, P.W.G.: A new method for the detection of peptides and similar compounds on paper chromatograms. Nature (London) 169, 922-923 (1952).

Schroeder, W.A.: Separation of peptides. In: Ion Exchanges in Organic and Biochemistry (eds. C. Calmon, T.R.E. Kressman), pp. 299-317. New York: Interscience Publ. Inc. 1957.

Siren, G., Blombäck, B., Alden, T.: Proteins in forest tree leaves. Dept. of Reforestation, Royal College of Forestry, Stockholm, 28, 1-22 (1970).

Smith, I.: Chromatographic and electrophoretic techniques, Vol. 1, Chromatography. London: Heinemann Medical Books Ltd. 1969.

Sobel, A.E., Hirschman, A., Besman, L.: A convenient microtitration method for estimation of amino acids. J. Biol. Chem. 161, 99-103 (1945).

Spackman, D.H., Stein, W.H., Moore, S.: Automatic recording apparatus for use in the chromatography of amino acids. Anal. Chem. 30, 1190-1206 (1958).

Stegemann, H.: Bestimmung von Aminosäuren mit dithionitreduziertem Ninhydrin. Hoppe-Seylers Z. Physiol. Chem. 319, 102-109 (1960).

Thompson, J.F., Morris, C.J., Smith, J.K.: New naturally occurring amino acids. Ann. Rev. Biochem. 38, 137-158 (1969).

Thorsteinson, A.J.: Host selection in phytophagous insects. Ann. Rev. Entomol. 5, 193-218 (1960).

Vetter, W.: Amino acids. In: Biochemical Applications of Mass Spectrometry (ed. G.R. Waller), pp. 387-404. New York, London, Sydney, Toronto: Wiley Interscience 1972.

Walton, H.F.: Principles of ion exchange. In: Chromatography, 2nd edition (ed. E. Heftmann), pp. 287-324. New York: Reinhold Publ. Corp. 1967a.
Walton, H.F.: Techniques and applications of ion exchange chromatography. In: Chromatography, 2nd edition (ed. E. Heftmann), pp. 325-342. New York: Reinhold Publ. Corp. 1967b.
Wehrmann, J.: Methodische Untersuchungen zur Durchführung von Nadelanalysen in Kiefernbeständen. Forstw. Cbl. 78, 77-97 (1959a).
Wehrmann, J.: Die Mineralstoffernährung von Kiefernbeständen (*Pinus silvestris*) in Bayern. Forstw. Cbl. 78, 129-149 (1959b).

CHAPTER 5

Photosynthesis, Respiration, and Dry Matter Production

W. ŻELAWSKI and R. B. WALKER

Introduction

A full description of methods used in studying CO_2-exchange and dry matter production is unnecessary, since a number of excellent reviews have been published on these subjects in recent years (Eckardt, 1968; Heath, 1969; Šetlík, 1970; Šesták et al., 1971; Duvigneaud, 1971). With the exception of the last-named of these, the treatments pertain to plants in general. Thus, emphasis will be placed here upon methodology pertinent to forest trees, advantages and disadvantages of the methods, potential errors inherent in their use, and some improvements developed by the authors and their coworkers. The chapter is directed especially to those who have just begun to work with CO_2-exchange or dry matter production, and to those who intend to use the methods occasionally in relation to tree-improvement work.

Definitions and Recommended Terminology

Terminology used in the study of photosynthesis and related processes unfortunately has never been established with sufficient precision. The short definitions below represent some refinement and modification of the terms recommended by Larcher (1969a) and Šesták et al. (1971), hoping to further their efforts toward uniformity of usage. Following each term is one or more symbols commonly used for it, and after the definition the units which are often used for the quantity are given.

Photosynthesis (Ph, F, P). The general term for the light-dependent biochemical and biophysical process involving CO_2 uptake, splitting of water with the release of O_2, and reduction of carbon to carbohydrate. None of the symbols listed is unique for this process. Assimilation is sometimes used as a synonym, but should be reserved for references to CO_2-incorporation or dry matter accumulation, implying a process of the whole organism, not just the photosynthesizing tissue.

Photosynthetic (or Respiratory) Activity. The rate of the process expressed per unit time and per any biological reference unit, e.g., whole plant, single assimilatory organ, single cell, chloroplast.

Photosynthetic (or Respiratory) Rates. The intensity of the process (usually net flux of CO_2 or O_2) expressed per unit time and per any type of reference unit (dry or fresh weight, surface area, volume, N-content, protein content, etc.).

Gross (or True) Photosynthetic Rate (P_G, Ph_G, P_T, TPS). The photosynthetic rate which would occur if there were no reduction because of CO_2 release and O_2 uptake from respiratory activity. Obviously the gross rate can only be estimated, because photosynthesis and light respiration take place simultaneously and can not be measured separately by any direct technique (usually $mg/dm^2/h$).

Net (or Apparent) Photosynthetic Rate (P_N, Ph_N, APS). The actual measured gas exchange resulting from the simultaneous uptake and release of both CO_2 and O_2. This term may be used in a general sense, but more often refers to the specific measurement of net mass flux of CO_2 into a leaf or branch (mg/dm^2/h or µg/cm^2/min; or frequently mg/dry-g/h).

Net Assimilation Rate (E, NAR). Although this term is sometimes used as a synonym of net photosynthetic rate, it should be reserved for the actual incorporation of CO_2 into the dry matter of the plant (g/dm^2/week) (see "Determination of Dry Matter Increase").

Photosynthetic Capacity (Ability). The rate of photosynthesis attained when all external factors are at or above optimum with the exception of the CO_2 concentration, which is at the natural level (ca. 0.032%).

Potential Photosynthetic Capacity. The rate of photosynthesis attained when all factors, including the CO_2 concentration, are at or above optimum.

Photosynthetic Performance. The actual rate of photosynthesis occurring at a particular time, usually more or less limited by one or more internal or external factors.

Assimilation Number. Photosynthesis per mg chlorophyll (e.g., mg CO_2/mg chlorophyll/h).

Photosynthetically Active Radiation (PhAR, PAR). Visible radiation in the spectral region from 400 to 700 nm (sometimes given as 380 to 720 nm).

Photosynthetic Efficiency. The efficiency of transforming solar energy (or usually photosynthetically active radiation) into the chemical energy of photosynthetic products (% of PhAR energy converted into chemical energy; or often moles of CO_2 fixed per Einstein of incident or of absorbed PhAR).

Dark Respiration (R_D). CO_2-evolution or O_2-uptake resulting from biological oxidation of organic substrates through the Krebs cycle. In green tissue, this may be depressed by light even at low irradiance (units same as for Ph_N).

Photorespiration or Light Respiration (R_L). Respiration occurring in illuminated green tissue which is different from dark respiration not only in rate but also as a biochemical process (Zelitch, 1971). Its rate apparently increases with increasing irradiance, but since measurement of this rate can be made only by one of several laborious indirect methods (Jackson and Volk, 1970), determinations on trees have been made only occasionally (units similar to those for Ph_N).

Production (P). The total amount of organic matter (biomass) accumulated on a unit area of ground surface (kg/ha; g/m^2; kcal/m^2).

Productivity ($\overline{P}$). The rate of production per unit time and per any reference unit, but usually per area of ground surface (e.g., g/m^2/day; kg/ha/year).

Photosynthetic Productivity and Respiratory Productivity. Various measures expressing the effectiveness of the process, e.g., percent of assimilated carbon incorporated into the yield, measures relating photosynthetic and/or respiratory rates to the production of organic material, etc.

Water Use Efficiency. The ratio of dry matter produced to the amount of water transpired (e.g., g dry matter/g H_2O). Sometimes the inverse of this ratio is utilized, the amount of water used to produce a unit of dry matter (e.g., g H_2O/g dry matter); this is called the water requirement, which can be calculated by multiplying the transpiration ratio (net H_2O efflux/net CO_2 influx) by 1.5 since dry matter produced is usually about 0.67 of the weight of the CO_2 absorbed (Levitt, 1972).

Reference Units and Bases for Expressing Rates

The variety of bases on which gas exchange rates may be expressed, and conversions between these are often difficult. Some advantages and disadvantages of the various methods of expression are given in the discussion below. Certainly a good understanding of these methods is necessary for handling and properly interpreting data.

Fresh Weight. The fresh weight of photosynthetic organs is less reliable than dry weight because of diurnal fluctuations and seasonal changes in water content, thus it cannot be recommended as a basis for expressing rates. Through the course of warm days the water content often drops appreciably as water deficit develops, and also toward the end of the growing season water content usually decreases. In both instances the reduced water content would cause photosynthetic rates based on fresh weight to be overestimated in comparison with those for tissue with higher water contents.

Dry Weight. The use of dry weight eliminates the difficulty mentioned above with fresh weights, but does not take into account the succulence, shape, content of photosynthesizing tissue, and distribution of this tissue in the leaf or other green organ. Further, stomatal distribution should be considered, because of its inherent influence on leaf resistance.

As a growing season progresses, the dry weight of leaves increases because of the absorption and retention of Ca and other immobile mineral elements. Also, with evergreen leaves, the ash content increases with age from year to year. Further, recently expanded leaves increase in non-photosynthetic components (especially cellulose, cutin and secondary products) during maturation. In addition, leaves may vary in weight during the course of a 24-h day as well as during the year (especially if evergreen) because of changes in the starch or other stored food content. Since any of these variations in weight may occur with little or no accompanying change in surface area or properties of the photosynthetic apparatus, expressions of photosynthetic rate on a weight basis have limited meaning. The meaning can be enhanced by including descriptive information on pertinent factors which may affect weight or the relation between weight and area. Despite its limitations, dry weight has been the most common basis for expressing gas exchange rates primarily because it is easy to determine, but also because it is related to biomass values, which are almost always given on a weight basis.

Surface Area. This is a frequently used basis for expressing rates, primarily because of its direct relation to light interception. Also, energy exchange, transpiration, and diffusion resistance of the leaf are all usually expressed on an area basis, so it is often consistent to express CO_2-exchange rates in the same terms.

Often the question arises whether the area used for calculations should represent a <u>one-side</u> or <u>two-side</u> (total) surface. The one-side area appears reasonable for dorsiventral leaves which receive radiation primarily on one side, and on which the stomata are often mostly on the lower surface. On the other hand, with vertically-oriented, fascicled, thick or cylindrical assimilatory organs which often have stomata on all of the surface, a better argument can be made for use of two-side (total) area. In any case, a clear statement should be made of the method of computation.

Measurement of the surface is not always easy, but there is a variety of photoplanimetric and other techniques available (Květ and Marshall, 1971). Often the projected area is used as the "one-side" area, but for greatest accuracy the leaves should be oriented in a plane and a correction applied for the area on the sides of the leaves, if the thickness is an appreciable proportion of the width. For complicated surfaces such as heather or coniferous branchlets or fascicles, the glass bead adherence technique (Thompson and Leyton, 1971) is often useful.

<u>Photosynthetic Apparatus and Chlorophyll</u>. Since only a portion of the leaf is functioning as the photosynthetic apparatus, it would be desirable to express photosynthetic rates in some kind of measure related to this active part. The volume of the assimilatory tissue might be a possibility, but determination of this volume is difficult, laborious, and not very accurate. The number or volume of chloroplasts are other good possibilities, but their determination is also difficult, and for many kinds of assimilatory organs not feasible because of breakage of chloroplasts during extraction.

Fortunately, the chlorophyll content is easy to calculate from spectrophotometric readings of an acetone extract from the chlorophyllous tissue (Šesták, 1971), and this content characterizes the amount of assimilatory tissue reasonably well. An expression of photosynthetic rate on the basis of chlorophyll content is sometimes called the <u>assimilation number</u>.

<u>N-Content or Protein Content</u>. These give a measure of the amount of protoplasm present, so expression of rates on such a basis is particularly suitable if the focus of the study is on respiratory or other enzymatic activities. However, it should be remembered that not all living cells of the leaf have chlorophyll.

Also, the expression of photosynthetic rate with reference to a particular enzymatic activity, e.g., carboxylation, can be useful but suitable only for detailed work in the laboratory.

<u>Morphological Basis</u>. Sometimes it may be useful to express CO_2-exchange rates per plant, per leaf or other organ, or even per cell. In such instances the use of the term "photosynthetic (respiratory) activity" seems especially appropriate. Such measures are obviously very sensitive to the size of the object, thus are useful only when sizes are uniform or when notation is made of size differences.

<u>Choice of the Basis for Expressing Rates</u>. Although this choice may sometimes be arbitrary, there are some principles which may aid in selection. Surface area is particularly appropriate if factors related to surface may be limiting the rate: especially the absorption of light, and also the entry of CO_2, which is strongly affected by stomatal distribution and leaf resistance. If light is limiting, chlorophyll content might seem to be an appropriate basis, but in

fact pigment levels are usually high enough that they have little or no effect on photosynthetic rate. On the other hand, if chlorophyll formation is being limited strongly by some factor such as temperature or mineral nutrition so that the leaves are chlorotic, chlorophyll content may make the best reference for calculation. Weight should always be recorded to link rate with biomass, but is usually less desirable than surface area as a basis for expressing rates. This is especially true if comparisons are to be made between varieties or species. When feasible, the necessary data should be taken for expression or conversion of rates in terms of all of the main bases: surface area, dry weight, and chlorophyll content. These data should be included in publications.

Photosynthetic Capacity and Photosynthetic Efficiency

Characteristic Photosynthetic Capacities of Trees. As a rule, comparisons between plants have been made in terms of maximum values or photosynthetic capacities. With special reference to forest trees, the most recent compilation of such capacities was made by Larcher (1969b); a summary of these data appears in Table 1.

Table 1. Summary of photosynthetic capacities of trees

Type of trees and their origin	Photosynthetic capacity ($mg/dm^2/h$ (one-side basis))
Deciduous	15 to 25
Evergreen broad-leaved from temperate and warm temperate regions	14 to 17
Evergreen broad-leaved from Pacific warm temperate regions	11 to 14
Tropical rain forest	18 to 22
Trees and shrubs from monsoonal and semi-arid regions	10 to 11
Coniferous	12 to 18

If leaves are thin or have a high water content, as in many crop plants and woody mesophytes, the dry weight of one dm^2 of foliage (one-side area) will often be less than one g, thus the mg/h values will be larger if expressed on a dry weight than on an area basis. In xeromorphic and needle leaves, however, one dm^2 (one-side area) often will weigh more than one g, making mg/h smaller in the dry weight than in the area calculation. If recalculated on a two-side (total) area basis, the values in Table 1 would drop to one-half of their magnitude. For example, if an average conifer exhibits a rate of 15 $mg/dm^2/h$, with a dm^2 of foliage having a dry weight of 1.5 g, the comparative calculations would be: $mg/dm^2/h$ (one-side area) = 15; $mg/dm^2/h$ (two-side area) = 7.5; mg/dry-g/h = 10.

Comparisons of Photosynthetic Capacity. As indicated in Table 1, reported photosynthetic capacities may differ appreciably between species or groups of species. However, interpretation of this variation is often difficult because of disparity in expression of rates and differences in the physiological state of the plant material (see "Reference Units and Bases for Expressing Rates").

An additional feature which makes comparisons of photosynthetic activities difficult is mutual shading of the assimilatory organs. This is particularly noticeable in some of the pines (Kramer and Clark, 1947; Kramer and Kozlowski, 1960), resulting in increases in photosynthetic performance with irradiance up to the intensity of full sunlight. Using an artificial radiant source, Ludlow and Jarvis (1971a) observed that net photosynthesis was substantially higher in *Picea sitchensis* at given irradiances if the light were diffuse rather than direct. Also Żelawski et al. (1973), using a special spherical source of high-intensity diffuse multi-directional light, reported net photosynthetic rates as high as 33 mg/dry-g/h in 5-week old *Pinus silvestris* seedlings, which is at least twice as high as values usually reported for this species (Larcher, 1969b).

Potential Photosynthetic Capacity. The potential photosynthetic capacity, expressing the rate when all factors including CO_2 are optimal, has limited application. The natural levels of CO_2 are ordinarily relatively low and stable, and well below optimal, if temperature and irradiance are favorable. However, the potential capacity is of interest for special purposes, such as the study of diffusion resistances or in evaluating the effect of additions of CO_2 to the air, as in greenhouses. For example, Koch (1969) reported a marked rise in photosynthetic capacity of several forest tree species when the CO_2 concentration was increased, with the maxima at about 5 to 8 times the normal atmospheric level, with declines at higher concentrations. Also, he noted a gradual decline in the rate of CO_2 uptake in *P. silvestris* with time at 0.16 vol % CO_2, attributing this decline to increasing stomatal resistance. Also using *P. silvestris* with an elevated CO_2 concentration (0.5 vol % in the gaseous phase), but measuring manometrically, Żelawski et al. (1974) observed the highest rates of photosynthesis ever reported for this species. Expressed in ml O_2/dry-g/h the rates were: for juvenile needles, in July as high as 27, and in October as high as 18; for secondary (fascicled) needles, in June as high as 19, and in October as high as 10 (1 ml $O_2 \triangleq$ 2 mg CO_2).

Photosynthetic Efficiency. Much interest exists in photosynthetic efficiency, because a more efficient plant in this sense will produce more dry matter from a given amount of incident radiation (i.e., PhAR). Sometimes incident radiation is corrected for reflection and transmission, so that calculations are on the basis of absorbed PhAR (Lemon, 1967). Although efficiencies are usually higher with elevated CO_2 concentrations, most are determined at the normal atmospheric level at which plants usually perform. If light is limiting the rate, the calculated efficiency will be higher, because the fraction of energy converted to dry matter is greater. This is true, even though photosynthetic performance and productivity increase on up to the optimal or "saturating" level of irradiance (Kok, 1960). Values for efficiencies can be calculated directly from the slope of the straight-line portion of the net photosynthesis vs. irradiance curve (Ludlow and Jarvis, 1971a). Also, regardless of the photosynthetic capacity of the plant, only if the foliage is dense enough to fully absorb the incident radiation, if below saturation, will the highest efficiencies be calculated. Further, if the radiation of the entire year is counted, evergreen plants in mild climates have an advantage over deciduous plants, because they can photosynthesize over a greater portion of the year.

With the limitations and to some extent artificialities which enter into the determination of photosynthetic efficiency, comparisons between species are difficult to make realistically. If efficiency is calculated in terms of net CO_2 uptake per unit of PhAR (e.g., mg/dm^2/h/

Einstein), the various problems previously discussed with respect to expressing rates may make comparisons difficult, but in any case differences in photosynthetic capacity will be very important in establishing the efficiencies. Most of these comparisons are made from results obtained during relatively short periods of measurement, so that influences of length of growing season and of cooler or duller periods of the year often are not considered. In available comparisons the crop plants, which often have photosynthetic capacities as much as twice those of forest trees (Larcher, 1969b; Zelitch, 1971), appear to be more efficient. However, Krueger and Ruth (1969) pointed out that conifer seedlings often photosynthesize at rates comparable to those of woody mesophytes.

When comparisons of the photosynthetic efficiency are made on the basis of the percent of the incident seasonal or annual PhAR which is converted to dry matter, the values of 2 to 2.5% reported for forest trees exceed those for crops such as rice and sugarcane (Hellmers, 1964). He mentioned several factors which may account for the higher efficiencies of forest stands in these comparisons. The forest canopies often are fully closed, resulting in full interception of the light. Also, the level of irradiance is frequently low for much of the year, which raises the calculated values, as mentioned above. Further, in forest canopies CO_2 levels build up at night from soil and understory respiration, thus stimulating the rate of photosynthesis the next day.

The photochemical and production efficiencies discussed above are based primarily on net photosynthesis, thus take biochemical efficiency into account only indirectly. Neglecting some subdivisions and variations, plants can be classified into one group with low photorespiration (maize, sugarcane, some members of the Chenopodiaceae) which are particularly efficient at higher temperatures and high irradiances, and another group including most crop and wild plants of temperate origin which have high rates of photorespiration (Jackson and Volk, 1970; Zelitch, 1971). Photorespiration in the latter group may be greater than net photosynthesis under some conditions, and certainly several times the rate of dark respiration. As far as is known, all forest tree species have this substantial photorespiration. However, under the relatively cool temperatures and moderate irradiances commonly experienced by most forest trees in the temperate zones, photorespiration does not prevent the plants from attaining productivities which are often high, probably as high as plants with low photorespiration would reach under similar conditions.

In plants having low photorespiration there is a very effective recycling of CO_2 released in the leaf from respiration back into photosynthesis, as indicated by the very low CO_2 compensation points. Another type of recycling of CO_2, quite unrelated to photorespiration, may occur if part of the CO_2 from stem respiration is carried upward to the leaves in the transpiration stream (Żelawski et al., 1970).

Photosynthetic efficiency is also reflected in the water use efficiency or water requirement (see "Definitions and Recommended Terminology"), although the latter also are dependent on diffusion resistance of the leaves and meteorological effects on transpiration as well as the sensitivity of the photosynthetic mechanism to water deficit. Forest tree species, with water requirements ranging from about 350 down to under 200 (Polster, 1950), compare favorably with the range of 250-350 recorded for the group of very efficient crop plants (e.g., maize, sugarcane) mentioned above (Black, 1973).

Genetic Variation

Variation in photosynthetic capacity between forest species was discussed earlier ("Photosynthetic capacity ... efficiency"), but often in forestry research and practice differences within a particular species are just as important. Numerous authors have shown that such varietal and provenancial differences do exist (Kozlowski and Keller, 1966; Żelawski, 1967, 1968; Larcher, 1969b).

Even though intraspecifc differences in photosynthetic capacity have been reported frequently, in any particular instance it is not easy to ascertain whether they are inherent in the forms studied, or may be attributable to more or less temporary nonuniformities. Such variation may arise from dissimilarities in phenology or stage of development (as in altitudinal ecotypes), in mineral nutrition, in proportion of foliage of different years, in succulence of foliage (as influenced by water relations), and the like. Also single measurements may fluctuate considerably, so means of a number of values, or comparisons made at more than one time, often will give a more valuable index of intraspecific variability.

A useful approach for studying genetic variation in photosynthesis is investigation of photosynthetic behavior in relation to environmental factors. In a number of studies genetic differences in response to temperature, irradiance, frost resistance, and water deficit have been reported (see reviews cited above). Since these reviews appeared, intraspecific variability has been reported in relation to a number of factors, viz.: shade tolerance and the ratio of photosynthesis to respiration in *P. silvestris* (Żelawski et al., 1968); influences of irradiance and CO_2 concentration on *P. silvestris* (Gordon and Gatherum, 1969); drought influences on *Pseudotsuga menziesii* (Zavitkovski and Ferrell, 1970); CO_2 concentration and temperature effects on photosynthesis and photorespiration of Populus clones (Luukkanen and Kozlowski, 1972); effects of temperature, conditioning temperatures, and age on photosynthesis and respiration in *P. menziesii* (Sorensen and Ferrell, 1973); differences in *P. sitchensis* correlated with leaf resistances (Ludlow and Jarvis, 1971a).

As discussed above in the section on photosynthetic efficiency, the rates of photorespiration and dark respiration as well as photosynthetic capacity are important in productivity. Genetic variation in respiration is a subject deserving more attention, not only because of its importance in productivity but also because of the many physiological processes dependent on the release of energy and on compounds produced in respiration. In this connection those processes involved in growth are the ones of most immediate interest.

Age and Stage of Development

Changes in photosynthetic activity during formation and development of foliage over the growing season are relatively well known (Kramer and Kozlowski, 1960). In a detailed study which supports this general statement Dickmann (1971) found that the photosynthetic rate of leaves of *Populus deltoides* seedlings was negative (i.e., evolution of CO_2 occurred) when the leaves were very small. When the leaves reached 1/20th of their full size, the rate became positive. The activity continued to increase until the leaves were fully expanded, stabilized for some time, then finally began to decrease late in the season. In the same species, Larson et al. (1972) showed that the percentage of

^{14}C fixed in the tip of the lamina declined with increasing age of the leaf, while the percentage in the base rose; this is evidence for uneven development of photosynthetic activity within a single leaf. In *P. menziesii*, Künstle (1971) found very low photosynthetic performance in the recently developed needles in June, but much higher values later in the season.

In seedlings of *P. silvestris*, Żelawski (1972) described in detail the changes which occurred in the shoots during their first growing season. Using the mean dry weight per needle (x) as a measure of the developmental stage of the shoot, he found the average photosynthetic rate, P_N (mg CO_2/dry g/h) to be equal to:

$$P_N = \frac{17.6}{x} + 5.7.$$

The photosynthetic rate of the shoot, an average value for all needles expanded at different times, decreased toward the end of the growing season by about 50%. This resulted both from declining performance of younger needles as they matured, and from the increasing influence of accumulated mature foliage. A further decline in photosynthetic rate of the juvenile (non-fascicled) needles occurred during their second growing season (Żelawski et al., 1971), and a similar decline also was observed during this second growing season in the photosynthetic rate of the secondary (fascicled) needles developing from the initials in the buds of the previous year (Żelawski et al., 1969, 1971), again with a decline of about 50% by the end of the growing season. Also, in these studies the photosynthetic capacity expressed on a dry weight basis was about twice as high in juvenile as in adult foliage. However, differences in structure and water content between the two types of foliage make the rates on a weight basis diverge unduly.

In evergreens, especially the conifers, there has been considerable interest in the effect of age of foliage on the photosynthetic capacity. The work of Freeland (1952) with a number of coniferous species established that the capacity is highest in fully expanded needles of the current growing season, then declines slowly. It commonly reaches about half of the original capacity after three years, but this needs to be verified for a particular species under the existing conditions. For example, Brix (1971) found that the 4-year old foliage from 24-year old *P. menziesii* trees still had about 60 to 70% of the original capacity. In a similar vein, Salo (1974) noted little if any decline in capacity of 1-year old Pseudotsuga foliage as compared with fully expanded needles of the current year. On the other hand, Ludlow and Jarvis (1971a) observed a sharp decline in photosynthetic capacity in *P. sitchensis* with age of foliage, and were able to correlate this decline with increasing leaf resistances. The age of the plant itself may have some further influences on photosynthetic capacity (Iwanoff and Kossowitch, 1929).

Also, in evergreens it is common to observe marked changes in photosynthetic activity during the calendar year (Pisek, 1960). Often there is a winter depression of photosynthetic performance followed by a springtime recovery. Exact measurement of the depression and recovery may be confounded by the influence of aging of the leaves. In mild climates the wintertime depression may be merely low performance because of reduced irradiance or temperature without photosynthetic capacity being reduced, but in colder climates capacity is also commonly depressed (Pisek, 1960).

Materials and Methods

Experimental Material

General Considerations. In the choice of experimental material, the principles covered in the two previous sections on genetic variation and stage of development are pertinent. Unless genetic variation itself is under study, the material should be chosen for genetic uniformity if feasible. Likewise unless stage of development is under consideration, material in a well-defined stage of development and stable metabolically such as fully expanded foliage of the current season should be selected. Especially to be avoided are very young shoots, or shoots with developing buds, since photosynthesis will be partially masked by high respiratory rates (Neuwirth, 1959). At least for conifers, the period of relatively stable rate occurs in the middle to late summer after the needles stop growing and continues until the late autumn or winter depression sets in (Żelawski and Góral, 1966b; Larcher, 1969b; Żelawski et al., 1969, 1971; Künstle, 1971; Żelawski, 1972; Salo, 1974). However, no definitive advice can be given on the best times for taking measurements, especially since summer droughts, early frosts or other influences may sharply affect activity.

Morphology and Position. To make sampling most representative, the morphological and physiological differentiation of leaves within the crown and along the shoots should be taken into account. Leaf size, internode length, pattern of branching, water deficits, morphological adaptation (see below), and proportions of different ages of foliage may all vary from one part of the crown to another, especially from the upper to the lower crown.

Even in a single shoot there may be a pattern of variation. For example, in *P. silvestris*, in which the leaf initials are set in the buds of the previous year but leaf development is more or less simultaneous in the spring, Żelawski et al. (1971) observed morphological and physiological differences in secondary (fascicled) needles along the shoots of second year seedlings. They found that rates of photosynthesis (dry weight basis) were lower in the upper than in the middle and basal parts of the shoot during the whole growing period except for the earliest stages of needle growth. Thus sampling from the middle or lower parts of these shoots was advisable for best representation of the material.

Normally chlorophyll contents of photosynthetic organs are high enough that there is no limitation of photosynthesis. However, if chlorophyll contents drop to low values, as in leaves chlorotic from mineral deficiency or disease, photosynthesis will be reduced, especially at higher irradiances (Kozlowski and Keller, 1966).

Chlorophyllous tissues other than leaves, e.g., bark of young shoots of *Populus tremuloides*, usually do not play any significant role in the uptake of CO_2 from the atmosphere. Rather, they recycle a part of the CO_2 released in dark respiration, and may make saving in the carbohydrate balance of the stem (Żelawski et al., 1970; Schaedle, 1975). Usually the gas exchange of twigs is measured simultaneously with that of the leaves, and their contribution or withdrawal neglected as being very small. This is usually justifiable, but the amount of the contribution should be checked for particular material being used.

Adaptations. A common visible adaptation is the development of thicker leaves with heavier cuticle and somewhat lower chlorophyll content in open exposures (so-called sun leaves), and thinner sometimes broader leaves with less cuticle and more chlorophyll in shaded locations (so-called shade leaves). With greater or lesser shading intermediate types may develop. There is ample evidence that these morphological manifestations are accompanied by differences in photosynthesis, with sun leaves showing higher capacities (Larcher, 1969b) and better performance at higher irradiances, but with shade leaves showing better performance at low irradiances (Kozlowski and Keller, 1966; Krueger and Ruth, 1969). Rates for shade leaves appear higher in relation to sun leaves if calculations are on a weight basis than on an area basis because the shade leaves are usually thinner and more succulent.

Adaptations to exposure to higher temperatures by increase in temperature optima for photosynthesis and respiration, and decreases in these optima through exposure to lower temperatures have been observed in a number of species (Mooney and Harrison, 1970; review by Pisek et al., 1973). Also adaptations to soil moisture regimes through changes in stomatal behavior, xeromorphy in leaves, and other possible changes have been reported in many instances (Kozlowski and Keller, 1966; Żelawski et al., 1969; Zavitkovski and Ferrell, 1970; Levitt, 1972).

Age of Plant and Use of Excised Branches. Seedlings of a few weeks up to a few years in age have been used frequently in studies on woody plants. Clearly, seedlings have advantages in bringing together material from widely separated regions, easy manipulation in experimental procedures, uniformity of replicate plants, and in early assessments of physiological attributes of provenances. On the other hand, they often differ morphologically from mature plants, are known to make better use of weak irradiances and to photosynthesize and respire at higher rates than older plants (Larcher, 1969b), and often offer too little tissue for concomitant determinations of chemical composition, water potential, and histological structure. An obvious alternative to the use of seedlings is work on intact larger plants in the open, but experimental work in the field is very demanding of labor and equipment (Koch et al., 1971). Because of these demands, the use in the laboratory of branches excised from larger plants growing in the field, has long been practiced.

Measurement of gas exchange on intact leaves or branches is inherently safer, because of less disturbance of physical and metabolic processes, than that on excised branches. Consequently, the use of excised material often has been avoided because of concern that the water balance could not be maintained properly, or that gas exchange might be disturbed by the accumulation of photosynthetic products (Neales and Incoll, 1968). However, Larcher (1963) made a detailed study of the CO_2 uptake of excised branches from several species and concluded 1. that there is often a change, up or down, in the photosynthetic rate immediately after excision, but that storage in the dark for a few hours with the cut stem in water restores the rate to about the same as before excising, and 2. there is no basic reason to avoid the use of excised branches for photosynthetic determinations. In recent studies with conifers, a number of workers have reported the use of excised branches (Brix, 1971; Ludlow and Jarvis, 1971a; Salo, 1974), and such material has long been used for photosynthetic studies by Pisek and his coworkers (Pisek, 1960).

Certainly careful precautions are necessary in collecting, storing, and handling excised material to ensure good results. Favorable water status is easier to maintain with conifers or other xerophytic types than with most broad-leafed species. Much larger branches should be collected than are intended for use, then smaller branches cut off for experimentation after the storage period, which should not exceed a few days. Also, excised material should not be used for long term experiments or those in which high rates of transpiration occur.

Conditions in Plant Chambers

Enclosing of plant material into a chamber for measurement of gas exchange, especially out-of-doors, tends to cause an increase in air and leaf temperatures which may be extreme in full sun, an increase in the humidity of the air because of transpiration in a volume with limited ventilation, and a depletion of CO_2 in the light because of photosynthesis. Each of these effects causes a measured photosynthetic rate to deviate from the rate in unenclosed leaves nearby. For ecological studies, it is often desired to simulate in the chamber the outside conditions as closely as possible, as these deviations are usually considered undesirable. For physiological studies, the deviations may not be important if they are within the normal physiological range for the experimental material and can be recorded accurately (Hellmuth, 1967). However, there is a great advantage in being able not only to measure but also control these variables. As a result a great deal of effort has been expended on development of gas-exchange chambers and accessory equipment for control and measurement of these factors (see reviews: Larcher, 1969a,b; Jarvis, 1970; Jarvis et al., 1971). One widely used commercial unit designed by Koch and Walz (1967) and manufactured by the Siemens Corporation can be operated in a mode to follow atmospheric temperature and relative humidity, or set to operate at chosen temperature and relative humidity (Lange et al., 1970). However, no matter how sophisticated and complicated, no equipment has fully overcome the problems caused by the inescapable necessity of enclosing the foliage.

In the laboratory, temperature control of gas-exchange chambers is easier than in the sun, although a filter to absorb the bulk of the infrared radiation of the light source is usually needed. Water at the desired temperature, circulating in a jacket around the chamber, is effective in temperature control of the air. The air, with its low heat capacity, quickly attains the temperature of the jacketed walls if there is good turbulent mixing. For physiological work, the leaf temperature should be used, as it may vary considerably from the air temperature, depending on irradiance and rate of transpiration. Thermocouples clipped to the bottom surface of the leaf probably give the most reliable measurements of leaf temperature, since thermocouples inserted into the leaf may be affected by traumatogenic respiration.

Humidity of the air in the chamber is much more difficult to control than temperature, but is very important in photosynthetic measurements because of its influence on stomatal aperture (Schulze et al., 1972). They worked with desert species, but Tranquillini (1963) also found conifer seedlings to be sensitive, with dry air reducing photosynthesis except at low flow rates at which transpiration could have raised the humidity appreciably. Working with *P. sitchensis*, Ludlow and Jarvis (1971a) observed marked increases in leaf resistance when water vapor pressure difference between the leaves and the air exceeded about 11 millibars. Although *P. sitchensis* may be exceptionally sensitive, it still appears wise to maintain as small a vapor pressure difference between leaves and air in the chamber as feasible. The vapor pressure difference is a better guide than relative humidity, because the latter changes with temperature. Simultaneous measurement of transpiration rate along with CO_2 exchange also provides a check on changes in leaf resistance. The humidity of the chamber may be controlled by saturating the incoming air at a particular temperature, somewhat lower than the chamber temperature, although this involves some inconvenience and necessitates drying the air stream before it enters the CO_2 analyzer.

Carbon dioxide concentration also deserves close attention in the chamber. If rates of air flow are high, in the range of several l/min or higher, laminar air streams without sufficient turbulence may cause "dead" spaces to form (Gaastra, 1959). Such an occurrence is favored by any obstruction to free circulation of air. This possibility, as well as the possibility of boundary layers increasing the leaf resistance makes the use of an internal circulating fan sufficient to produce an air movement of at least 0.5 m/sec imperative. If air flow rates are low, in the range of 0.5 to 2 l/min, and there is a sizable amount of rapidly photosynthesizing material in the chamber, depletion of CO_2 may be sufficient to depress the rate appreciably if irradiance and temperature are near optimal levels. If the depletion does not exceed 10% (20-30 µl/l), which can be achieved by adjusting either the flow rate or the amount of plant material in the chamber, the effect on the measured rate will be small and can be neglected. Effects of artificially increased CO_2 concentrations have already been discussed in "Photosynthetic Capacity and Photosynthetic Efficiency".

Scanty information is available on the influence of the rate of air movement in the chamber. The ordinary air passage through the chamber gives very slow rates of air movement; a flow rate of 33 l/min through a small chamber of 10.2 cm^2 cross-sectional area gave an air velocity of only 0.56 m/sec (Tranquillini, 1964). In experiments on the influence of wind, using field-grown plants in the Patscherkofel climatized wind tunnel, Tranquillini (1969) and Caldwell (1970) found reductions in stomatal aperture and gas exchange to begin above 1 to 2 m/sec even for the sensitive *Rhododendron ferrugineum*. Only by excessive stirring with internal fans would such air velocities be reached in gas exchange chambers of customary size with a few l/min air flow through the system.

Irradiance

There is no ideal artificial light source for photosynthetic measurements, because in no one type are light quality, attainable intensity, and proportion of photosynthetically active radiation to infrared very similar to sunlight. Hence a variety of types are in use depending on cost, availability, and preference: xenon arc, fluorescent tubes, incandescent bulbs, mercury vapor lamps with luminophor enriching the spectrum, and others. Often exact spectra and specifications of these sources are difficult to obtain from manufacturers.

If a unidirectional light source is used, intensity decreases in proportion to the square of the distance, so accurate recording of the latter is important. This may be a problem if all foliage is not in a plane at the same distance from the source. Also the source may vary in intensity over the area it irradiates; this is particularly true near the ends of fluorescent tubes.

Mutual shading of assimilatory organs is another factor contributing significantly to the total irradiance received by the plant material (Kramer and Clark, 1947). Multidirectional irradiance is more similar to the diffuse sunlight often occurring under natural conditions.

Surrounding the plant chamber with a number of smaller lamps provides better distribution of irradiance than a single one or several stronger sources.

Another way to provide multidirectional diffuse irradiance is by multiple reflection of the light beam in a photometric sphere (Żelawski et al., 1973). The white painted mat inner surface reflects light and irradiates the plant chamber in the center (Fig. 1). The plant is protected from direct irradiance from the source, and theoretically receives equal energy from all directions. In practice, the distribution of irradiance within the sphere is interfered with somewhat by the presence of the plant material or equipment in the chamber, but this is insignificant at irradiances above the saturation level. However, because of this effect, an experimental object must be limited to a space approximating 1/10th of the diameter of the sphere.

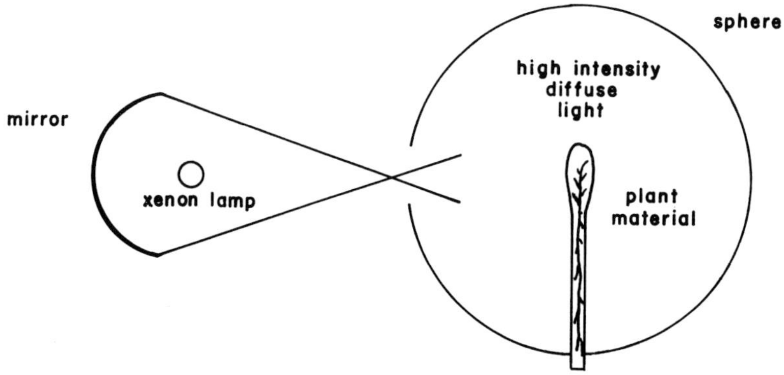

Fig. 1. Diagram of a spherical system for illuminating material with high intensity diffuse radiation (Żelawski et al., 1973). The radiation from the Xenon lamp is directed by the mirror into the sphere, where it is reflected in all directions from the walls

Removal of the excess heat from radiant sources is usually achieved with water jackets or transparent filter basins filled with running water, or sometimes the lamps are bathed by the water. Optical filters which cut off infrared radiation and interference mirrors which reflect this radiation backwards are useful only at very high irradiances because photosynthetically active radiation is also reduced by such devices.

Various units are in use for expressing irradiance, and sometimes confusion may occur, particularly if notation is not made as to whether total radiation or only photosynthetically active radiation is being indicated. If total radiation should be used (as in studies of transpiration, influences on leaf temperature, and energy balance) units of energy or power as calories or ergs per unit area and time or watts per unit area are appropriate. However, for photosynthesis, which involves photosynthetically active radiation (PhAR), the preferred way to express irradiance is in units of photon flux density such as micro Einsteins/m^2/sec. If calories, ergs, or watts are used for irradiance, a clear statement should be made about the spectral band being measured, and its relation to PhAR kept in mind. Illuminance values such as foot-candles or lux have often been used to express light intensity, since they can be easily measured with a photoelectric light meter. They have the advantage of including little if any far-red or infra-red radiation, but cannot be precisely converted to energy or flux units. Nonetheless, ready and sufficiently accurate conversion can be made of illuminance readings to PhAR flux values by making a calibration curve for the illuminance meter against a PhAR flux meter (such as that made by Lambda Instruments Corp., Lincoln, Nebraska, USA). Approximate equivalents for the most common units used for PhAR are given in Table 2.

Table 2. Approximate PhAR equivalents

0.45 cal $cm^{-2}min^{-1}$ [a]	7000 ft-c
1500 μEinsteins $m^{-2}sec^{-1}$	75000 lux
315 W m^{-2}	3.1 x 10^5 ergs $cm^{-2}sec^{-1}$

[a] 1.0 cal $cm^{-2}min^{-1}$ total solar radiation, commonly reached on sunny summer days in the temperate zone, ≅ 0.45 cal $cm^{-2}min^{-1}$ in PhAR. The langley (ly) is sometimes used to designate cal $cm^{-2}min^{-1}$.

Measuring the Uptake and Evolution of Carbon Dioxide

The various physical, physico-chemical, and chemical methods for the determination of CO_2 and O_2 and measuring the rates of photosynthesis and respiration have been reviewed recently in detail (see review by Čatský et al., 1971; Janáč et al., 1971). Thus, in the limited space here emphasis will be on methods more commonly used, with only brief comment on the others. However, some mention will be made of certain older methods, usually more laborious than the new ones, but which may be used for special purposes or if more elaborate equipment is unavailable.

Conductimetric and Chemical Methods. In the conductimetric method, CO_2 is absorbed in deionized water, followed by measuring the electrical conductivity, which is proportional to the CO_2 concentration. This method has advantages in field measurements because of its simplicity of design and operation, insensitivity to water vapor, and the possibility of continuous multi-channel work.

Colorimetric and electrometric measurements are based on the proportionality between the pH of a dilute bicarbonate solution and the partial pressure of CO_2 in the air which is in equilibrium with the solution. The precision of the determination of pH limits the accuracy of this method, but by using a portable pH meter or even colorimetric pH indicators it is possible to measure concentrations of CO_2 even under field conditions without the necessity of a power supply.

Chemical determination of CO_2 based on titration of excess hydroxide after precipitation of the carbonate is an old method which is laborious and not very accurate, because of its numerous steps. However, if decrease in conductivity of the hydroxide solution is measured instead of the titer, the method may be applicable for special purposes under laboratory conditions. There are several modifications of the conductimetric method to avoid bubbling between the electrodes, which often disturbs the measurements.

Katharometer Method. Gas analysis by the katharometer (diaferometer or hot wire) method is based on differences in thermal conductivity of gases. Although this method is not widely used yet for measuring CO_2 concentrations, it could be advantageous because of high sensitivity, short time response, linearity of the calibration curve, and relative simplicity of the instrument. A major disadvantage is the simultaneous sensitivity to water vapor and to O_2, but this can be circumvented by drying the air and by taking into consideration the ratio

$$\frac{\Delta CO_2}{\Delta O_2}$$

for calculations.

Infra-red Gas Analysis. Although in special instances one or more of the methods mentioned above may be used, the one most often utilized in gas exchange research is that based on the absorption of infra-red radiation by CO_2. A wide variety of infra-red gas analyzers (IRGA) is commercially available, with many of them suitable for physiological work. The method is advantageous because of its high sensitivity, stability, possibility of continuous recording, and feasibility of using multiple channels. Its disadvantages are in the cost and complexity of the equipment, possible interference from non-physiological CO_2 sources, some interference from water vapor, and the necessity of a stable power supply. For use in the field special caravans are commonly built using mobile generators for the power source (e.g., Neuwirth, 1965; Lange et al., 1970).

Systems for CO_2 analysis using infra-red analyzers are commonly set up either as closed or open. The differences in arrangement of components for these two types are shown in Fig. 2a and b, and a short discussion of the characteristics and applications of each follows below.

In a closed system the air circulates between the analyzer and the plant chamber without contacting the outside atmosphere (Fig. 2a). Time is measured for a particular depletion or buildup in the CO_2 concentration to occur in the circulating air: e.g., from 350 to 300 µl/l. A closed system is advantageous mainly because of its accuracy with small amounts of material or in case of very low rates. By choosing the proper relationship between the volume of the system and the amount of tissue, it is possible to make short runs of the order of a few minutes or even seconds. Measuring the time for a given CO_2 consumption at various levels of CO_2 permits the determination of a photosynthetic CO_2 curve and of the CO_2 compensation point. The major disadvantage of the closed system is that the measurement reflects the influence of a continuously changing CO_2 concentration. This influence can be minimized by always measuring the time for the same amount of change in CO_2 level for runs which are to be compared. However, this does not circumvent the difficulty that continuous change in the CO_2 level demands replenishment of this gas. This is usually done by a manual operation which hampers automatic recording. An alternate system is semi-closed with frequent automatic injections of CO_2 to maintain its level nearly constant; in this case the amount of CO_2 injected over a given time establishes the rate of photosynthesis (Roždestvenskii et al., 1965; Jarvis et al., 1971).

A closed system must be absolutely airtight, and the internal volume, including tubing, pump, water vapor trap, and any other components must be determined precisely, as this volume is used in the calibra-

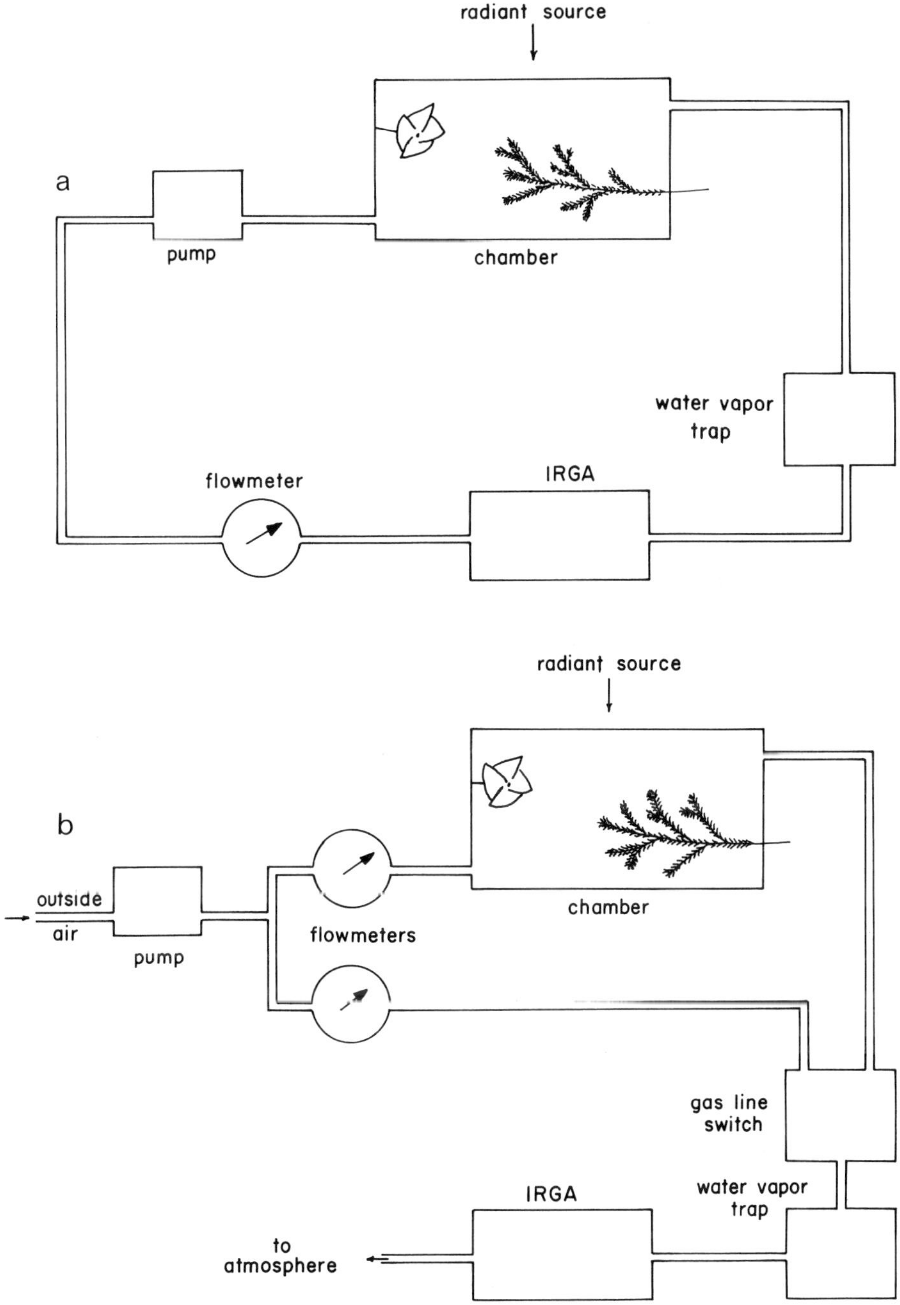

Fig. 2a and b. (a) The components for infra-red CO_2 analysis arranged in a closed system. (b) The components for infra-red CO_2 analysis arranged in an open system

tion. Also, the air must circulate properly, especially to give good mixing of added CO_2; particularly careful attention must be paid to the mixing when the volume of the chamber is small in relation to the to the rest of the system. Introduction of a water vapor trap ahead of the analyzer may dry the air in the system enough to enhance transpiration and possibly increase leaf resistance (see Experimental Material). This effect can be prevented or eliminated by passing the air through an humidifying bottle before it enters the chamber, but this adds another component to the system and may cause condensation in the apparatus at times.

In the open type of system, the difference in CO_2 concentrations of the air entering and leaving the chamber is measured (Fig. 2b). This difference may be measured either by alternately switching from the ingoing to the outgoing air (absolute mode of IRGA), or by passing the ingoing reference air and the outgoing air through the two cells of the analyzer simultaneously (differential mode of IRGA). The relationship between the CO_2 concentration and instrument reading is usually not quite linear, which should be corrected for by careful calibration (see "Calibration Techniques"). The differential mode is more sensitive, because usually 50 to 100 μl/l CO_2 is extended over the full scale of the instrument; this is of particular advantage in measuring respiration in which volumes evolved are often small. However, in the differential mode, sensitivity and reproducibility are affected by the level of CO_2 in the "background" or comparison air, so this level should either be kept constant or monitored with an auxiliary absolute CO_2 measurement. Also, in the differential mode, maintaining equal or at least constant flow rates through the reference and sample tubes is essential. The absolute mode, in which the whole scale of the instrument is often calibrated to 200 to 400 μl/l CO_2 is particularly suited to monitoring fluctuations of CO_2 in the atmosphere, which may be substantial, and for special applications such as determination of CO_2 compensation points when calibrated from "zero". In either mode, precise determination of the flow rate through the plant chamber is necessary for calculation of the photosynthetic or respiratory rate. Open systems are better than closed systems when the time course of a process is being investigated, because continuous recording can be used and several samples can be measured in parallel by the use of a multichannel gas line switch. The semi-closed systems with continuous replenishment of CO_2 are probably the best of all, but are more complex.

Measuring the Evolution and Uptake of Oxygen

Measurement of changes in O_2 is inherently more difficult than that of CO_2 because the amount of O_2 is small compared with the large natural background in the air. Also methods for determination of O_2 are less sensitive than those for CO_2. The development of more sensitive O_2 analyzers based on paramagnetic or katharometric principles has reduced this limitation (Jarvis et al., 1971). Despite these improvements, measurement of O_2 changes around the aerial parts of plants is seldom carried out. On the other hand, water plants or root systems immersed in nutrient solution can be studied in this way, because the solubility of O_2 in water is low and the concentrations are thus much lower than in the air. The Clark electrode, which measures the partial pressure of O_2 using polarography (Clark et al., 1953), seems to be the most convenient for this purpose.

Another possibility for measuring oxygen exchange by aerial parts of plants is the use of a manometric technique, with a Warburg or comparable apparatus. Although better suited for the study of algal suspensions or other small aquatic forms, the technique can be adapted for use with leaves (Semikhatova et al., 1971). In use with leaves or other aerial organs, a number of limitations and precautions must be kept in mind. In the first place, the Warburg vessels are so small that only pieces of organs, e.g., leaf discs, can be contained, thus wounding may affect the gas exchange rates. This disadvantage may be partially overcome by using small containers other than the Warburg vessels, such as plexiglass boxes designed for pine needles (Łotocki and Żelawski, 1973). For small plants, it would even be possible to construct vessels large enough for the entire organism. A second series of difficulties is concerned with CO_2 supply, which must come from buffer solutions, with concentrated carbonate buffer recommended for leaves in order to have an adequate CO_2 concentration in the air above the buffer (Warburg and Krippahl, 1960). This means that the leaf material must be suspended or otherwise kept from contact with the potentially damaging buffer solution. Irradiating at levels sufficient to avoid limitations by light may then give problems in control of leaf temperatures in the vessels, and even of water loss from the tissue and increase in leaf resistance.

Despite these potential problems, the manometric approach has some advantages. Dark respiration is relatively easy to measure, with the carbonate buffer, of course, omitted, and the respiratory CO_2 absorbed in concentrated hydroxide in the center wells of the vessels; this absorption decreases the pressure in the closed system, making possible the reading of O_2 consumption on the manometer. Even in the measuring of photosynthesis, if careful attention is given to maintaining stable and known conditions in the vessels, manometry offers the advantage of simultaneous comparison of a number of samples under uniform conditions of temperature, humidity, CO_2 concentration, and irradiance.

Measurements Using Carbon Isotopes

Methodology for use of ^{14}C, the main isotope employed, has been described in detail (Vosnesenskiĭ et al., 1971). This isotope has been used widely for studying photosynthetic products, and their translocation, biochemical pathways, and accumulation.

The total amount of ^{14}C absorbed may be used to express relative photosynthesis in counts per minute. For calculating rates of photosynthesis in absolute units (e.g., mg/h) the specific radioactivity of the $^{14}CO_2$ in the exposure, the efficiency of counting, and possibly isotopic discrimination effects have to be taken into account. To attain absolute values from the relative $^{14}CO_2$ measurements themselves, the relative $^{14}CO_2$ fixation may be calibrated against determinations by the infra-red technique (Zalenskiĭ et al., 1955).

Often closed systems have been used to expose plant material to $^{14}CO_2$, to avoid radioactive contamination possibilities. In such cases a calculated amount of $^{14}CO_2$ is released into the system by dripping of acid into labeled carbonate, usually with relatively high specific activity to permit sufficient labeling of the material in a short exposure. Sometimes continuous monitoring of radioactivity in the closed circuit has been used to calculate absolute $^{14}CO_2$ absorption from the initial slope of the absorption curve and the specific activity of the $^{14}CO_2$. Also come investigators have attempted to distinguish between gross and net photosynthesis by simultaneously recording the uptake of $^{14}CO_2$ and total CO_2 in a closed system. The idea originated with Lister et al. (1961), but they could not develop a sufficiently sensitive assay of changes in radioactivity because of limitations of the counters available at that time. More recently the use of a combination of several G-M tubes (Steward et al., 1971), ionization chambers (Ludwig and Canvin, 1971), or scintillation counters (Pawlak and Nalborczyk, 1973) has made possible increases in counting from about 10% up to 40 to 60%. Thus, now measuring small amounts of $^{14}CO_2$ is feasible along with more precise simultaneous determination of CO_2 uptake or evolution.

In closed systems, long exposure times can lead to error in the determination because of reduction of the photosynthetic rate with diminishing CO_2 concentration. Releasing the CO_2 in increments aids in maintaining a more even level of CO_2, but at the same time lowers the degree of utilization of the applied radioactivity. Also, with longer exposure times, labeled assimilates are translocated out of the leaves (the organs usually analyzed) into other parts of the plant. Error from this effect increases with time of exposure and with the rate of translocation.

The limitations of closed systems mentioned above have led to the development of a variety of methods for exposing plant organs to a continuous stream of air labeled with $^{14}CO_2$, with provision for trapping the radioactivity in a suitable absorbent at the exit from the system. With such systems (Voznesenskiĭ et al., 1971), short exposures (a few seconds or a minute) are suitable for labeling early products in the photosynthetic pathway and for determination of rates of "true" photosynthesis. On the other hand, long exposures (a few minutes up to hours) will result in labeling of many compounds of the metabolic pathways, so that the rate of CO_2 uptake may represent net photosynthesis or may indicate a resultant of uptake and evolution from substrates which have been labeled in the course of the experiment. A detailed description of a method using the short exposure time in the field, embodying a number of modifications from previously described systems for field use, has recently appeared (Tieszen et al., 1974); (Fig. 3).

In any methods using $^{14}CO_2$, determination of radioactivity in the plant material is an important consideration. Self absorption of the low-energy β-particles prevents determination except in internal gaseous counters, by wet combustion followed by trapping in ethanolamine, or dissolving in organic bases and measuring with the liquid scintillation technique. Tieszen et al. (1974) describe an in-vial combustion procedure said to be rapid and dependable. In any case, methods for radioactivity analysis are time-consuming, and the plant material, often in short supply because of need of other types of analyses, is destroyed.

The $^{14}CO_2$ methods have important special uses as outlined above, and are of particular utility when chemical aspects are being investigated. However, for most determinations of photosynthetic rates, more convenient and precise methods are available.

Other isotopes of carbon have received limited use. The ^{11}C isotope is inconvenient because it has a half-life of only about 20 min. Applications of the non-radioactive isotope ^{13}C (along with the stable isotope ^{18}O), are limited by the complicated procedures and costly equipment involved in the mass spectrographic analyses. Only in special cases, as in the identification of water as the source of the O_2 evolved in photosynthesis, has this methodology been used in the past.

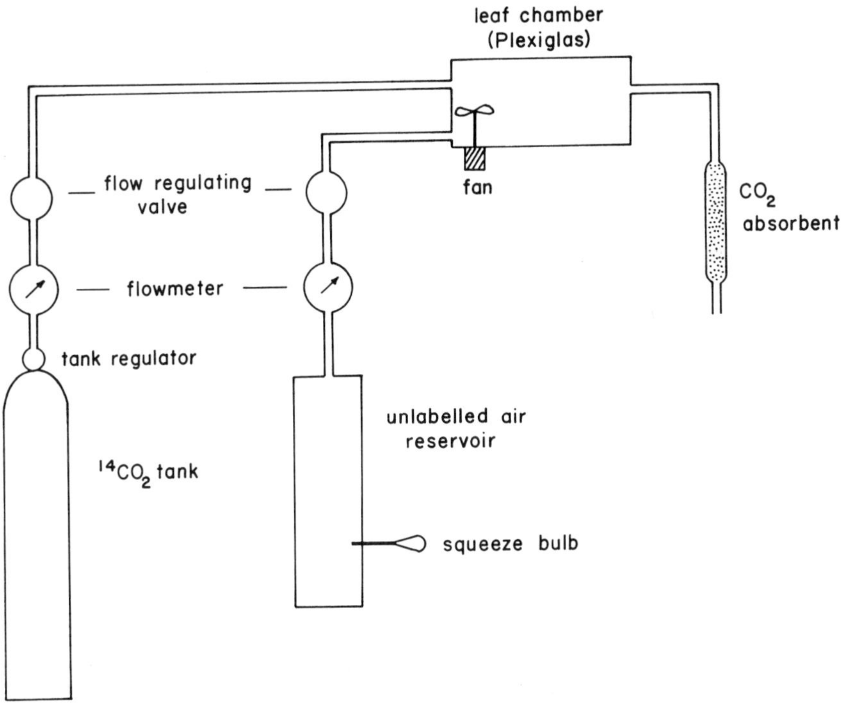

Fig. 3. Field apparatus for measurement of photosynthesis using $^{14}CO_2$ (Tieszen et al., 1974). The reservoir of unlabeled air is used for flushing the system; the CO_2 absorbent prevents $^{14}CO_2$ from escaping into the atmosphere

Presently it finds use in distinguishing between plants with the C_3 or C_4 pathways of photosynthesis because of isotopic discrimination (Bender et al., 1973), and in some studies of photorespiration (Ludlow and Jarvis, 1971b).

Determination of Dry Matter Increase

Sometimes determination of the accumulation of organic biomass over a given period is considered a measure of photosynthesis. In an opposite way, rates of photosynthesis measured on branches may be extrapolated to values for the net production of a forest stand (Helms, 1970). However, as mentioned earlier, high rates of photosynthesis are not always accompanied by high rates of growth or dry matter production. Indeed, correlation between these rates might be expected in seedlings, and was, in fact, found with Douglas-fir seedlings (Brix, 1967), but cannot be expected to be easy to demonstrate in older forest plants. Respiration of non-green parts and its influences on growth, translocation and storage, lags in utilization of photosynthate, and seasonal and hormonal controls on both terminal and diameter growth insure that the relationship between photosynthetic rates and growth in large plants will be complex.

Increase in leaf area as a measure of growth has long been in use, but useful refinements and applications to tree species are still being made (Harms, 1971). Likewise gravimetric measurements on matched leaves or parts of leaves have long been used as indicators of photosynthetic rate and of growth. However, only in recent years has the control of temperature, irradiance, and other factors made this method more accurate and useful. The details of the procedure and equipment are given in the review by Šetlik and Šesták (1971). The method is applicable to various short term studies, and to longer term experiments if the material maintains its activity with time. It is possible to standardize the procedures so that they may be used for mass testing, as in surveys for genotypic differences. There are possible criticisms of the method in terms of traumatogenic effects of cutting the samples, gaseous diffusion features of the leaf discs, and buildup of assimilates and its effect on rates.

More applicable for productivity studies, especially in plant stands, is the method of growth analysis (Květ et al., 1971; Ondok, 1971; Ledig, 1974). In this method, rates of dry matter increase, various characteristics of organs, and allometric relations between characteristics of these organs are measured. The inclusion of mathematical methods makes possible investigation of the dynamics of productivity and modelling of the seasonal course of dry matter accumulation. There are limitations in working with larger trees or with limited numbers of trees, however, because destructive sampling is called for at each experimental date. Also, there is only the possibility to compare populations, as individual variations are effectively lost in the destructive sampling method. However, considerable effort has gone into development of methods for indirect assessment of the primary values (dry weight, leaf area, etc.) used in growth analysis (Ondok, 1971), which should enhance the potential use of such methods in forest science.

The use of growth analysis, its mathematical background, and formulas and calculations are given in depth in the reviews cited above. Thus, here only a short presentation of characteristics measured, definitions, and symbols is warranted. The principal characteristics measured or estimated are:

Crop growth rate or primary productivity (C, CGR): the biomass increment per unit of ground area occupied by the stand.

Relative growth rate (R, RGR): (= efficiency index) the biomass increment per unit of existing biomass.

$$\mathrm{RGR} = \frac{dW}{dt} \cdot \frac{1}{W}$$

where time (t) is usually in weeks, W is total dry weight of the plant in g (usually measured in weekly harvests), so RGR is commonly in the units of g/g/week (see Fig. 4).

Net assimilation rate (NAR, E): (= unit leaf rate) the biomass increment per unit of time and per unit of any measure of the size of the assimilatory organs (usually leaf surface area, weight, or chlorophyll content). The method of calculation of the NAR depends on whether the allometric relation between the increase in total biomass and the increase in leaves is linear or quadratic.

Leaf area ratio (F, LAR): the ratio between leaf area and total dry weight.

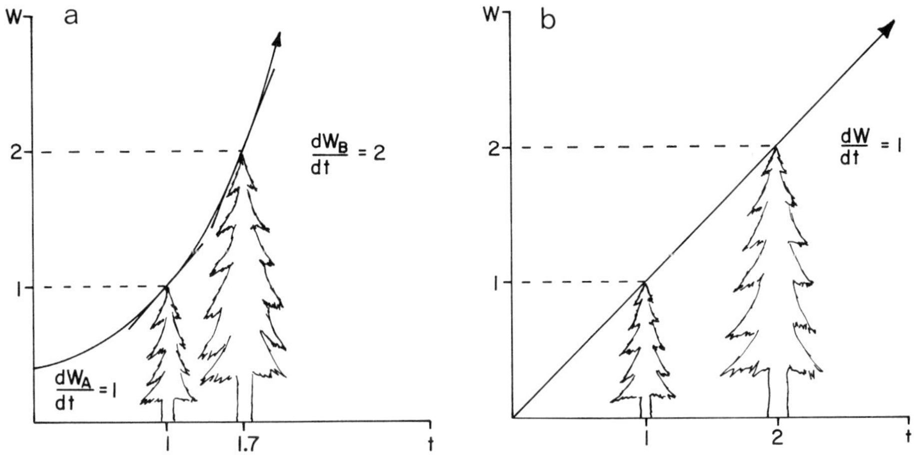

Fig. 4a and b. Illustration of the concept of Relative Growth Rate from Ledig (1974). In (a) the absolute growth rate is twice as great in the larger plant as in the smaller plant (note slopes), but the Relative Growth Rate remains constant, because it is calculated in relation to the amount of growing tissue. On the other hand, in (b) the absolute growth rate for the two plants is the same, but this means that the Relative Growth Rate has declined

Specific leaf area ($\frac{A}{W_L}$, SLA): the ratio of leaf area (A) to leaf dry weight (W_L) i.e., the surface area of a unit weight of an assimilatory organ.

Leaf weight ratio ($\frac{W_L}{W}$, LWR): the ratio between leaf weight and total dry weight.

Note: these values are interrelated as follows:

$$F = \frac{A}{W} = \frac{A}{W_L} \cdot \frac{W_L}{W} \qquad R = \frac{NAR \cdot F}{\frac{1}{W} \quad \frac{dW}{dt}} = \frac{1}{A} \; \frac{dW}{dt} \; \frac{A}{W}$$

Leaf area index (LAI, L): the area (A) of assimilatory organs per unit of ground area (P), i.e., LAI = $\frac{A}{P}$. Crop growth rate (C) and net assimilation rate (NAR) are related as follows:

$$C = \frac{NAR \cdot L}{\frac{dW}{dt} \cdot \frac{1}{P}} = \frac{dW}{dt} \cdot \frac{1}{A} \cdot \frac{A}{P}$$

Leaf area duration (LAD, D) and biomass duration (BMD, Z) are characteristics integrating the leaf area over a certain period of time. The ratio of leaf area duration and biomass duration ($\frac{D}{Z}$) is a mean value of leaf area ratio (F).

Growth analysis has found considerable application in research on crop plants. It has only rather recently been used with forest trees, and then primarily with seedlings. However, using allometric analysis, it can now be extended to larger trees and to forest canopies. The method has more potential for comparisons of both phenotypes and genotypes of trees in young stages of development than has been as yet exploited.

Calibration Techniques

Methods of calibration used in various techniques of photosynthetic research have been described in detail (Semikhatova and Chulanovskaya, 1965; Janáč and Čatský, 1971). Thus, here there is no need to consider these methods except for a few general precautions. Most of this section will be devoted to less familiar approaches for calibration of both closed and open differential modes for IRGAs, and for volume determinations (Szaniawski et al., 1976). IRGAs are available with a longer and a shorter cell in series, which is an advantage in calibrating (Legg and Parkinson, 1968).

General Precautions. The most common way of calibrating infra-red CO_2 analyzers is with the use of high-pressure mixtures of CO_2 in nitrogen or air. Such mixtures are available commercially, with the CO_2 stated to be at a certain concentration ± about 1 or 2 vpm[1]. However, the experience of various laboratories has indicated that these values often are not very accurate. In the worst instances, they may be grossly in error (Janáč and Čatský, 1971), but more often the errors may be only a few vpm. Nonetheless, an error of only 1 or 2 vpm in both the zero and span gases in calibration for the differential mode will give serious errors in the subsequent readings, especially if the differentials are small. Commercial compressed nitrogen often contains appreciable CO_2, which must be removed by an absorbent before such gas can be used for the zero adjustment of an analyzer in the absolute mode. Further, the concentration of CO_2 in tanks may gradually change during storage, because of slow diffusional mixing, surface adsorption, or sedimentation at the bottom of the tank.

Preparation of mixtures by low pressure volumetric or flow meter methods requires several stages of dilution and is rarely accurate enough, especially for calibrations in the differential mode. Commercially available mixing pumps (Janáč and Čatský, 1971) can be quite precise, but the equipment necessary to give even two stages of dilution costs about 2/3 as much as the IRGA itself.

Special Calibration Using Carbonate. Żelawski and Góral (1966a) described a method for calibration of closed circuit systems based on the release of CO_2 from dilute carbonate solution and H_2SO_4 into CO_2-free air. Precise preparation of the carbonate solution and careful determination of the volume of the whole circuit are particularly important for success with this method. The carbonate is heated for two h at 250°C, weighed, then dissolved in freshly boiled redistilled or triple-distilled water. For the range of 0 to 500 vpm CO_2, the solution is prepared so that one ml yields enough CO_2 to increase the concentration by 100 vpm in the closed system. Since dilute bicarbonate solutions avidly absorb CO_2 from the air, they must not be in contact with the air either during preparation or during the calibration procedure.

The internal volume of the closed system is determined with a water manometer. First an over-pressure of several hundred mm of water is developed. This excess air is released carefully, and its volume is measured precisely. The internal volume of the system can then be calculated from the proportion:

---

[1]vpm (volume per million) is equal to µl/l: e.g., normal air at 0.032 vol % CO_2 has 320 vpm (= 320 µl/l). These are preferred over ppm (parts per million) for volumetric use.

$$\frac{P}{p} = \frac{X}{v}$$

in which P = barometric pressure, p = over-pressure created in the system, X = the unknown volume of the system, and v = the volume of the air introduced to create the over-pressure. The barometric pressure must be measured near the location of the system being calibrated. Using a range of over-pressure and replicating determinations, it is possible to attain an accuracy of 0.1% (Żelawski and Góral, 1966a).

Calibration of IRGAs in the differential mode is also possible with a modified method using bicarbonate solutions (Szaniawski et al., 1976).

Manometric Determination of Volumes of Vessels. Sometimes the determination of the volumes of Warburg vessels or other containers being used for manometric studies of gas exchange is not feasible with water or mercury methods (Umbreit et al., 1972). For example, plastic containers may not tolerate the treatment with strong acid needed to remove the last traces of the mercury. In such cases, the volume may be easily measured manometrically. The method is based on the well-known fact that the pressure multiplied by the volume gives a constant value in a closed system.

A convenient method is a modification of the old one of Vogler (Semikhatova and Chulanovskaya, 1965). The volume of a closed system (V) is changed by a decrease or increase in the level of the manometer fluid (e.g., 150 mm), and the corresponding increase or decrease of pressure is measured in the open part of the manometer. This manipulation is carried out twice, making sure that adequate changes are recorded: first with the empty vessel (pressure change is P_1), and second with some known volume (W) of water or other substance in the vessel to diminish its volume (pressure change is P_2). The unknown volume (V) of the system is then calculated by the formula:

$$V = \frac{WP_2}{P_2 - P_1}$$

Data in the literature indicate that this method is accurate only to within about 5%, which is not sufficiently good for the determination of vessel constants for manometry. However, with several repetitions using alternately over-pressure and under-pressure in the same vessel, an accuracy can be attained which is about the same as that achieved with the mercury calibration. Special attention must be paid to the cleanliness of the manometer capillaries, but the method is very convenient and rapid.

Another alternative for determination of vessel and manometer volumes and constants is use of a moderately priced special device for this purpose (e.g., Gilmont Instruments, Great Neck, New York, USA). These calibrators are said to be convenient and accurate.

Conclusion

This review should not be regarded as a full manual of methods ready for application, but rather as an introduction and guide to these techniques. Considerations involved in choosing the proper method and in the correct interpretation of results have been emphasized to stimulate a more critical approach than is sometimes adopted. The present status of methodology and instrumentation makes possible precise physical measurements, but the physiological responses are not always evaluated adequately by the techniques applied. It is essential to become aware of the limitations of each procedure to be used, and to draw only those conclusions from the results which these limitations permit. For example, individual one-time measurements of photosynthesis should not be used to guide a tree-breeding program. Rather, the application of gas exchange studies to such a program should involve an extended study of the relation between the gas exchange and productivity and growth. Certainly the study of photosynthesis and respiration can add substantially to information about forest species, and the techniques for such study can be learned by the forest geneticist for use in appropriate applications.

Acknowledgement. Appreciation is expressed to Dr. David J. Salo for critical reading of the manuscript and valuable suggestions, and to Gail Salo for preparation of the figures.

References

Bender, M.M., Rouhani, I., Vines, H.M., Black, C.C., Jr.: $^{13}C/^{12}C$ ratio changes in Crassulacean acid metabolism plants. Plant Physiol. 52, 427-430 (1973).

Black, C.C., Jr.: Photosynthetic carbon fixation in relation to net CO_2 uptake. Ann. Rev. Plant Physiol. 24, 253-286 (1973).

Brix, H.: An analysis of dry matter production of Douglas-fir seedlings in relation to temperature and light intensity. Can. J. Botany 45, 2063-2072 (1967).

Brix, H.: Effects of nitrogen fertilization on photosynthesis and respiration in Douglas-fir. Forest Sci. 17, 407-414 (1971).

Caldwell, M.M.: The effect of wind on stomatal aperture, photosynthesis, and transpiration of *Rhododendron ferrugineum* L. and *Pinus cembra* L. Cbl. ges. Forstwesen 87, 193-201 (1970).

Čatský, J., Lake, J.V., Begg, J.E., Voznesenskiĭ, V.L.: Physicochemical measurement of p CO_2 and chemical determination of carbon dioxide. In: Plant Photosynthetic Production (eds. Z. Šesták, J. Čatský, P.G. Jarvis), pp. 198-237. The Hague: Dr. Junk N.V. Publishers 1971.

Clark, L.C., Jr., Wolf, R., Grander, D., Taylor, Z.: Continuous recording of blood oxygen tensions by polarography. J. Appl. Physiol. 6, 189-193 (1953).

Dickmann, D.I.: Photosynthesis and respiration by developing leaves of cottonwood (*Populus deltoides* Bartr.). Botan. Gaz. 132, 253-259 (1971).

Duvigneaud, P. (ed.): Productivity of forest ecosystems (Proc. Brussels Symposium), 707 pp. Paris: UNESCO 1971.

Eckardt, F.E. (ed.): Functioning of terrestrial ecosystems at the primary production level (Proc. Copenhagen Symposium), 516 pp. Paris: UNESCO 1968.

Freeland, R.: Effect of age of leaves upon the rate of photosynthesis in some conifers. Plant Physiol. 27, 685-690 (1952).

Gaastra, P.: Photosynthesis of crop plants as influenced by light, carbon dioxide, temperature, and stomatal diffusion resistance. Mededel. Landbouwhogs. Wageningen 59, 1-68 (1959).

Gordon, J.C., Gatherum, G.E.: Effect of environmental factors and seed source on CO_2 exchange of Scotch-pine seedlings. Botan. Gaz. 130, 5-9 (1969).

Harms, W.R.: Estimating leaf-area growth in pine. Ecology 52, 931-934 (1971).
Heath, O.V.S.: The physiological aspects of photosynthesis, 310 pp. Palo Alto: Stanford Univ. Press 1969.
Hellmers, H.: An evaluation of the photosynthetic efficiency of forests. Quart. Rev. Biol. 39, 249-257 (1964).
Hellmuth, E.O.: A method of determining true values for photosynthesis and respiration under field conditions. Flora 157B, 265-286 (1967).
Helms, J.: Summer net photosynthesis of Ponderosa pine in its natural environment. Photosynthetica 4, 243-253 (1970).
Iwanoff, L.A., Kossowitch, N.L.: Über die Arbeit des Assimilationsapparates verschiedener Baumarten. I. Die Kiefer (*Pinus silvestris*). Planta 8, 427-464 (1929).
Jackson, W.A., Volk, R.J.: Photorespiration. Ann. Rev. Plant Physiol. 21, 385-432 (1970).
Janáč, J., Čatský, J.: Calibration of IRGA. In: Plant Photosynthetic Production (eds. Z. Šesták, J. Čatský, P.G. Jarvis), pp. 152-158. The Hague: Dr. W. Junk N.V. Publishers 1971.
Janáč, J., Čatský, J., Jarvis, P.G., Brown, K.W., Eckardt, F.E., Fock, H., Schaub, H., Björkman, O., Gauhl, E., Pieters, G.A.: Infra-red gas analyzers and other physical analyzers. In: Plant Photosynthetic Production (eds. Z. Šesták, J. Čatský, P.G. Jarvis), pp. 111-197. The Hague: Dr. W. Junk N.V. Publishers 1971.
Jarvis, P.G.: Characteristics of the photosynthetic apparatus derived from its response to natural complexes of environmental factors. In: Prediction and Measurement of Photosynthetic Productivity (Proc. IBP/PP Meeting, Třeboň), pp. 353-367. Wageningen: PUDOC 1970.
Jarvis, P.G., Čatský, J., Eckardt, F.E., Koch, W., Koller, D.: General principles of gasometric methods and the main aspects of installation design. In: Plant Photosynthetic Production (eds. Z. Šesták, J. Čatský, P.G. Jarvis), pp. 49-110. The Hague: Dr. W. Junk N.V. Publishers 1971.
Koch, W.: Untersuchungen über die Wirkung von CO_2 auf die Photosynthese einiger Holzgewächse unter Laboratoriumsbedingungen. Flora B 158, 402-438 (1969).
Koch, W., Walz, H.: Kleinklimaanlage zur Messung des pflanzlichen Gaswechsels. Naturwissenschaften 12, 321-322 (1967).
Koch, W., Lange, O.L., Schulze, E.-D.: Ecophysiological investigations on wild and cultivated plants in the Negev desert. I. Methods: A mobile laboratory for measuring CO_2 and water vapour exchange. Oecologia (Berl.) 8, 296-309 (1971).
Kok, B.: Efficiency of photosynthesis. In: Encyclopedia of Plant Physiology (ed. W. Ruhland), Vol. V/1, pp. 566-633. Berlin-Göttingen-Heidelberg: Springer 1960.
Kozlowski, T., Keller, T.: Food relations of woody plants. Botan. Rev. 32, 293-382 (1966).
Kramer, P.J., Clark, W.S.: A comparison of photosynthesis in individual pine needles and entire seedlings at various light intensities. Plant Physiol. 22, 51-57 (1947).
Kramer, P.J., Kozlowski, T.T.: Physiology of trees, 642 pp. New York: McGraw-Hill Book Co. 1960.
Krueger, K.W., Ruth, R.H.: Comparative photosynthesis of red alder, Douglas-fir, Sitka spruce, and western hemlock seedlings. Can. J. Botany 47, 519-527 (1969).
Künstle, E.: Der Jahresgang des CO_2-Gaswechsels von einjährigen Douglasientrieben in einem 20-jährigen Bestand. Allg. Forst u. J. Ztg. 142, 105-108 (1971).
Květ, J., Marshall, J.K.: Assessment of leaf area and other assimilating plant surfaces. In: Plant Photosynthetic Production (eds. Z. Šesták, J. Čatský, P.G. Jarvis), pp. 517-555. The Hague: Dr. Junk N.V. Publishers 1971.

Květ, J., Ondok, J.P., Necas, J., Jarvis, P.G.: Methods of growth analysis. In: Plant Photosynthetic Production (eds. Z. Šesták, J. Čatský, P.G. Jarvis), pp. 343-391. The Hague: Dr. W. Junk N.V. Publishers 1971.

Lange, O.L., Schulze, E.-D., Koch, W.: Evaluation of photosynthesis measurements taken in the field. In: Prediction and Measurement of Photosynthetic Productivity (Proc. IBP/PP Meeting, Třeboň), pp. 339-352. Wageningen: PUDOC 1970.

Larcher, W.: Die Eignung abgeschnittener Zweige und Blätter zur Bestimmung des Assimilationsvermögens. Planta 60, 1-18 (1963).

Larcher, W.: Physiological approaches to the measurement of photosynthesis in relation to dry matter production by trees. Photosynthetica 3, 150-166 (1969a).

Larcher, W.: The effect of environmental and physiological variables on the carbon dioxide gas exchange of trees. Photosynthetica 3, 167-198 (1969b).

Larson, P.R., Isebrands, J.G., Dickson, R.E.: Fixation patterns of ^{14}C within developing leaves of eastern cottonwood. Planta 107, 301-314 (1972).

Ledig, F.T.: Concepts of growth analysis. In: Proc. 3rd North Amer. Forest Biol. Workshop, Colorado State Univ. College of Forestry, Ft. Collins, Colorado (eds. C.P.P. Reid, G.H. Fechner), pp. 166-182 (1974).

Legg, B.S., Parkinson, J.K.: Calibration of infra-red gas analyzers for use with CO_2. J. Sci. Instr. 1 (Ser. 2), 1003-1006 (1968).

Lemon, E.: Aerodynamic studies of CO_2 exchange between the atmosphere and the plant. In: Harvesting the Sun (eds. A. San Pietro, F.A. Greer, T.J. Army), pp. 263-290. New York: Academic Press 1967.

Levitt, J.: Responses of plants to environmental stresses, 697 pp. New York: Academic Press 1972.

Lister, G.R., Krotkov, G., Nelson, C.A.: A closed-circuit apparatus with an infra-red CO_2-analyzer and a Geiger tube, for continuous measurements of CO_2 exchange in photosynthesis and respiration. Can. J. Botany 39, 581-591 (1961).

Ludlow, M.M., Jarvis, P.J.: Photosynthesis in Sitka spruce (*Picea sitchensis*). I. General characteristics. J. Appl. Ecol. 8, 925-953 (1971a).

Ludlow, M.M., Jarvis, P.J.: Methods for measuring photorespiration in leaves. In: Plant Photosynthetic Production (eds. Z. Šesták, J. Čatský, P.G. Jarvis), pp. 294-315. The Hague: Dr. W. Junk N.V. Publishers 1971b.

Ludwig, L.J., Canvin, D.T.: An open gas exchange system for the simultaneous measurement of the CO_2 and $^{14}CO_2$ fluxes from leaves. Can. J. Botany 49, 1299-1313 (1971).

Luukkanen, O., Kozlowski, T.T.: Gas exchange in six Populus clones. Silvae Genetica 21, 220-229 (1972).

Łotocki, A., Żelawski, W.: Effect of ammonium and nitrate source of nitrogen on productivity of photosynthesis in Scots pine (*Pinus silvestris* L.) seedlings. Acta Soc. Botan. Pol. 42, 599-605 (1973).

Mooney, H.A., Harrison, A.T.: The influence of conditioning temperature on subsequent temperature-related photosynthetic capacity in higher plants. In: Prediction and Measurement of Photosynthetic Productivity (Proc. IBP/PP Meeting, Třeboň), pp. 411-417. Wageningen: PUDOC 1970.

Neales, T.F., Incoll, L.D.: The control of leaf photosynthesis rate by the level of assimilate concentration in the leaf: a review of the hypothesis. Botan. Rev. 34, 107-125 (1968).

Neuwirth, G.: Der CO_2-Stoffwechsel einiger Koniferen während des Knospenaustriebes. Biol. Zbl. 78, 559-584 (1959).

Neuwirth, G.: Aufbau und Arbeitsweise eines Feldlabors zur Messung von CO_2-Assimilation und Transpiration in Pflanzenbeständen. Biol. Plant. 7, 212-217 (1965).

Ondok, J.P.: Indirect estimation of primary values used in growth analysis. In: Plant Photosynthetic Production (eds. Z. Šesták, J. Čatský, P.G. Jarvis), pp. 392-411. The Hague: Dr. W. Junk N.V. Publishers 1971.

Pawlak, M., Nalborczyk, E.: Scintillation chamber for the low radioactivity measurement of the $^{14}CO_2$. Patent No. P 164164 (1973).

Pisek, A.: Immergrüne Pflanzen. In: Encyclopedia of Plant Physiology (ed. W. Ruhland), Vol. V/2, pp. 376-414. Berlin: Springer 1960.

Pisek, A., Larcher, W., Vegis, A., Napp-Zinn, K.: The normal temperature range. In: Temperature and Life (eds. H. Precht, J. Christophersen, H. Hensel, W. Larcher), pp. 102-194. Berlin: Springer 1973.

Polster, H.: Die physiologischen Grundlagen der Stofferzeugung im Walde, 96 pp. München: Bayer. Landwirtsch. Verlag 1950.

Rozdestvenskii, V.J., Culkin, V.G., Kleshnin, A.F.: Avtomaticheskoe podderzhanie postoiannogo diapazona konzentracji CO_2 v ustanovkach dla izucenia fotosinteza rastienii. Fizjologia rastienii 12, 178-181 (1965).

Salo, D.J.: Factors affecting photosynthesis in Douglas-fir. Ph.D. Diss., Univ. of Washington, Seattle (Diss. Abstr. Intl. Sec. B, 35, 3765B-3766B, 1974), 149 p.

Schaedle, M.: Tree photosynthesis. Ann. Rev. Plant Physiol. 26, 101-115 (1975).

Schulze, E.-D., Lange, O.L., Buschbom, U., Kappen, L., Evenari, M.: Stomatal responses to changes in humidity in plants growing in the desert. Planta 108, 259-270 (1972).

Semikhatova, O.A., Chulanovskaya, M.V.: Manometricheski metody izucheniya dykhaniya i fotosinteza rastenii. Moskva-Leningrad: Nauka 1965.

Semikhatova, O.A., Chulanovskaya, M.V., Metzner, H.: Manometric methods of plant photosynthesis determination. In: Plant Photosynthetic Production (eds. Z. Šesták, J. Čatský, P.G. Jarvis), pp. 238-256. The Hague: Dr. W. Junk N.V. Publishers 1971.

Šesták, Z.: Determination of chlorophylls a and b. In: Plant Photosynthetic Production (eds. Z. Šesták, J. Čatský, P.G. Jarvis), pp. 672-701. The Hague: Dr. W. Junk N.V. Publishers 1971.

Šesták, Z., Jarvis, P.G., Čatský, J.: Criteria for the selection of suitable methods. In: Plant Photosynthetic Production (eds. Z. Šesták, J. Čatský, P.G. Jarvis), pp. 1-48. The Hague: Dr. W. Junk N.V. Publishers 1971.

Šetlik, J. (ed.): Prediction and measurement of photosynthetic productivity, 632 pp. (Proc. IBP/PP Technical Meeting, Třeboň.) Wageningen: PUDOC 1970.

Šetlik, J., Šesták, Z.: Use of leaf tissue samples in ventilated chambers for long term measurements of photosynthesis. In: Plant Photosynthetic Production (eds. Z. Šesták, J. Čatský, P.G. Jarvis), pp. 316-342. The Hague: Dr. W. Junk N.V. Publishers 1971.

Sorensen, F.C., Ferrell, W.K.: Photosynthesis and growth of Douglas-fir seedlings when grown in different environments. Can. J. Botany 51, 1689-1698 (1973).

Steward, F.C., Graven, G.H., Weerasinghe, P.R., Bidwell, R.G.S.: Effects of prior environmental conditions on the subsequent uptake and release of carbon dioxide in the light. Can. J. Botany 49, 1999-2007 (1971).

Szaniawski, R., Wierzbicki, B., Żelawski, W.: Calibration of the differential infra-red CO_2 analyzer by use of diluted carbonate solution. Photosynthetica (in press) (1976).

Thompson, F.B., Leyton, L.: Method for measuring the leaf surface area of complex shoots. Nature 229, 572 (1971).

Tieszen, L.L., Johnson, D.A., Caldwell, M.M.: A portable system for the measurement of photosynthesis using 14-carbon dioxide. Photosynthetica 8, 151-160 (1974).
Tranquillini, W.: Die Abhängigkeit der Kohlensäureassimilation junger Lärchen, Fichten und Zirben von der Luft- und Bodenfeuchte. Planta 60, 70-94 (1963).
Tranquillini, W.: Blattemperatur, Evaporation und Photosynthese bei verschiedener Durchströmung der Assimilationsküvette. Ber. dtsch. Bot. Ges. 77, 204-218 (1964).
Tranquillini, W.: Photosynthese und Transpiration einiger Holzarten bei verschieden starkem Wind. Cbl. ges. Forstwesen 86, 35-48 (1969).
Umbreit, W.W., Burris, R.H., Stauffer, J.F.: Manometric and biochemical techniques (5th ed.). Minneapolis: Burgess Publishing Co. 1972.
Voznesenskiĭ, V.L., Zalenskiĭ, O.V., Austin, R.B.: Methods of measuring rates of photosynthesis using carbon - 14 dioxide. In: Plant Photosynthetic Production (eds. Z. Šesták, J. Čatský, P.G. Jarvis), pp. 276-293. The Hague: Dr. W. Junk N.V. Publishers 1971.
Warburg, O., Krippahl, G.: Weiterentwicklung der manometrischen Methoden (Carbonatgemische). Zeitsch. Naturf. 15, 364-467 (1960).
Zalenskiĭ, O.V., Semikhatova, O.A., Voznesenskiĭ, V.C.: Metody primeneniya radioaktivnogo ugleroda C^{14} dlya izucheniya fotosinteza. Moskva-Leningrad: Izd. Akd. Nauk SSSR (1955).
Zavitkovski, J., Ferrell, W.K.: Effect of drought upon rates of photosynthesis, respiration, and transpiration of seedlings of two ecotypes of Douglas-fir. II. Two-year-old seedlings. Photosynthetica 4, 58-67 (1970).
Żelawski, W.: Variation in the photosynthetic efficiencies of trees with special reference to Scots pine (*Pinus silvestris* L.). 14th IUFRO-Congress, Munich, pp. 515-535 (1967).
Żelawski, W.: Vergleichende Untersuchungen über Wachstum und Assimilationsaktivität einiger polnischer Kiefernprovenienzen. Tagungsber. dtsch. Akad. Landwirtschaftwiss. 100, 221-232 (1968).
Żelawski, W.: Uptake and evolution of carbon dioxide and accumulation of organic substance in Scots pine (*Pinus silvestris* L.) seedlings. Bull. Acad. Polen. Sci. Ser. Biol. 20, 747-753 (1972).
Żelawski, W., Góral, I.: Gazowy analizator (IR) w badaniach z zakresu fizjologii i ekologii roslin. Wiad. Botan. 10, 75-85 (1966a).
Żelawski, W., Góral, I.: Seasonal changes in the photosynthesis rate of Scots pine (*Pinus silvestris* L.) seedlings grown from seed of various provenances. Acta Soc. Botan. Pol. 35, 587-598 (1966b).
Żelawski, W., Kinelska, J. Łotocki, A.: Influence of shade on productivity of photosynthesis in seedlings of Scots pine (*Pinus silvestris* L.) during the second vegetation period. Acta Soc. Botan. Pol. 37, 505-518 (1968).
Żelawski, W., Kucharska, J., Kinelska, J.: Relationship between dry matter production and carbon dioxide absorption in seedlings of Scots pine (*Pinus silvestris* L.) in their second vegetation season. Acta Soc. Botan. Pol. 40, 243-256 (1971).
Żelawski, W., Kucharska, J., Łotocki, A.: Productivity of photosynthesis in Scots pine (*Pinus silvestris* L.) seedlings grown under various soil moisture conditions. Acta Soc. Botan. Pol. 38, 143-155 (1969).
Żelawski, W., Łotocki, A., Nalborczyk, E.: Productivity of photosynthesis in Scots pine (*Pinus silvestris* L.) seedlings during their first and second growing seasons after X-irradiation of seed. USDA-Final Rep., pp. 15-33 (1974).
Żelawski, W., Riech, F.P., Stanley, R.G.: Assimilation and release of internal carbon dioxide by woody plant shoots. Can. J. Botany 48, 1351-1354 (1970).

Żelawski, W., Szaniawski, R., Dybczynski, W., Piechurowski, A.: Photosynthetic capacity of conifers in diffuse light of high illuminance. Photosynthetica 7, 351-357 (1973).

Żelitch, I.: Photosynthesis, photorespiration, and plant productivity, 347 pp. New York: Academic Press 1971.

CHAPTER 6

Analyses of Monoterpenes of Conifers by Gas-Liquid Chromatography

A. E. SQUILLACE

Introduction

Monoterpenes are C_{10} chemical compounds with carbon skeletons consisting of two isopentane (isoprene) units joined together and are a major component of the steam-volatile part of oleoresin produced in tissues of most conifers. Most of them are monoterpene hydrocarbons ($C_{10}H_{16}$). But often the volatiles in oleoresin include oxygenated monoterpenes, sesquiterpenes, and n-hydrocarbons and are often studied along with the monoterpene hydrocarbons. In the literature, these chemical are variously termed "monoterpenes", "terpenes", "volatile oils", "essential oils" or "turpentine". For the chemistry of these compounds see Simonsen and Owen (1953, 1957) and de Mayo (1959).

During the past 20 years interest in monoterpene composition of conifers has greatly increased. This increase resulted mainly from the development of gas-liquid chromatography (GLC) about 20 years ago and from the success of subsequent studies employing GLC in several disciplines. Prior to the development of GLC, monoterpene composition was determined by fractional distillation, which was a relatively slow process. Large samples of oleoresin were required and this usually necessitated bulking of samples from several trees and severely restricted the number of samples that could be analyzed. The early research was largely limited to commercial purposes and to taxonomic studies at a species level (Mirov, 1958a, 1961, 1963). The development of GLC greatly increased the speed and accuracy of chemical analyses and permitted the use of small samples. Study of variation among trees within species and more extensive sampling of the species' ranges became feasible. The adaptability of GLC to study of monoterpene composition not only stimulated more comprehensive taxonomic studies (Zavarin, 1968b; Turner and Flake, 1974) but also aroused the interest of scientists in other disciplines.

Genetics studies have shown that monoterpene composition varies greatly among trees within species and is strongly inherited (Hanover, 1966a, b, c, 1971; Squillace and Fisher, 1966; Hilton, 1968; Franklin and Snyder, 1971; Squillace, 1971; Baradat et al., 1972; Rockwood, 1973a). Opportunities were revealed to breed for improved composition of turpentine (Squillace and Fisher, 1966) and to use monoterpenes in indirect selection for other commercially important traits (Hanover, 1966b, 1971).

Proof of strong genetic control also stimulated study of genetic relationships among trees growing in different portions of a species' range. Normally, in studies of this nature, geneticists must collect seed from native trees growing in various portions of the species' range and plant them in a common environment. But when the trait is subject to relatively little environmental effect, the researcher can study the trees in their native habitats (Hanover and Furniss, 1966; Rudloff, 1973a; Gansel and Squillace, in press). Racial patterns of monoterpene composition in some species are now known to such an extent that the geographic source of seed of plantations can be identified by sampling monoterpene composition. Knowledge of such racial patterns also opens up the possibility of using monoterpene composition for seed certification.

High versus low levels of some monoterpenes are controlled by single genes (Hanover, 1966c; Squillace, 1971; Baradat et al., 1972; Rockwood, 1973a). Such traits are very useful as gene markers for identification of hybrids (Hanover and Wilkinson, 1969) and relatives (Chudnyi, 1971) and for studies of the degree of natural inbreeding and of population structure.

Several recent studies have shown relationships between monoterpene composition and insect resistance (Heikkenen and Hrutfiord, 1965; Smith, 1965, 1966b; Berton and Chararas, 1967; Rudinsky et al., 1970; Smelyanets and Chursin, 1972; Overhulser et al., 1972; Wilkinson, 1972; Hanover, 1975), disease resistance (Polozhentsev et al., 1970; Federov and Manukov, 1972; Krupa and Nylund, 1972; Krupa et al., 1973; Rockwood, 1973b, 1974), and animal damage (Santamour et al., 1963; Radwan, 1972). Such findings can greatly aid in development of pest resistant strains of trees.

Thus, study of monoterpene composition offers many opportunities for better understanding and management of forest trees. This chapter is designed to provide some guidelines for scientists beginning such studies. Although the characterization of monoterpenes by GLC is rather simple, there are many pitfalls that the inexperienced investigator can encounter, from sampling techniques to analyses and interpretation of data. In the examples to be cited, emphasis will be on slash pine (*Pinus elliottii* Engelm.), because of the author's experience. A bibliography of selected recent literature, cross-referenced by species is appended.

Material and Methods

Sampling

Choice of trees for studying monoterpene composition and the sampling intensity to be used depend greatly upon the objective of the investigator.

1. If the objective is to characterize a species, the investigator must recognize that individuals often vary greatly both within stands and between stands in different portions of the species' range. As a minimum, one must sample several trees in each of several portions of a species' range. The most reliable way to characterize a species would be to sample trees over its entire range in what amounts to a study of geographic variation.

2. If the objective is to study geographic variation, one can either sample planted trees in a provenance test or wild trees in their native habitat. Studies conducted to date suggest that plantation or environmental effects on monoterpene composition are relatively small (Hanover, 1966a; Squillace and Fisher, 1966). Sampling in a provenance plantation has the advantages of accessibility and ease of sample collection. Sampling of trees in their native habitats allows better coverage of the species' range than is available in most provenance

tests, but collection of samples is more costly. The range of interest to the investigator should be sampled completely. The more provenances sampled the greater the likelihood of discovering the true patterns of variation. In slash pine, a grid of sampling points spaced approximately 50 miles apart was satisfactory (Gansel and Squillace, in press). Sample size (number of trees per provenance or stand) is best determined from preliminary studies; a minimum of 15-20 trees is recommended. The trees should, of course, be selected randomly within provenances or stands.

3. If the objective is to study inheritance, clones or families are required. The content of some monoterpenes seems to be controlled by single gene pairs. To study this kind of inheritance, control-pollinated families and their parents must be sampled. Offspring of self-fertilization would be especially desirable. The number of parents or families to sample is difficult to specify because it will depend upon the variety of phenotypes found. Usually 20-25 individuals per family will be adequate. More may be required if evidence suggests that several genes rather than one are involved.

Source of Oleoresin

A source of oleoresin must be chosen by the investigator. Oleoresin occurs in various organs such as roots, stems, branches, foliage, catkins, and cones. Likewise, it occurs in various tissues such as xylem, phloem, and cortex. Variation in composition occurs among both tissues and organs in most investigated species. For an outline of resin duct systems, see Pauly (1963).

In choosing a source, the investigator may be restricted by his objectives. For example, if he is interested in breeding trees for superior composition of oleoresin, he will likely wish to sample stem xylem oleoresin. But if he is interested in taxonomic relationships, population structure, etc., it may be preferable to sample cortical or needle oleoresin. In some species, oleoresin of these tissues often contains more major monoterpenes (those comprising more than a few percent of the total) than oleoresin from xylem tissue. For example, slash pine oleoresin in stem or branch xylem can contain major amounts of α-pinene, β-pinene, and β-phellandrene only. Oleoresin from cortical tissue, on the other hand, can contain major amounts of myrcene and/or limonene as well. Major amounts of myrcene or limonene have never been found in xylem oleoresin of this species. Bernard-Dagan et al. (1971) likewise found more major constituents in cortical vs. xylem oleoresin in *Pinus pinaster*. Greater numbers of major constituents, of course, offer greater opportunities for genetic variation.

The investigator must also decide whether to obtain exuded oleoresin or to collect tissues containing oleoresin and extract it. Many conifers exude oleoresin from wounded stems, branch tips, or needles, and wounding is perhaps the simplest and most common procedure. In species that do not exude resins, elaborate extraction procedures involving distillation may be required, especially if yields per unit weight or volume of tissue are desired. In some species, wood increment cores, needles, or buds contain sufficient oleoresin that extraction with a solvent such as pentane or ethyl ether may be sufficiently reliable to determine composition. Concentration of the extract may be necessary in some cases. Another alternative is to inject solid tissue samples directly into gas chromatographs specially adapted for this purpose (Rudloff, 1965; Roberts, 1968b).

Factors to Consider in Sample Collection

Tree Age. Monoterpene composition may change greatly during juvenile stages. Funes et al. (1971, 1973) found large changes in needle oil composition of *Pinus pinaster* up to the age of 65 days, but it remained constant thereafter. Slash pines having genes for high myrcene or β-phellandrene in cortical oleoresin may not contain high amounts of these chemicals until about 1-1/2 to 2 years of age, or longer, if grown under sub-optimal conditions (Fisher and Squillace, 1966). The same is true of stem xylem oleoresin up to about 5 years of age. Hanover (1971) found greater amounts of β-pinene in old than in juvenile (4 years) *Pinus monticola* trees. Age effects seem to be rather small beyond the juvenile stage in both slash pine (Squillace, 1971) and *Pinus ponderosa* (Smith, 1964a). Unless the investigator purposely wishes to study age effects, he should sample beyond the juvenile stage or restrict sampling to trees of the same age.

Season. Most investigators have reported relatively small effects of season on monoterpene composition of stem xylem or cortical oleoresin. But appreciable effects have been found in foliar oleoresin of some species, such as *Pinus sylvestris* (Juvonen, 1966) and *Juniperus pinchotii* (Adams, 1970). When seasonal effects are unknown, collections should preferably be made within a relatively short period of the year.

Position on Stem - Xylem Oleoresin. In some species, sampling height has little effect on composition of stem xylem oleoresin (Blight and McDonald, 1964). But appreciable effects have been noted in slash pine (Roberts, 1970) and *Pinus taeda* (Rockwood, 1972). Franklin (in press) further showed that distance from live crown was associated with variation in composition in slash pine and recommended sampling at base of crown to minimize such effects. Most people have sampled within a few feet of the ground and this is, of course, required when characterizing composition of exuded oleoresin for commercial purposes. Horizontal position on stem seems to have no effect according to several studies (Bannister et al., 1962; Blight and McDonald, 1964; Smith, 1964b). The crown of a grafted tree can affect the composition of oleoresin below the graft union (Mirov, 1945). But rootstocks do not affect composition of oleoresin in the scion (Hanover, 1966a; Squillace and Fisher, 1966; Chudnyi, 1971; Schmidtling, 1974).

Position in Crown and Age of Tissue - Cortical Oleoresin. In sampling needles or cortical tissue in branch tips, elevational and horizontal position in crown is not very important (Hanover, 1966a; Rudloff, 1967a, 1972a; Pauly and Rudloff, 1971; Rockwood, 1972). However, needle age is important in some species (Juvonen, 1966; Maarse and Kempner, 1970; Bernard-Dagan et al., 1971; Zavarin et al., 1971; Poltavchenko and Rudakov, 1972). Age of cortical tissue was also important in some species (Hanover, 1966a; Roberts, 1970). In several species of *Abies*, composition of oleoresin in cortical blisters varied with vertical position of blister on the stem (Zavarin, 1968a; Thor and Barnett, 1974). These findings suggest that in sampling needles, cortical tissue, or cortical blisters one should hold age of tissue constant.

Damaged Tissues. Appreciable changes in monoterpene composition can be caused by mechanical wounds (Roberts, 1968a), insect attacks (Bernard-Dagan et al., 1971), or diseases (Federov and Manukov, 1972; Krupa et al., 1973). Hence, damaged trees should be avoided unless such damage is being studied.

Sample Storage Duration. Time of sample storage (up to several months) seems to have only minor effects (Smith, 1964b; Squillace, 1971). But some monoterpenes, such as myrcene, α-terpinene, and β-phellandrene, will polymerize during prolonged storage and, hence, prolonged storage should be avoided.

Collection, Preparation, and Storage of Samples

Exuded stem xylem oleoresin is usually collected by inserting close-fitting screwcap vials into holes drilled approximately 1 cm into the xylem. In many pines, several milliliters of oleoresin can be obtained within a day. Such large amounts of oleoresin are preferable if the proportions of monoterpenes relative to other constituents of oleoresin are desired. If only the relative proportions of monoterpenes are desired, samples as little as 1 or 2 microliters of oleoresin are satisfactory and usually requires only a small wound.

For collection of oleoresin from cortical tissue, one procedure is to excise branches about 1 cm from the tip. In many species, a droplet of oleoresin sufficient for collection will exude within a few minutes. Sometimes excising several branch tips will be necessary. In doing so, one must be careful to see that only cortical oleoresin is obtained. If the incision is made too far below the branch tip, some xylem oleoresin may exude along with cortical oleoresin. Such a mixture, of course, is undesirable. Cortical ducts are near the periphery of the branch while xylem ducts are near the center - both can be observed with a hand lens. Another procedure is to make shallow cuts into the phloem of 1-year-old branch internodes (Hanover, 1966b), but here again care is needed to avoid collecting xylem oleoresin.

A convenient way to collect the oleoresin is to use a micropipette, preferably with an aspirator. The pipettes containing oleoresin can be placed into vials to which a solvent such as pentane, ethyl ether, or acetone is then added (approximately 2 parts solvent to 1 part oleoresin). Tightly sealed vials should then be stored in a refrigerator or freezer. For routine collections of many samples, it is highly desirable to keep sample size approximately constant. This will permit adding the same amount of solvent to each vial, maintaining relatively constant concentrations which are conveniently chromatographed. Unfortunately, some trees do not exude sufficient oleoresin, and in such cases one should decrease the proportion of solvent added to avoid low concentrations.

For species that do not exude oleoresin readily, one can excise several buds or branch terminals, split them lengthwise, and place into vials containing a solvent such as pentane or ethyl ether (Schmidtling, 1974). Finely chopped leaf or wood tissue handled in this manner also may be satisfactory in some species. In such samples, it may be necessary to concentrate the sample before analyzing by GLC.

When amounts of monoterpenes per unit weight or volume of tissue are desired, more elaborate collection and extraction procedures may be required (Rudloff, 1961, 1964; Juvonen, 1966; Zavarin et al., 1971; Pauly et al., 1973).

Operation of Gas Chromatograph

Detailed techniques of GLC operation are best learned by personal instruction from an experienced operator. Short courses are often given by manufacturers of GLC. Textbooks such as Burchfield and Storrs (1962) and Dal Nogare and Juvet (1962) should be referred to as well as publications of Zubyk and Conner (1960), Rybinski and Rozmej (1968), Rudloff (1969, 1974), Baerheim Svendsen and Karlsen (1971), and Bardyshev and Kulikov (1971). A few suggestions follow.

1. Selection of columns is critical. A 4.8 x 610 mm copper column packed with Carbowax 20M on a Chromosorb W, 60-80 mesh support works well with oleoresin of several pine species of the southeastern United States. However, other columns are preferable in some species (for example, see Rudloff, 1961; Burchfield and Storrs, 1962; Klouwen and Ter Heide, 1962; Blight and McDonald, 1963; Bardyshev and Kulikov, 1972). In selecting columns, one should seek the advice of investigators working with the species of interest or with closely related species. Most manufacturers of chromatographic equipment provide literature on column selection.

2. It is helpful at the start to prepare a composite of standards of the monoterpenes one expects to find in the species being studied. Standards should first be checked out individually for impurities. The composite chromatogram of Fig. 1 was prepared by starting out with a single monoterpene and then adding others one at a time and labeling each peak successively until all were incorporated.

It is also helpful to run the composite at a series of different column temperatures to determine retention times of each chemical at various column temperatures and to determine the optimum column temperature to use for separations of interest (Fig. 2). Note that the presence or absence of all chemicals in the composite can be readily detected at a column temperature of 120°C. When peaks overlap severely, it may be necessary to run samples at two or more temperatures to permit reliable quantitative determinations. When overlapping is not severe, the procedure described by Mori (1972) can be used to increase accuracy of quantitative determinations.

3. When very weak concentrations of the sample are run, be cautious of impurities in the solvent. Weak concentrations require high detector sensitivity, and under such conditions impurities in the solvent may appear. When it is necessary to run at a very high sensitivity, one should run the solvent alone to check for impurities.

Quantitative Determinations

In addition to detecting monoterpenes present in a sample (qualitative determinations), the investigator is usually interested in the relative amounts, or in the yields of monoterpenes present (quantitative determinations). There are three basic types of quantitative determinations commonly employed, each requiring a different procedure. They are arranged below in order of increasing amount of information obtained, as well as in order of increasing amount of effort required:

1. Content of each monoterpene relative to other monoterpenes present in the sample - "total monoterpene basis".

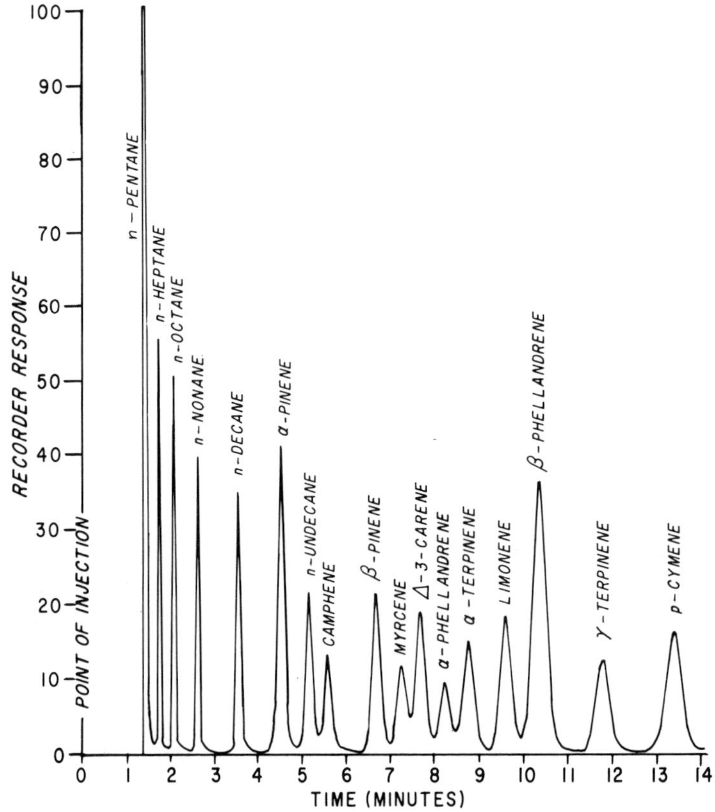

Fig. 1. Chromatogram of some standard monoterpenes and n-hydrocarbons. (Column: 4.8 x 610 mm, copper; support: Chromosorb W, 60-80 mesh; substrate: Carbowax 20M; column temperature: 120°C)

2. Content of each monoterpene relative to all other constituents of oleoresin - "oleoresin basis".

3. Content of each monoterpene per unit weight (or volume) of tissue - "tissue basis".

These types of determinations each measure a different parameter. They will often be correlated to some degree in samples from different trees, because the total monoterpene content of oleoresin and yield of oleoresin per unit of tissue often vary less than the relative proportions of monoterpenes. If total monoterpene content and yields are constant, all determinations would be perfectly correlated. But since total monoterpene content and yields are seldom constant, perfect correlation is rare and one cannot "extrapolate" from one type of determination to the other. For example, if a tree has a high proportion (in comparison to other trees) of α-pinene on a total monoterpene basis, it does not necessarily have a high proportion of α-pinene on an oleoresin basis, nor does it necessarily have a high content of α-pinene per unit weight of tissue. Extrapolations in the opposite direction would be equally invalid. Advantages and disadvantages of each of these types of quantitative determinations will be discussed along with computational procedures.

Fig. 2. Retention times (relative to α-pinene) for some monoterpenes and n-hydrocarbons at various column temperatures. (Column: 4.8 x 610 mm, copper; support: Chromosorb W, 60-80 mesh; substrate: Carbowax 20M)

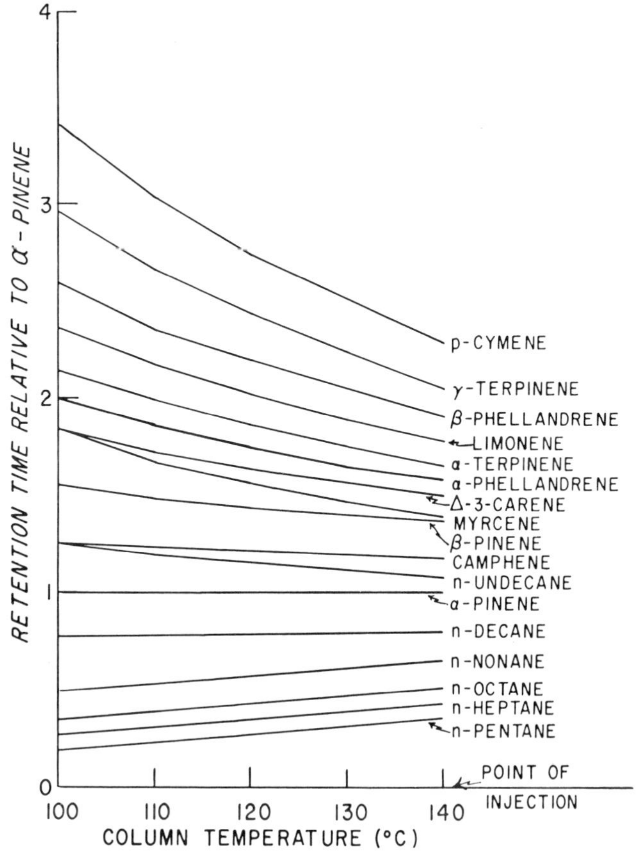

Total Monoterpene Basis. Under this procedure, the investigator merely determines the area under the peak of each monoterpene and divides it by the total area under all monoterpene peaks in the sample (normalization). Usually, the proportions, which necessarily add to 1, are expressed in percent and these then necessarily add to 100%.

This is the simplest and most rapid procedure for determining relative monoterpene composition. The sample of oleoresin can be as little as 1 or 2 μl. Since such small amounts of oleoresin can readily be obtained from many tissues by wounding or through use of solvents, the procedure has wide application. The weight or volume of the sample need not be determined, and an internal standard is not necessary. The method is accurate, being subject only to errors in determining peak areas, assuming no errors in detector response. It is especially suited to genetic studies in which many trees must be sampled. Relative composition on a total monoterpene basis is strongly inherited and, as noted by Zavarin and Cobb (1970) and Smith (in press), it is not strongly associated with variation in total monoterpene content of oleoresin.

One disadvantage of the method is that the percentages of monoterpenes are not independent because they must total 100% for each sample. This constraint tends to cause negative correlations between pairs of monoterpenes and to mask biological correlations which may be present

(Mosimann, 1962). Methods of computing expected correlations due to constraint and of adjusting for them will be discussed later, as well as problems inherent in analyses of percentage data.

Oleoresin Basis. To determine the amounts of each monoterpene relative to total oleoresin, the weight (or volume) of the sample of oleoresin is determined, and a known weight (or volume) of a standard is added to the sample. After chromatographing, the peak area for the standard is obtained along with peak areas for each of the sample monoterpenes. The percent of each monoterpene on an oleoresin basis (P_m) is then obtained from the following formula (assuming the standard and sample monoterpene have equal molecular weights):

$$P_m = \frac{A_m W_s}{A_s W_o} \times 100 \quad \text{where:}$$

A_m = area under the sample monoterpene peak

A_s = area under the standard peak

W_s = weight (or volume) of the standard

W_o = weight (or volume) of the oleoresin sample

For an alternative procedure involving the development of calibration curves from an internal standard with synthetic mixtures, see Dal Nogare and Juvet (1962). These authors also discuss "absolute methods" that directly compare peak heights or areas of an unknown component with calibration curves prepared from pure samples. The latter procedure requires precise control of operating conditions from sample to sample.

The total percentage of all monoterpenes in the sample is often called "percent turpentine". Note that monoterpene content on an oleoresin basis can readily be converted to content on a total monoterpene basis by dividing the former by percent turpentine. The relationship between these kinds of measurement is depicted graphically for a hypothetical sample below.

	Total oleoresin				
	Monoterpenes			Resin acids and neutrals	
Monoterpene composition (%) on a	A	B	C	D	Sum
Total monoterpene basis:	40	40	20		100
Oleoresin basis:	10	10	5	75	100
	25 = percent turpentine				

The relative merits of determining and expressing monoterpene composition on a total monoterpene vs. oleoresin basis are debatable. Obviously, the latter procedure requires more work, since both the sample and the standard must be weighed or their volumes determined. Also, the sample size must be reasonably large for accurate determinations, although as little as 10 μl may be sufficient (Hanover, 1966a). Even such sizes cannot be obtained in many cases.

Some investigators favor the oleoresin basis over the total monoterpene basis because of the belief that data from the former are subject to very little constraint. As can be seen in the above diagram, both types of determinations are subject to constraint since in both cases all constituents necessarily add to 100. The magnitude of the constraint depends on the means of the constituents. Since the means of the monoterpenes are always less on the oleoresin basis, the expected correlations due to constraint are likewise less. But between-tree variation in percent turpentine tends to be relatively small compared to composition (see, for example, Mirov, 1963; Zavarin and Cobb, 1970; Smith, in press) causing a greater degree of actual constraint for the monoterpenes expressed on an oleoresin basis than one would expect on the basis of their means. As a result, observed correlations between constituents would tend to be similar for both types of determinations, and this was found to be true in a group of samples in which both types of determinations were made (Table 1). Thus, the advantage of reduced constraint under the oleoresin basis is often overrated.

Table 1. Observed and expected correlation coefficients between monoterpene content of exuded stem xylem oleoresin determined on a total monoterpene basis vs. oleoresin basis for 114 slash pines[a]

Constituents	Total monoterpene basis		Oleoresin basis	
	Observed	Expected	Observed	Expected
α-pinene vs β-pinene	0.68	0.49	0.53	0.07
α-pinene vs β-phellandrene	0.75	0.60	0.72	0.08
β-pinene vs β-phellandrene	0.04	0.39	0.00	0.06

[a]Data from an office report, Southeastern Forest Experiment Station, Olustee, Florida.

Another advantage of the oleoresin basis procedure is that it permits studying percent turpentine, which can be important commercially, in addition to monoterpene composition. This additional trait, when strongly inherited, can also be of value in taxonomic and similar investigations. Although both within- and between-species variation in percent turpentine have been often reported, the inheritance of this trait seems to have received little attention. In a slash pine study, percent turpentine was found to be inherited but not as strongly as the percentages of individual components (Table 2). Heritability of turpentine percent was also less than heritability of most individual monoterpenes in *P. monticola* (Hanover, 1966b). The relatively low genetic effect on percent turpentine may be partially due to errors associated with measurements of weights or volumes of samples and standards.

Some investigators have expressed the opinion that data on an oleoresion basis reflect genetic effects better than data on a total monoterpene basis. However, the data of Table 2 do not support this view and indicate the reverse may be true. More comparative data on the extent of genetic control are needed. Obviously, if an investigator is interested in studying correlations between monoterpenes and other constituents of oleoresin, he must necessarily obtain data on an oleoresin basis. But in studies requiring extensive sampling, such as chemosystematic studies, the total monoterpene basis would likely be preferable.

Table 2. Components of variance obtained from data on a total monoterpene vs. oleoresin basis in the same experiment[a]

Source of variation	df	Components of variance as proportion of total variance			
		α-pinene	β-pinene	β-phellandrene	Total monoterpenes (percent turpentine)
Total monoterpene basis:					
Clones	5	0.95	0.94	0.94	--
Ramets/clones	108	0.05	0.05	0.05	--
Samples/ramets/clones	114	0.00	0.00	0.01	--
Oleoresin basis:					
Clones	5	0.91	0.91	0.91	0.25
Ramets/clones	108	0.07	0.08	0.06	0.53
Samples/ramets/clones	114	0.02	0.01	0.03	0.22

[a]Duplicate samples (about 1 cc of each) of stem xylem oleoresin were obtained from each of 19 ramets in each of 6 slash pine clones. Data are from an office report in files of the Southeastern Forest Experiment Station, Olustee, Florida.

Tissue Basis. Under this procedure the investigator determines the weight (or volume) of monoterpenes per unit weight (or volume) of tissue. Usually the volatiles are extracted from a known weight or volume of tissue by steam distillation or solvent extraction and the amount of total oil per unit of tissue determined. (For detailed procedures used in making such determinations see Drew and Pylant, 1966; Juvonen, 1966; Adams, 1969; Bernard-Dagan et al., 1971; Zavarin et al., 1971; Rudloff, 1972b). Then the oil is chromatographed to determine the relative composition of monoterpenes. For an alternative procedure involving the introduction of an internal standard, see Maarse and Kepner (1970).

Most investigators express the total yield of oil on a tissue basis and the monoterpene composition on a total monoterpene basis. But some investigators express composition on a tissue basis; this is accomplished by multiplying total yield per unit of tissue by composition on a total monoterpene basis.

When monoterpene composition is expressed per unit of tissue, the values are affected by both relative composition and yield. Relative composition is strongly controlled by genetic factors, often by single genes, as shown by several investigations. Yield of monoterpenes may also be controlled by genetic factors, but usually to a lesser degree, because usually many genes are involved. Yield is also more subject to environmental effects (Rudloff, 1972b; Powell and Adams, 1973), and since more measurements are involved, it is more subject to error. Thus, monoterpene composition expressed on a tissue basis is likely to be less strongly inherited than composition expressed on a total monoterpene basis. Hence, although it may be desirable to express composition on a tissue basis for some purposes (especially commercial), expression on a total monoterpene basis would likely be preferable for chemosystematic studies and other studies requiring highly heritable traits as has been recommended by Powell and Adams (1973).

Usually the yields of monoterpenes are expressed per unit of dry weight of tissue. In these cases, one must recognize that yields expressed in this manner become dependent upon dry weight density of the tissues. Likewise, if the yields are expressed on a green weight basis, the green density (which may be highly reflective of water content) becomes a factor affecting the results. These considerations suggest that yields should preferably be determined on a volume rather than a weight basis, if volume can be accurately measured.

Handling and Interpretation of Data

Correlations Among Constituents. Investigators frequently compute correlations among the monoterpene proportions in a group of samples. The purpose may be to obtain clues as to biosynthetic pathways, to predict the influence of genetic selection for one constituent on other constituents, or for other reasons. As noted earlier, when monoterpenes are expressed on a relative basis (either total monoterpene or oleoresin basis), correlations due to constraint are present. Constraint correlations may be misleading and can be produced in data by taking proportions of originally uncorrelated measurements. Meaningful biological relations may or may not be present. That constraint correlations are negative can readily be seen by supposing that only two monoterpenes (A and B) are present in all samples. When the proportion of A is high, B must necessarily be low, and vice versa. In this case, the expected correlation due to constraint is -1. If three or more monoterpenes are involved, the expected correlations due to constraint will not be so strong, but can be very appreciable.

Expected correlations (r) due to constraint are predictable and depend only upon the means of the two monoterpenes of interest. They can be computed with the following equation from Mosimann (1962):

$$r_{AB} = -\sqrt{\frac{\bar{A}\bar{B}}{(1-\bar{A})(1-\bar{B})}} \quad \text{where}$$

r_{AB} = the expected correlation between monoterpenes A and B, whose mean proportions are $\bar{A}$ and $\bar{B}$, respectively.

If desired, one can test whether or not the observed correlation found between two chemicals is significantly different from the expected correlation due to constraint. More usefully, one can determine a "net measure of distance" of the observed correlation from that due to constraint (James E. Mosimann, personal communication). Patterns of such distances hopefully will reflect biological relationships. The procedure is as follows and is exemplified in Table 3.

1. Compute the mean proportion of each monoterpene and the observed correlations between all pairs of constituents.

2. From the means, compute the expected correlations due to constraint, using Mosimann's formula.

Table 3. Tests of significance of differences between observed and expected correlations, and calculation of "net distances"

Pairs of monoterpenes[a]	Correlations		Corresponding values of $\underline{z}$[b]				Net distance[e]
	Observed	Expected	Observed	Expected	Difference[c]	t[d]	
α-pinene vs β-pinene	0.13	0.60	0.13	0.69	0.82	9.7**	0.68
α-pinene vs myrcene	0.59	0.16	0.68	0.16	0.52	6.1**	0.48
α-pinene vs β-phellandrene	0.58	0.29	0.66	0.30	0.36	4.2**	0.34
β-pinene vs myrcene	0.42	0.23	0.45	0.23	0.22	2.6**	0.22
β-pinene vs β-phellandrene	0.71	0.40	0.89	0.42	0.47	5.5**	0.44
Myrcene vs β-phellandrene	0.20	0.11	0.20	0.11	0.31	3.6**	0.29

[a] Determined from cortical oleoresin of 142 slash pine trees. Mean proportions were 0.30, 0.46, 0.06, and 0.16 for α-pinene, β-pinene, myrcene, and β-phellandrene, respectively

[b] A transformation of r. See Snedecor and Cochran (1967, p. 185)

[c] Actual $\underline{z}$ - expected $\underline{z}$

[d] $t = \frac{\text{Difference}}{1/\sqrt{n-3}}$, with ∞ d.f. See Snedecor and Cochran (1967, p. 185)

[e] Net distance = value of r corresponding to the $\underline{z}$ difference

** = Significant at the 1% level

3. Determine z transformations (Snedecor and Cochran, 1967) corresponding to the observed and expected correlations.

4. Compute the difference between observed and expected z values.

5. Transform the z differences into values of r to represent "net distances".

6. If desired, compute "t" and determine significance.

All of the t-values arrived at in step 6 above are significant. Should a t-value for a pair of proportions, say A and B, not be significant, one should also check the correlations of each proportion with the remainder, 1-A-B. Neither of these correlations should be significant if the correlation of A and B is due to constraint.

The net distances found in Table 3 suggest at least one pattern. The four monoterpenes fall into two groups, with α-pinene and β-pinene in one and myrcene and β-phellandrene in the other. Net distances between members of the same group are positive while those between members of different groups are negative. One biological implication might be that myrcene and β-phellandrene are often produced at the expense of α-pinene and β-pinene.

Statistical Analyses. Frequently analyses of variance are run on monoterpene percentage (or proportion) data to test for significance of differences among families or for other purposes. In doing so, one must be cautious of the fact that such data are often non-normally distributed and/or heterogeneous, making analyses of variance tenuous, as noted by Coyne and Keith (1972) and Thor and Barnett (1974). Where strong biomodality occurs, the data can be classified into high and low categories (or more categories if several modes occur), and the data can then be analyzed by chi-square.

Phenotypic Classification for Monoterpene Content. Relative amounts of monoterpenes frequently show either bimodal or skewed frequency distributions and these are often taken as evidence of control by few genes with large effects. Two precautions are:

1. A skewed distribution alone (without two or more modes) does not necessarily mean control by few genes. Skewness in percentage data is common and may be merely due to the existence of limits 0 and 100%.

2. Because of the constraint present in relative composition data, the frequency distributions for the various monoterpenes present in a group of samples are not likely to be independent. This fact should be considered when evaluating evidence for bimodality and when setting criteria for phenotypic classification.

A group of 2290 slash pines sampled for cortical monoterpenes will serve as an example. Myrcene showed clear bimodality and single-gene inheritance was readily proved for this monoterpene on the basis of parent and progeny data (Squillace, 1971). α-pinene showed a tendency toward bimodality in the frequency distribution for all trees (Fig. 3C). But closer examination of the data showed that the tendency toward bimodality for α-pinene was false, being due to the fact that trees with high myrcene tended to have low α-pinene (Fig. 3A), while trees with low myrcene tended to have high percentages of α-pinene (Fig. 3B). Note that there is no clear evidence of bimodality in Fig. 3A and B. The small mode on the left of Fig. 3B and the extended

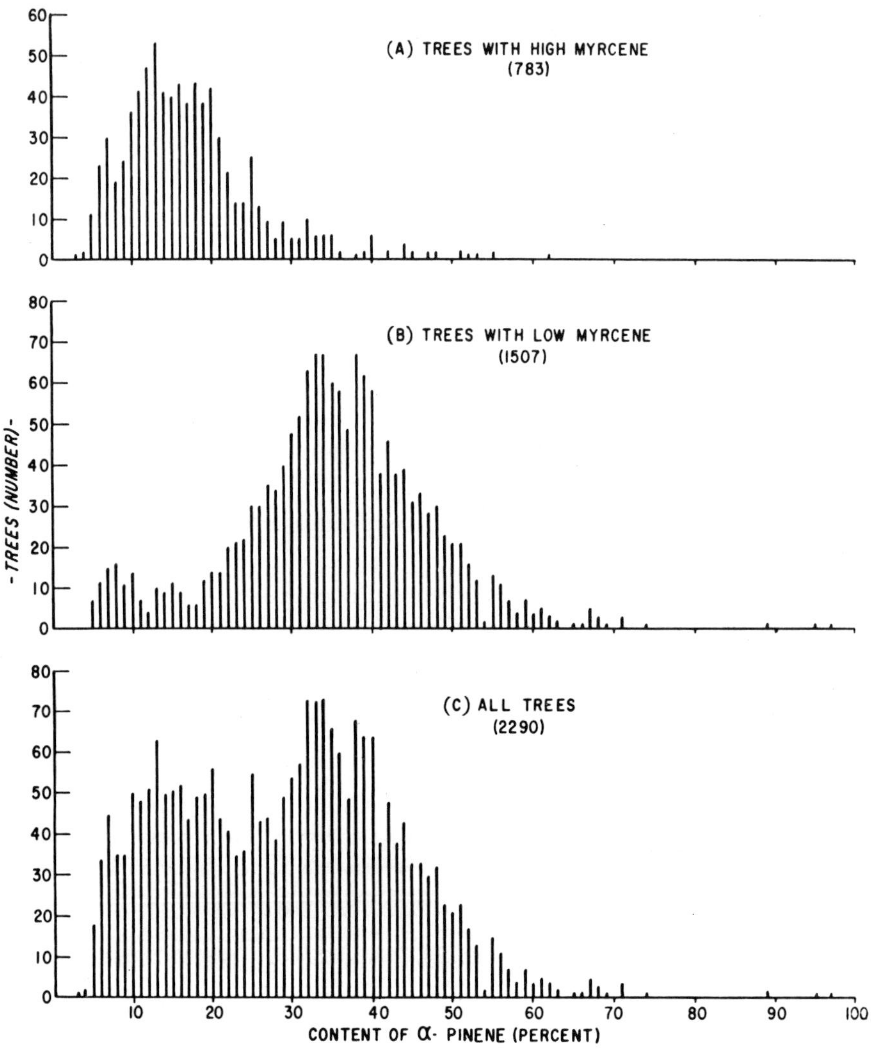

Fig. 3A-C. Frequency distribution for α-pinene content (total monoterpene basis) in cortex of slash pines having high myrcene (A), in those having low myrcene (B), and in all trees (C)

tails on the right of both figures were shown to be largely due to an interaction with limonene - trees with high limonene tended to have low α-pinene while those with low limonene tended to have high α-pinene.

Thus, it is advisable to subdivide data into combinations of high and low classes for all monoterpenes showing clear modes. This procedure may also permit refining classification criteria, as exemplified by an interaction between β-pinene and limonene in the same data discussed above. When trees were classified into high and low limonene groups, bimodality for β-pinene occurred in both groups (Fig. 4). But the nodes appeared at different points on the horizontal axes; this result suggested that trees be classified for β-pinene according to the presence or absence of high limonene, as indicated in the upper

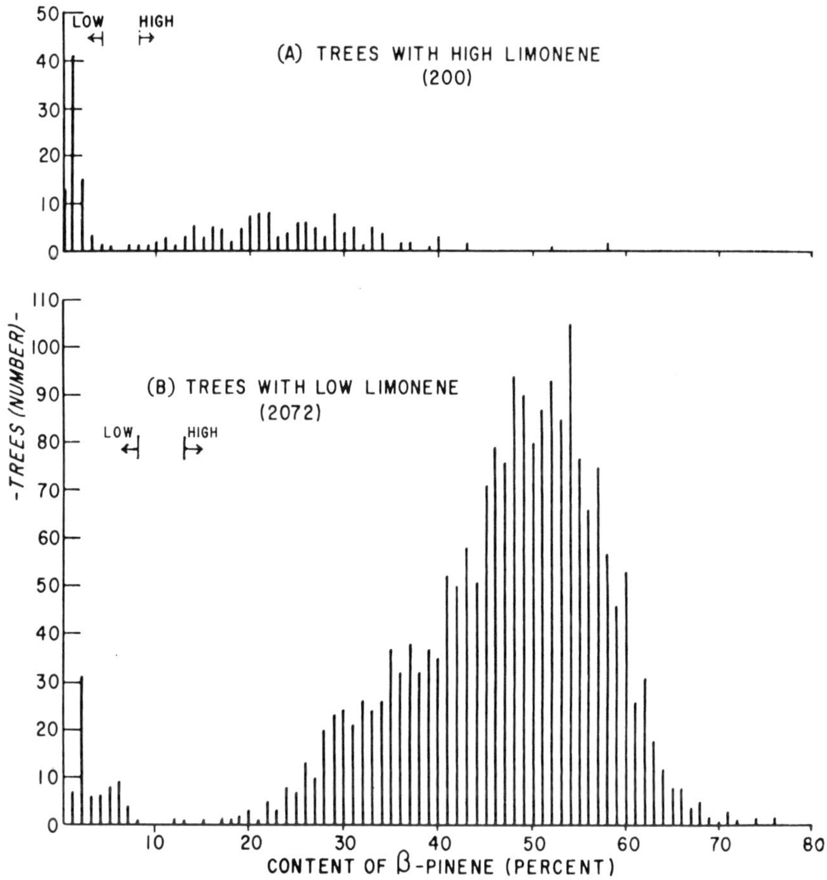

Fig. 4A and B. Frequency distributions for β-pinene content (total monoterpene basis) of slash pines having high limonene (A) and low limonene (B)

left portions of Fig. 4A and B. Further details of the techniques used for slash pine are given in a manuscript in press by Gansel and Squillace.

Determining Mode of Inheritance. Knowledge of the mode of inheritance of monoterpenes can be of great value in breeding for specific monoterpene composition, in searching for gene markers and for other purposes. Suggestions in regards to determining mode of inheritance follow.

1. Normal frequency distributions suggest, but do not prove, multigenic control. Conclusions of such control should be supported by family tests, since strong environmental effects can also cause normal distributions.

2. Bimodal (or trimodal) distributions suggest control by few genes. But analyses of parents and their progenies are required to determine the exact number of genes involved and the presence or absence of dominance. For examples of test procedures, see Hanover (1966c), Squillace (1971), Baradat et al. (1972), and Rockwood (1973a).

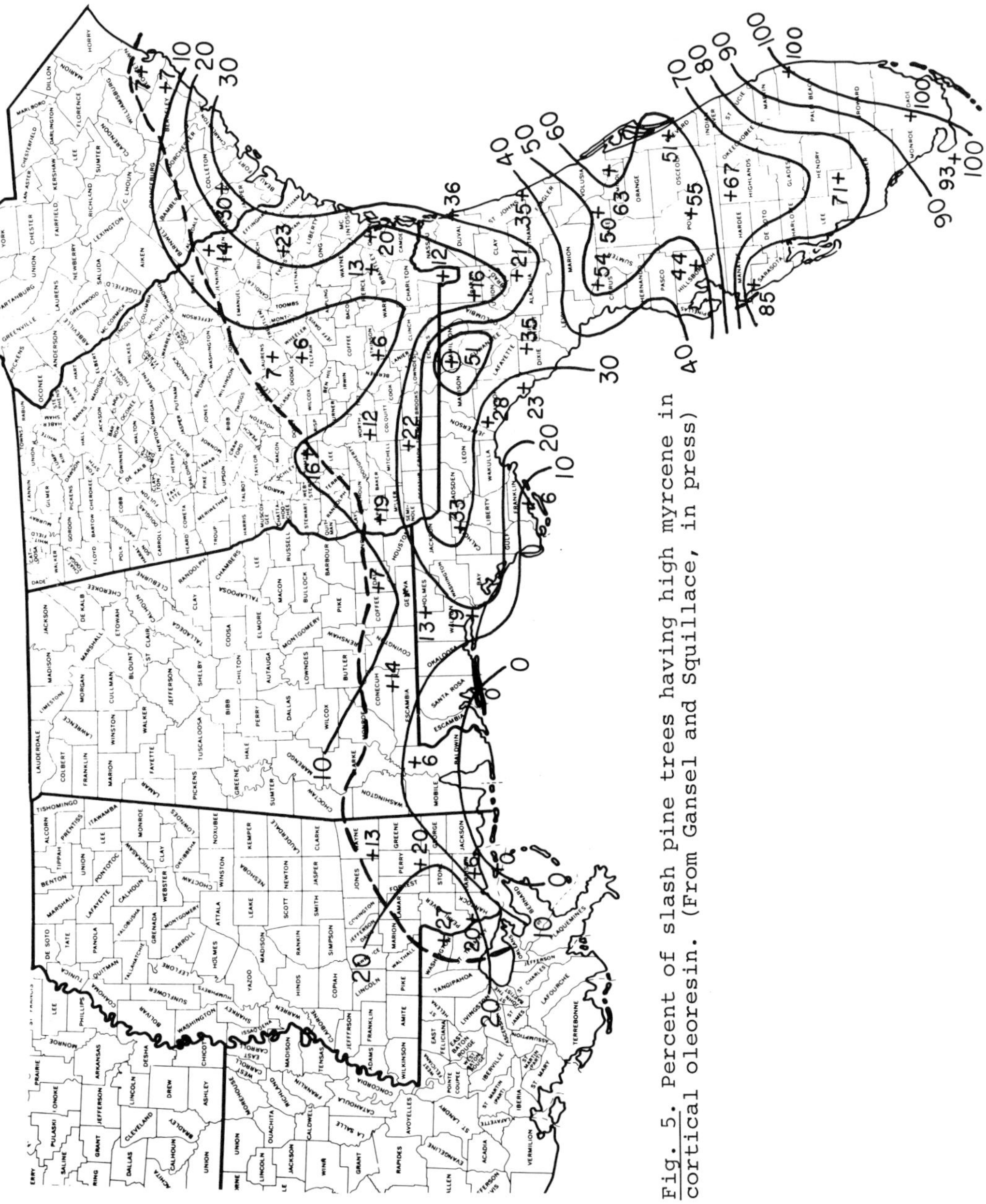

Fig. 5. Percent of slash pine trees having high myrcene in cortical oleoresin. (From Gansel and Squillace, in press)

3. In cases where two modes occur, the following rule is helpful in determining which phenotype is dominant: if segregation occurs when "high parents" are mated, then high is dominant over low; if segregation occurs when "low parents" are mated, then low is dominant.

Determining Geographic Patterns. In studying geographic patterns of variation, usually the mean values obtained for each geographic source are employed. However, when bimodality is evident in the frequency distribution of the data, it is preferable to use the proportion of "high" or "low" trees at each source (Gansel and Squillace, in press). It is also desirable to plot values representing each source on a map and look for patterns. Statistical analyses alone can sometimes be misleading. Isograms should be drawn when warranted (Fig. 5). For further procedures in distinguishing and outlining clines, see Flake et al. (1969, 1973), Adams and Turner (1970), and Wilkinson et al. (1971).

Determining Biosynthetic Pathways. Many investigators use observed correlations between monoterpenes as clues to biosynthetic pathways. Zavarin (1970) postulated that positively correlated monoterpenes should be biosynthetically closer to each other than those to which they are strongly negatively correlated. As noted earlier, one can determine "net distances" between observed and constraint correlations. These may also be helpful in determining biosynthetic pathways. The "neutrality" concept of Conner and Mosimann (1969) may also be useful in such investigations. For reviews of biosynthesis of monoterpenes, see Sandermann (1962), Loomis (1967), Hanson (1967), Poltavchenko and Rudakov (1969), Banthorpe et al. (1972), and Juvonen and Hiltunen (1974).

Summary

Composition of the monoterpenes in oleoresin of conifers is strongly inherited and is often controlled by single genes with major effects. Composition can be accurately and rapidly determined by gas-liquid chromatography. Because of these unique features, study of monoterpene composition offers excellent opportunities to increase our knowledge of genetic relationships between and within species of trees and to aid in the development and utilization of superior trees.

Although determination of monoterpene composition by GLC is a simple process, the inexperienced investigator can encounter many problems and pitfalls, from sampling techniques to analysis and interpretation of data. These have been discussed and some guidelines presented. A survey of literature showed that although monoterpene composition has been studied in many forest tree species, most studies were exploratory, involving few trees. There is a need for comprehensive geographic variation studies and these should be conducted while truly native stands still exist.

Species Cross Reference

(Selected references on monoterpene composition of conifers, published since 1958. Earlier reports on pines are included in Mirov, 1961. For references on history and economics of turpentine, see Weiner and Pollock, 1974).

References

1. Adams, R.P.: Chemosystematic and numerical studies in natural populations of Juniperus. Ph.D. Thesis, University of Texas, Austin (1969).
2. Adams, R.P.: Seasonal variation of terpenoid constituents in natural populations of *Juniperus pinchotii* Sudw. Phytochemistry 9, 397-402 (1970).
3. Adams, R.P., Turner, B.L.: Chemosystematic and numerical studies of natural populations of *Juniperus ashei*. Taxon 19, 728-751 (1970).
4. Akimov, Y.A., Kuznetsov, S.I.: Evaluation of the essential oil from the vegetative organs of species and forms of Cedrus. Rastit. Resur. 8, 562-565 (1972). For. Abstr. (Oxford) 34, 3156 (1973).
5. Akimov, Y.A., Kuznetsov, S.I.: Essential oil composition of *Pinus hamata*. Rastit. Resur. 10, 413-418 (1974).
6. Akimov, Y.A., Lishtvanova, L.N., Nilov, G.I.: Composition of the essential oil of *Pinus pithyusa*. Rastit. Resur. 8, 99-103 (1972). For. Abstr. (Oxford) 33, 5537 (1972).
7. Akimov, Y.A., Podgornyi, Y.K.: Variation in the content and composition of the essential oils of *Pinus pithyusa*. Rastit. Resur. 9, 235-242 (1973). For. Abstr. (Oxford) 35, 576 (1974).
8. Ames, G.R., Chandra-Rajan, M., Mathews, W.S.A.: Geographical variations in the composition of *Pinus merkusii* turpentines. Trop. Sci. 11, 170-176 (1969).
9. Andersen, N.H., Syrdal, D.D.: Terpenes and sesquiterpenes of *Chamaecyparis nootkatensis* leaf oil. Phytochemistry 9, 1325-1340 (1970).
10. Anderson, A.B., Riffer, R., Wong, A.: Monoterpenes, fatty and resin acids of *Pinus edulis* and *Pinus albicaulis*. Phytochemistry 8, 1999-2001 (1969a).
11. Anderson, A.B., Riffer, R., Wong, A.: Monoterpenes, fatty and resin acids of *Pinus contorta* and *Pinus attenuata*. Phytochemistry 8, 2401-2403 (1969b).
12. Anderson, A.B., Riffer, R., Wong, A.: Monoterpenes, fatty and resin acids of *Pinus ponderosa* and *Pinus jeffreyi*. Phytochemistry 8, 873-875 (1969c).
13. Anderson, A.B., Riffer, R., Wong, A.: Monoterpenes, fatty and resin acids of *Pinus lambertiana* and *Pinus monticola*. Phytochemistry 8, 869-872 (1969d).
14. Anderson, A.B., Riffer, R., Wong, A.: Chemistry of genus Pinus. VII. Monoterpenes, fatty and resin acids of *Pinus monophylla* and *Pinus quadrifolia*. Holzforsch. 24, 182-184 (1970).
15. Anderson, A.B., Riffer, R., Wong, A., Ynalvez, L.A.: Composition of Mindoro pine gum oleoresin. Wood Sci. 1, 12-14 (1968).
16. Baboshin, B.K., Sidorov, R.I., Rudakov, G.A., Nikolaeva, Z.K., Ivanova, L.S.: Study of the composition of terpenic hydrocarbon mixtures by gas-liquid chromatography. Gidroliz. i Lesokhim. Prom. 16, 14-15 (1963).
17. Badam, L., Mednikov, F.A.: Chemical composition of the resinous substances obtained in trial tapping of conifers growing in the forests of the Mongolian Peoples Republic. For. Abstr. (Oxford) 35, 2062 (1974).
18. Baerheim Svendsen, A., Karlsen, J.: Gas chromatographic analysis of lower terpenes in plant material. Pharmacogn. Phytochem., 1st Inter. Congr. 1970, 17-40 (Pub. 1971).
19. Bailey, L.F.: Leaf oils from Tennessee Valley conifers. J. Forestry 46, 882-889 (1948).
20. Bamgiottii, A.M., Vincieri, F.F., Coran, S.A.: Terpenoid hydrocarbons from *Pinus pinea*. Riv. Ital. Essenze, Profumi, Piante Off., Aromi, Saponi, Cosmet., Aerosol 54, 875-877 (1972).
21. Bamgiottii, A.M., Vincieri, F.F., Cosi, F.: Monoterpene and sesquiterpene hydrocarbons of *Pinus mugo*. Phytochemistry 11, 1455-1460 (1972)

22. Bannister, M.H., Brewerton, H.V., McDonald, I.R.C.: Vapour-phase chromatography in a study of hybridism in Pinus. Sven. Paperstidn. 62, 567-573 (1959). For. Abstr. (Oxford) 21, 121 (1960).
23. Bannister, M.H., Williams, A.L., McDonald, I.R.C., Forde, M.B.: Variation of turpentine composition in five population samples of *Pinus radiata*. New Zealand J. Sci. 5, 486-495 (1962).
24. Banthorpe, D.V., Charlwood, B.V., Francis, M.J.O.: The biosynthesis of monoterpenes. Chem. Rev. 72, 115-155 (1972).
25. Banthorpe, D.V., Davies, H.ff.S., Gatford, C., Williams, S.R.: Monoterpene patterns in Juniperus and Thuja species. Planta Med. 23, 64-69 (1973).
26. Baradat, P., Bernard-Dagan, C., Fillon, C., Marpeau, A., Pauly, G.: Les terpénés du Pin maritime: aspects biologiques et génétiques. II. Hérédité de la teneur en monoterpènes. Ann. Sci. For. 29, 307-334 (1972).
27. Bardysev, I.I., Bulgakov, A.N., Udarov, B.G.: Composition of the mixtures of monoterpenes produced by certain climatic ecotypes of subspecies of *Pinus sylvestris*. Lesn. Zh. 16, 159-160 (1973). For. Abstr. (Oxford) 35, 7400 (1974).
28. Bardysev, I.I., Bulgakov, A.N., Udarov, B.G., Epshtein, Y.V., Gordon, L.V.: Composition of turpentines from soft resin obtained in different regions of the country. Gidroliz. Lesokhim. Prom. 25, 21-22 (1972).
29. Bardysev, I.I., Cerces, H.A.: *Pinus nigra* var. *caramanica* resin - a promising raw material for the rosin and turpentine industry. For. Abstr. (Oxford) 22, 3906 (1961a).
30. Bardysev, I.I., Cerces, H.A.: Chemical composition of oil of *Picea excelsa*. For. Abstr. (Oxford) 22, 3907 (1961b).
31. Bardysev, I.I., Dontsova, E.P.: Presence of α-terpinene in the turpentines of various conifers. Khim. Prirod. Soed. 5 (2), 123-124 (1969).
32. Bardysev, I.I., Dontsova, E.P., Bulgakov, A.N., Pertsovskii, A.L.: Chemical composition of turpentines from domestic softwoods. Dokl. Akad. Nauk BSSR 13 (8), 705-707 (1969).
33. Bardysev, I.I., Kulikov, V.I.: Optimum conditions for separating mixtures of terpene hydrocarbons by gas-liquid chromatography. Zh. Anal. Khim. 26 (9), 1857-1858 (1971).
34. Bardysev, I.I., Kulikov, V.I.: Gas-liquid chromatography of terpene hydrocarbons. Gidroliz. Lesokhim. Prom. 5, 17-20 (1972).
35. Bardysev, I.I., Mal'tsev, G.T.: Composition of the volatile oils distilled from the oleoresin and twigs of *Pinus nigra* var. *caramanica* trees and grafts taken from them. Lesn. Zh. 14, 164-167 (1971). For. Abstr. (Oxford) 33, 3561 (1972).
36. Bardysev, I.I., Papanov, G.J., Percovskij, A.L.: Chemical composition of the balsams of *Pinus sylvestris* and *P. nigra* growing in Bulgaria. Dokl. AN BSSR 14, 539-540 (1970). For. Abstr. (Oxford) 32, 5161 (1971).
37. Bardysev, I.I., Roman, L.W.: Investigation of the composition of the turpentine and the properties of rosin from *Pinus pithyusa* var. *stankewiczii* Suk., and *Pinus stankewiczii* Fom. For. Abstr. (Oxford) 22, 2591 (1961).
38. Bardysev, I.I., Udarov, B.G., Bulgakov, A.N., Kulagin, V.P.: Chemical composition of the turpentines from oleoresin of individual pines collected in the normal way or once per season. Lesn. Zh. 16, 118-120 (1973). For. Abstr. (Oxford) 35, 7976 (1974).
39. Bardysev, I.I., Vedeneev, K.P.: Chemical composition of turpentines from individual Scotch pine trees. Sb. Tr., Tsent. Nauch.-Issled. Proekt. Inst. Lesokhim. Prom. 19, 32-33 (1968).
40. Bardysev, I.I., Zen'ko, R.I., Gorbacheva, I.V., Vasil'kova, G.I., Karachun, T.P.: Properties and chemical composition of turpentines separated from oleoresin from Scotch pines growing in various regions of the Soviet Union. Gidroliz. Lesokhim. Prom. 22, 17-18 (1969).

41. Bernard-Dagan, C.: Les essences du Pin maritime: leur répartition dans les divers organes, nature et évolution des monoterpénes. Bull. Soc. Botan. France (Paris, 1966), 14 pp. (1968).
42. Bernard-Dagan, C., Fillon, C., Pauly, G., Baradat, P., Illy, G.: Les terpénes du Pin maritime: aspects biologiques et génétiques. I.- Variabilité de la composition monoterpénique dans un individu, entre individus et entre provenances. Ann. Sci. For. 28, 223-258 (1971).
43. Berton, A., Chararas, C.: Investigations into the concentration of vapours given off by the terpene constituents of certain Abies species and their attractiveness to *Pityokteines curvidens*. C. R. Acad. Agr. France 53, 1238-1243 (1967). For. Abstr. (Oxford) 29, 6029 (1968).
44. Blight, M.M., Brewerton, H.V., McDonald, I.R.C., Bannister, M.H.: β-pinene content of the *Pinus radiata* population of Kaingaroa forest. New Zealand J. Sci. 7, 457-459 (1964).
45. Blight, M.M., McDonald, I.R.C.: Note on ℓ-sabinene from *Pinus muricata*. New Zealand J. Sci. 6, 229-231 (1963).
46. Blight, M.M., McDonald, I.R.C.: Sample reproducibility in Pinus essential oil studies. New Zealand J. Sci. 7, 212-220 (1964).
47. Burchfield, H.P., Storrs, E.E.: Biochemical applications of gas chromatography, 680 pp. New York: Academic Press 1962.
48. Burdon, R.D., Zabkiewicz, J.A.: Identical and non-identical seedling twins in *Pinus radiata*. Can. J. Botany 51, 2001-2004 (1973).
49. Burley, J., Kemp, R.H., Keeble, B., Mathews, W.S.: Preliminary study of chemical variation in turpentine of *Pinus oocarpa*. In: Tropical Provenance and Progeny Research and International Cooperation (eds. J. Burley, D.G. Nikles), pp. 344-350. Oxford: Commonw. For. Inst. 1973.
50. Cabotage, E.V., Ynalvez, L.A., Riffer, R., Anderson, A.B.: Composition of Benguet pine gum oleoresin. Forest Prod. J. 17, 41-44 (1967).
51. Cabotage-Salud, E.: Naval stores. Forpride Dig. 1, 5-6 (1972).
52. Cameron, D.W., Sutherland, M.D.: Some constituents of *Pinus sylvestris* oil. Perfum. Essen. Oil Rec. 50, 200-203 (1959).
53. Chararas, C.: (The reactions of *Ips sexdentatus* to terpene exhalations from various conifers). C. R. Acad. Agr. France 53, 226-231 (1967). For. Abstr. (Oxford) 29, 4264 (1968).
54. Chararas, C., Berton, A.: New method of analysis of terpene vapors of *Pinus maritima* and behavior of *Blastophagus piniperda*. Rev. Pathol. Vegetale Entomol. Agr. France 40, 235-243 (1961).
55. Chararas, C., Berton, A.: Odoriferous constituents of terpene vapors from various tree species and their action towards *Ips sexdentatus*. C. R. Acad. Sci., Paris, 264 D, 1471-1474 (1967).
56. Chaudhary, S.S., Nazir, B.N., Handa, K.L.: Chemical investigations of several Indian pine needle oils. Indian Oil Soap J. 25, 306-312 (1960).
57. Cheng, Y.S., Rudloff, von, E.: The volatile oil of the leaves of *Chamaecyparis nootkatensis*. Phytochemistry 9, 2517-2527 (1970).
58. Chirkova, N.A., Pentegova, V.A.: The sesquiterpenes of *Abies sibirica*. Izv. Sibir. Otd. Akad. Nauk SSSR, Ser. Khim. Nauk 125 (1962).
59. Chudnyi, A.V.: Method of determining the clonal identity of pine grafts. Lesn. Khoz. 6, 45-46 (1971). For. Abstr. (Oxford) 33, 194 (1972).
60. Chudnyi, A.V.: Link between the composition of turpentine oils and the growth of Scotch pine and Siberian pine trees. Novoe Lesokhim., pp. 162-165 (1973). For. Abstr. (Oxford) 35, 5707 (1974).
61. Chudnyi, A.V., Dokuchaeva, M.: Change of the chemical composition of balms in relation to hybridization and vegetative multiplication of some species of the genus Pinus. Sin. Prod. Kanifoli Skipidara, 94-103 (1970).

62. Chudnyi, A.V., Krangauz, R.A., Gundaeva, E.I.: Effect of oleoresin (composition) on the resistance of Scots pine to *Fomes annosus*. Lesn. Khoz. 7, 60-61 (1972). For. Abstr. (Oxford) 34, 4654 (1973).
63. Chudnyi, A.V., Prokazin, E.P.: Geographical variation in the composition of turpentine oils of *Pinus sylvestris* on the territory of the USSR. Rastit. Resur. 9, 494-503 (1973). For. Abstr. (Oxford) 35, 3418 (1974).
64. Cobb, F.W., Jr., Kristic, M., Zavarin, E., Barber, H.W., Jr.: Inhibitory effects of volatile oleoresin components on *Fomes annosus* and four Ceratocystis species. Phytopathology 58, 1327-1335 (1968).
65. Cobb, F.W., Jr., Zavarin, E., Bergot, J.: Effect of air pollution on the volatile oil from leaves of *Pinus ponderosa*. Phytochemistry 11, 1815-1818 (1972).
66. Connor, R.J., Mosimann, J.E.: Concepts of independence for proportions with a generalization of the Dirichlet distribution. J. Am. Stat. Assoc. 64, 194-206 (1969).
67. Cori, O.: Terpene biosynthesis: utilization of neryl pyrophosphate by an enzyme system from *Pinus radiata* seedlings. Arch. Biochem. Biophys. 135, 416-418 (1969).
68. Costa, A.F., Do Vale, J.C.: Materials for the study of aromatic Portugese plants. Volatile oil of *Juniperus phoenicea*. Bol. Esc. Farm. Univ. Coimbra 13/14, 57-67 (1958). For. Abstr. (Oxford) 21, 2569 (1960).
69. Couchman, F.M., Rudloff, von, E.: Gas liquid chromatography of terpenes. Part XIII. The volatile oil of the leaves of *Juniperus horizontalis* Moench. Can. J. Chem. 43, 1017-1021 (1965).
70. Coyne, J.F., Keith, G.C.: Geographic survey of monoterpenes in loblolly and shortleaf pines. USDA Forest Serv. Res. Pap. SO-79, 12 pp. (1972).
71. Cvrkal, H.: Biochemical diagnosis of *Picea abies* in smoke-affected areas. Sborn. Csl. Akad. Zemed. Ved. (Lesn.) 5, 1033-1048 (1959). For. Abstr. (Oxford) 23, 2169 (1962).
72. Cvrkal, H.: The use of gas/liquid chromatography to distinguish physiological varieties of conifers. Lesson 6, 757 (1960). (A.B.I.P.C. 32:1420 (1962)).
73. Cvrkal, H., Janák, J.: Anwendung der Gaschromatographie zur Identifizierung einiger Terpene aus ätherischen Ölen von Nadelbäumen (*Coniferae*). Collect. Czech. Chem. Communs. 24, 1967-1974 (1959).
74. Dal Nogare, S., Juvet, R.S., Jr.: Gas-liquid chromatography theory and practice, 450 pp. New York: Interscience Publ. 1962.
75. Deryuzhkin, R.I., Krasnoboyarova, L.V., Latysh, V.G.: (Effect of various factors on the composition of essential oils of *Larix sibirica*). Lesn. Zh. 14, 110-114 (1971). For. Abstr. (Oxford) 34, 2569 (1973).
76. Drew, J., Pylant, G.D., Jr.: Turpentine from the pulpwoods of the United States and Canada. TAPPI 49, 430-438 (1966).
77. Edwards, E.E.L., Howard, G.E.: Destructively distilled turpentine from Cyprus. Trop. Sci., Lond. 2, 97-98 (1960). For. Abstr. (Oxford) 22, 1219 (1961).
78. Erdtman, H., Kubota, T.: The chemistry of the natural order Cupressales. XXXVII. Monoterpenes from (the essential oil of) the bark of *Juniperus communis* L. For. Abstr. (Oxford) 23, 4511 (1962).
79. Fedorov, N.I., Manukov, E.N.: Composition of gum turpentine and wood turpentine of Scots pine trees infected with *Fomes annosus*. Lesoved. Les. Khoz. 5, 128-133 (1972). For. Abstr. (Oxford) 34, 690 (1973).
80. Fisher, G.C.: Composition of turpentine from *Pinus khasya*. Chem. Ind. 44, 1761 (1963).
81. Fisher, G.C., Squillace, A.E.: GC charts terpenes in conifers and terpene variations among species. Chem. Eng. News 44, 27 (1966).
82. Flake, R.H., Rudloff, von, E., Turner, B.L.: Quantitative study of clinal variation in *Juniperus virginiana* using terpenoid data. Proc. Natl. Acad. Sci. 64, 487-494 (1969).

83. Flake, R.H., Rudloff, von, E., Turner, B.L.: Confirmation of a clinal pattern of chemical differentiation in *Juniperus virginiana* from terpenoid data obtained in successive years. In: Recent Advances in Phytochemistry, Vol. 6 (eds. V.C. Runckles, T.J. Mabry), pp. 215-227 (1973).
84. Flake, R.H., Turner, B.L.: Volatile constituents, especially terpenes, and their utility and potential as taxonomic characters in populational studies. In: Chemistry in Botanical Classification (eds. G. Bendz, J. Santesson), pp. 123-128. New York: Academic Press 1973.
85. Forde, M.B.: Inheritance of turpentine composition in *Pinus attenuata* x *radiata* hybrids. New Zealand J. Botany 2, 53-59 (1964).
86. Forde, M.B., Blight, M.M.: Geographical variation in turpentine of Bishop pine. New Zealand J. Botany 2, 44-52 (1964).
87. Franklin, E.C.: Within-tree variation of monoterpene composition and yield in slash pines of varying size. Forest Sci. (in press).
88. Franklin, E.C., Snyder, E.B.: Variation and inheritance of monoterpene composition in longleaf pine. Forest Sci. 17, 178-179 (1971).
89. Funes, A., Sanchez-Medina, F., Mayor, F.: (Terpene composition and content in *Pinus pinaster*). Montes 27, 17-25 (1971). For. Abstr. (Oxford) 33, 7200 (1972).
90. Funes, A., Sanchez-Medina, F., Mayor, F.: Terpene composition of *Pinus pinaster* seedlings and plants. Phytochemistry 12, 1391-1394 (1973).
91. Gallagher, M.J., Sutherland, M.D.: Terpenoid chemistry. IV. The turpentine of *Araucaria cunninghamii* Ait. Aust. J. Chem. 13, 367-371 (1960). For. Abstr. (Oxford) 23, 1375 (1962).
92. Gansel, C.R., Squillace, A.E.: Geographic variation of monoterpenes in cortical oleoresin of slash pine. Silvae Genetica (in press).
93. Gerhold, H.D., Plank, G.H.: Monoterpene variations in vapors from white pines and hybrids. Phytochemistry 9, 1393-1398 (1970).
94. Gilmore, A.R.: Variation in monoterpene composition of loblolly pine as related to geographic source of seed. In: 11th South. Forest Tree Improv. Conf. Proc., pp. 128-132 (1971).
95. Goryaev, M.I., Dzhalilov, D.R.: Analysis of the essential oil of Kzakh juniper. For. Abstr. (Oxford) 22, 3908 (1961a).
96. Goryaev, M.I., Dzhalilov, D.R.: The essential oil of needles of Turkestan juniper (*Juniperus turkestanica*). For. Abstr. (Oxford) 22, 3909 (1961b).
97. Green, C.L., Keeble, B., Burley, J.: Gum turpentine analysis of some *Pinus kesiya*, *P. caribaea*, and *P. oocarpa* provenances. Preprint, 10th Commonw. Forestry Conf., Oxford, Sept., 17 pp. (1974).
98. Greenwald, S.M.: Environmental effects on the growth and monoterpene production of *Pinus taeda* and *Ocimum basilicum*. Diss. Abstr. Int. B. 33 (1), 5695-5696 (1973).
99. Gregonis, D.E., Portwood, R.D., Davidson, W.H., Durfee, D.A., Levinson, A.S.: Volatile oils from foliage of coast redwood and big tree. Phytochemistry 7, 975-981 (1968).
100. Hancock, W.V., Swan, E.P.: The petroleum ether soluble extractives of British Columbia coastal and interior-type Douglas fir. Phytochemistry 4, 791-798 (1965).
101. Hannus, K., Pensar, G.: Silvichemicals in technical foliage. I. Water steam distilled oil from pine material. Pep. Puu 55, 509-517 (1973).
102. Hanover, J.W.: Environmental variation in the monoterpenes of *Pinus monticola* Dougl. Phytochemistry 5, 713-717 (1966a).
103. Hanover, J.W.: Genetics of terpenes. I. Gene control of monoterpene levels in *Pinus monticola* Dougl. Heredity 21 (1), 73-84 (1966b).
104. Hanover, J.W.: Inheritance of 3-carene concentration in *Pinus monticola*. Forest Sci. 12, 447-450 (1966c).
105. Hanover, J.W.: Genetics of terpenes. II. Genetic variances and interrelationships of monoterpene concentrations in *Pinus monticola*. Heredity 27 (2), 237-245 (1971).

106. Hanover, J.W.: Factors affecting the release of volatile chemicals by forest trees. Mitteilungen der forstlichen Bundes-Versuchsanstalt Wein 97, 625-644 (1972).
107. Hanover, J.W.: Physiology of tree resistance to insects. Ann. Rev. Entomol. 20, 75-95 (1975).
108. Hanover, J.W., Furniss, M.M.: Monoterpene concentration in Douglas-fir in relation to geographic location and resistance to attack by the Douglas-fir beetle. USDA Forest Serv. Res. Pap. 6, 23-28 (1966).
109. Hanover, J.W., Wilkinson, R.C.: A new hybrid between blue spruce and white spruce. Can. J. Botany 47, 1693-1700 (1969).
110. Hanson, J.R.: The biosynthesis of the monoterpenes. Perfum. Essent. Oil Rec. 58, 787-795 (1967).
111. Heikkenen, H.J., Hrutfiord, B.F.: *Dendroctonus pseudotsugae*: a hypothesis regarding its primary attractant. Science 150 (3702), 1457-1459 (1965).
112. Hillis, W.E., Ishikura, N.: The extractives of the mycorrhizas and roots of *Pinus radiata* and *Pseudotsuga menziesii*. Aust. J. Biol. Sci. 22, 1425-1436 (1969).
113. Hilton, R.L.: Genetic variation and interrelationships of the cortical monoterpenes, foliar mineral elements, and growth characteristics of eastern white pine. Ph.D. Diss., Mich. State Univ., 127 pp. (1968).
114. Hirose, Y.: Genetic interpretation of sesquiterpenoids in the essential oils. Pap. No. 28, Div. of Agric. Food Chem. 58th Nat. A.C.S. Meeting, N.Y. (1969).
115. Hodges, J.D., Lorio, P.L., Jr.: Comparison of oleoresin composition in declining and healthy loblolly pines. USDA Forest Serv. Res. Note SO-158, 4 pp. (1973).
116. Hodges, J.D., Lorio, P.L., Jr.: Moisture stress and composition of xylem oleoresin in loblolly pine. Paper No. 70, Proc. 3rd North Amer. Forest Biol. Workshop (1974).
117. Hörster, H.: Variability of the oils of *Juniperus communis*. I. The composition of the oils in leaves of different age. Planta Med. 23, 353-362 (1973). For. Abstr. (Oxford) 35, 2063 (1974).
118. Hörster, H.: Comparison of the monoterpene fractions of *Juniperus drupacea* and *Juniperus oxycedrus*. Planta Med. 26, 113-118 (1974).
119. Hörster, H., Csedo, C., Racz, G.: Gas-chromatographic analysis of the volatile oil from juniper leaves (*Juniperus communis*) harvested in Romania. Rev. Med. (Tirgu-Mures, Rom.) 20, 79-83 (1974).
120. Hrivnak, J., Mahdalik, M., Varadiova, E., Sojak, L.: Analysis of monoterpenes from the needles of *Pinus sylvestris* and *Picea excelsa* using capillary gas chromatography. Holzforsch. Holzverwert. 25, 24-26 (1973).
121. Hunt, R.S., Rudloff, von, E.: Chemosystematic studies in the genus Abies. I. Leaf and twig oil analysis of alpine and balsam firs. Can. J. Botany 52, 477-487 (1974).
122. Iconomou, N., Valkanas, G., Buechi, J.: Comparison of gum turpentines of *Pinus halepensis* and *Pinus brutia* grown in Greece. J. Chromatogr. 16, 29-33 (1964).
123. Ishikawa, M., Tsuchiya, T.: Gas chromatography of natural essential oils. Shika Zairyo Kenkyusho Hokoku 2, 401-408 (1962).
124. Joye, N.M., Jr., Proveaux, A.T., Lawrence, R.V.: Composition of pine needle oil. J. Chromatogr. Sci. 10, 590-592 (1972).
125. Juvonen, S.: Über die terpenbiosynthese-beeinflussenden Faktoren in *Pinus sylvestris* L. Acta Botan. Fenn. 71, 1-92 (1966).
126. Juvonen, S.: Über die Unterschiede in der Zusammensetzung des Kiefernnadelöles von verschiedenen Zweigwirteln desselben Baumes. Farm. Aikak. 78, 46-53 (1969).
127. Juvonen, S.: Terpene biosynthesis studies with C-14 labeled compounds in *Pinus silvestris* L. (Biogenetic studies, I.) Farm. Aikak. 79, 207-210 (1970a). For. Abstr. (Oxford) 33, 5768 (1972).

128. Juvonen, S.: Über die Harzgänge und das ätherische Öl der Kiefernnadel. Farm. Aikak. 79, 107-112 (1970b).

129. Juvonen, S.: Gaschromatographisch-massenspektrometrische Untersuchung von Kiefernnadelöl (*Pinus silvestris* L.). Farm. Aikak. 79, 137-145 (1970c).

130. Juvonen, S., Hiltunen, R.: Über das Vorkommen von 3-Caren-armen und -reichen Chemotypen bei *Pinus silvestris* L. in Finnland. Biogenetische Studien II. Farm. Aikak. 81, 137-145 (1972).

131. Juvonen, S., Hiltunen, R.: Neue Erkenntnisse über die Terpenbiosynthese in *Pinus silvestris* L. Farm. Aikak. 83, 71-82 (1974).

132. Juvonen, S., Lako, H.: Über die Unterschiede in der Zusammensetzung des ätherischen Öles verschiedener Zweige derselben Fichte. Farm. Aikak. 76, 126-130 (1967).

133. Karapandzhc, D.M.: Chemical composition of the essential oils from *Pinus nigra* Arn. needles. Glas. Hem. Drus., Beograd 31, 469-472 (1966a).

134. Karapandzhc, D.M.: Chemical composition of the essential oils from the needles of *Pinus heldreichii* Christ. Glas. Hem. Drus., Beograd 31, 473-476 (1966b).

135. Karlsen, J., Baerheim Svendsen, A.: Direct gas-liquid chromatography of volatile oil constituents in plant material. I. The monoterpene hydrocarbons of Norwegian spruce needle oil. Medd. Norsk Farm. Selskap 28, 85-90 (1966).

136. Karnik, M.G., Bhatia, K., Lal, J.: Seasonal variation in pine needle oil from *Pinus roxburghii*. Indian For. 92, 521-522 (1966). For. Abstr. (Oxford) 28, 1606 (1967).

137. Kepner, R.E., Maarse, H.: Changes in composition of volatile terpenes in Douglas fir needles during maturation. J. Agr. Food Chem. 18, 1095-1101 (1970).

138. Klouwen, M.H., Ter Heide, R.: Studies on terpenes. I. A systematic analysis of monoterpene hydrocarbons by gas-liquid chromatography. J. Chromatogr. 7, 297-310 (1962).

139. Kolesnikova, R.D., Latysh, V.G., Popova, N.I., Deryuzhkin, R.I., Krasnoboyarova, L.V.: Preparative chromatographic separation of certain larch essential oil components. Prirod. Soed. 8, 612-615 (1972).

140. Kowalski, J.: Studies on the essential oil of Polish silver fir. For. Abstr. (Oxford) 23, 6054 (1962).

141. Krupa, S., Andersson, J., Marx, D.H.: Studies on ectomycorrhizae of pine. IV. Volatile organic compounds in mycorrhizal and nonmycorrhizal root systems of *Pinus echinata* Mill. Eur. J. Forest Pathol. 3, 194-200 (1973).

142. Krupa, S., Nylund, J.E.: Studies on ectomycorrhizae in pine. III. Growth inhibition of two root pathogenic fungi by volatile organic constituents of ectomycorrhizal root systems of *Pinus sylvestris* L. Eur. J. Forest Pathol. 2, 88-94 (1972).

143. Latting, J., Scora, R.W.: Chemosystematics of *Pseudotsuga* in California. Amer. J. Botany 61, 34 (1974). Plant Breeding Abstr. 44, 7245 (1974).

144. Lee, C.J., Snajberk, K., Zavarin, E.: Chemical composition of the cortical essential oil from *Abies balsamea*. Phytochemistry 13, 179-183 (1974).

145. Lester, D.T.: Geographic variation in leaf and twig monoterpenes of balsam fir. Can.J. Forest Res. 4, 55-60 (1974).

146. Levinson, A.S., Lemoine, G., Smart, E.C.: Volatile oil from foliage of *Sequoiadendron giganteum*: change in composition during growth. Phytochemistry 10, 1087-1094 (1971).

147. Lewis, K.R.: Terpenes of Virginia pine (*Pinus virginiana*), with investigations of their attractancy to the southern pine beetle (*Dendroctonus frontalis*). Diss. Abstr. Int. B 32 (1), 7090 (1972).

148. Lishtvanova, L.N., Akimov, Y.A.: Composition of pine needle essential oil studied by capillary chromatography. Tr. Vses. Nauch.-Issled. Inst. Efirnomaslich. Kul't. 3, 54-56 (1971).

149. Loomis, W.D.: Biosynthesis and metabolism of monoterpenes. In: Terpenoids in Plants (ed. J.B. Pridham), pp. 59-82. New York: Academic Press 1967.

150. Lotan, J.E., Joye, N.M., Jr.: Some variation of terpenes in Montana lodgepole pine. USDA Forest Serv. Res. Note INT-120, 4 pp. (1970).

151. Maarse, H., Kepner, R.E.: Changes in composition of volatile terpenes in Douglas fir needles during maturation. J. Agr. Food Chem. 18, 1095-1101 (1970).

152. Manjarrez, A., Guzman, A.: Composition of turpentines obtained from the State of Michoacan (Mexico). Bol. Inst. Quim. Univ. Nal. Auton. Mex. 16, 20-24 (1964).

153. Martinek, A.: Study of monoterpenes from essential oils of mountain pine by combined gas and thin-layer chromatography. Planta Med. 17, 201-208 (1969).

154. Mayo, de, P.: Mono- and sesquiterpenoids, 320 pp. New York: Interscience Publ. Inc. 1959.

155. McGimpsey, J.R., Murray, J.: Essential oils of New Zealand Podocarpaceae. II. *Podocarpus spicatus*. J. Appl. Chem. 10, 340-344, (1960). For. Abstr. (Oxford) 22, 2594 (1961).

156. Melin, E., Krupa, S.: Studies on ectomycorrhizae of pine. II. Growth inhibition of mycorrhizal fungi by volatile organic constituents of *Pinus silvestris* (Scots pine) roots. Physiol. Plant. 25, 337-340 (1971).

157. Mirov, N.T.: Effect of the crown on the composition of oleoresin in pines. J. Forestry 43, 345-348 (1945).

158. Mirov, N.T.: Distribution of turpentine components among species of the genus Pinus. In: The Physiology of Forest Trees (ed. K.F. Thimann), pp. 251-268. New York: Ronald Press 1958a).

159. Mirov, N.T.: Composition of gum turpentine of pines. XXX. A report on *Pinus serotina*, *P. tenuifolia* and *P. yunnanensis*. J. Am. Pharm. Assoc. 47, 410-413 (1958b). For. Abstr. (Oxford) 20, 1258 (1959).

160. Mirov, N.T.: Composition of gum turpentines of pines. USDA Tech. Bull. 1239, 158 pp. (1961).

161. Mirov, N.T.: Chemistry and plant taxonomy. Lloydia 26, 117-124 (1963).

162. Mirov, N.T., Frank, E., Zavarin, E.: Chemical composition of *P. elliottii* var. *elliottii* turpentine and its possible relations to taxonomy of several pine species. Phytochemistry 4, 563-568 (1965).

163. Mirov, N.T., Iloff, P.M., Jr.: Chemical composition of gum turpentine of pines. XXIX. A report on *Pinus ponderosa* from five localities: Central Idaho, central Montana, southeastern Wyoming, northwestern Nebraska and central eastern Colorado. Am. Pharm. Assoc. J. Sci. Ed. 47, 404-409 (1958).

164. Mirov, N.T., Zavarin, E., Bicho, J.G.: Composition of gum turpentine of pines. *Pinus nelsonii* and *Pinus occidentalis*. J. Pharm. Sci. 51, 1131-1135 (1962).

165. Mirov, N.T., Zavarin, E., Snajberk, K.: Chemical composition of the turpentines of some eastern Mediterranean pines in relation to their classification. Phytochemistry 5, 97-102 (1966).

166. Mirov, N.T., Zavarin, E., Snajberk, K., Costello, K.: Further studies of turpentine composition of *Pinus muricata* in relation to its taxonomy. Phytochemistry 5, 343-355 (1966).

167. Monteoliva, M., Funes, A., Garcia-Peregrin, E., Zofra, M.F., Henares, M.: Biosintesis di terpenos en pinos. Ars Pharmaceutica 10, 285-347 (1969).

168. Mori, Y.: A practical quantitative method for unresolved gas chromatographic peaks. I. Two overlapping peaks. J. Chromatogr. 66, 9-18 (1972).

169. Mosimann, J.E.: On the compound multinomial distribution, the multivariate β-distribution, and correlations among proportions. Biom. 49 (1 and 2), 65-82 (1962).
170. Motl, O., Herout, V., Sorm, F.: On terpenes. CXII. The composition of the oil from *Juniperus oxycedrus* L. berries. Coll. Czech. Chem. Communs. 25, 1656-1661 (1960).
171. Motl, O., Paknikak, S.K.: On the terpenes. CXCIII. Composition of oil from *Cupressus funebris* leaves. Coll. Czech. Chem. Communs. 33, 1934-1942 (1968). For. Abstr. (Oxford) 32, 3501 (1971).
172. Murray, J.: Essential oils of New Zealand Podocarpaceae. III. *Podocarpus nivalis* and *Dacrydium laxifolium*. J. Appl. Chem. 10, 366 (1960). For. Abstr. (Oxford) 22, 3911 (1961).
173. Nikles, D.G.: Comparative variability and relationship of Caribbean pine (*Pinus caribaea* Mor.) and slash pine (*Pinus elliottii* Engelm.). Ph.D. Thesis, N.C. State Univ., Sch. of Forest Resour., 201 pp. (1966).
174. Ogilvie, R.T., Rudloff, von, E.: Chemosystematic studies in the genus Picea (Pinaceae). IV. The introgression of white and Engelmann spruce as found along the Bow River. Can. J. Botany 46, 901-908 (1968).
175. Ognyanov, I., Tsankova, E.: The composition of Bulgarian *Pinus nigra* var. *austriaca*. Herba Hungarica, Budapest 5, 71-74 (1966). For. Abstr. (Oxford) 28, 6687 (1967).
176. Ognyanov, I., Tsankova, E.: Sesquiterpene compounds in Bulgarian coniferous trees. Meshdunar. 4th Kongr. Efirnym Maslam (Mater.), 1, 252-255 (1968).
177. Ognyanov, I., Vlahov, R., Tsankova, E.: Composition of Bulgarian essential oil from *Pinus silvestris* L. leaves. Compt. rend. l'Acad. Bulg. Sci. 17, 483-486 (1964). For. Abstr. (Oxford) 26, 1562 (1965).
178. Oh, H.K., Sakai, T., Jones, M.B., Longhurst, W.M.: Effect of various essential oils isolated from Douglas fir needles upon sheep and deer rumen microbial activity. Appl. Microbiol. 15, 777-784 (1967).
179. Okay, M.: Untersuchung über türkische Kiefernnadelöle. Comm. Fac. Sci. Univ. Ankara 11, 1 p. (1963-64).
180. Overhulser, D., Hrutfiord, B., Gara, R.I.: The sitka spruce weevil-host relationship. Abstr. 2nd N.Am. For. Biol. Workshop, p. 28 (1972).
181. Pauly, G.: Les canaux sécretéurs du Pin maritime. In: Recueil des Travaux du Laboratoire de Biologie Végétale de la Faculté des Sciences de Bordeaux. Vol. IV, pp. 1-33 (1963).
182. Pauly, G., Gleizes, M., Bernard-Dagan, C.: Identification des constituants de l'essence des aiguilles de *Pinus pinaster*. Phytochemistry 12, 1395-1398 (1973).
183. Pauly, G., Rudloff, von, E.: Chemosystematic studies in the genus Pinus: the leaf oil of *Pinus contorta* var. *latifolia*. Can. J. Botany 49, 1201-1210 (1971).
184. Peloquin, R.L.: Geographic variation of the monoterpenes of *Pinus ponderosa*. M.S. Thesis, Stanford Univ., Palo Alto, Calif., 31 pp. (1964).
185. Pentegov, A.P., Cirkova, M.A.: Utilization of *Abies sibirica* needles for the production of oil and (synthetic) camphor. For. Abstr. (Oxford) 21, 3882 (1960).
186. Pentagova, V.A., Dubovenko, Z.V., Vol'skii, L.N., Vasilink, S.M., Chirkova, M.A., Schmidt, E.N.: Sesquiterpene hydrocarbons of the oleoresins of some Siberian conifers. Izv. Sibir. Otd. Akad. Nauk SSSR, Ser. Khim. Nauk. 2, 114-118 (1968). For. Abstr. (Oxford) 31, 1709 (1970).
187. Pigulevskii, G.V., Maksimova, A.M.: Akad. Nauk. SSSR 5, 66 (1961).
188. Polozhentsev, P.A., Chudnyi, A.V., Zolotov, L.A.: Certain peculiarities in the composition of gum turpentine in ailing pine trees. Lesn. Khoz. 22, 8, 60-62 (1969).

189. Polozhentsev, P.A., Zolotov, L.A., Latysh, V.G.: Composition and toxicity of pine oleoresin in sites of root rot. Izv. Vyssh. Ucheb. Zaved., Les. Zh. 13, 3-6 (1970).

190. Poltavchenko, Y.A., Rudakov, G.A.: Use of gas-liquid chromatography for studying the genesis of monoterpenes in conifer family pines. Izv. Nauch.-Issled. Inst. Nefte-Uglekhim. Sin. Irkutsk. Univ. 11 (pt. 1), 39-40 (1969). Chem. Abstr. 78, 2056g.

191. Poltavchenko, Y.A., Rudakov, G.A.: Essential oils of coniferous trees from the Baikal region and the problem of the origin of terpenes. Sintetich. Prod. iz Kanifoli i Skipidara, p. 82-94 (1970). For. Abstr. (Oxford) 33, 7199 (1972).

192. Poltavchenko, Y.A., Rudakov, G.A.: Biosynthesis of monoterpenes in *Pinus sibirica* and *P. pumila*. Biol. Nauk. 7, 95-100 (1972). For. Abstr. (Oxford) 34, 2655 (1973).

193. Poltavchenko, Y.A., Tkach, T.N., Tkach, V.S., Rudakov, G.A.: Changes in the distribution of monoterpenes in *Pinus sylvestris*. Biol. Nauk. 11 (10), 71-76 (1968). For. Abstr. (Oxford) 30, 3506 (1969).

194. Powell, R.A., Adams, R.P.: Seasonal variation in the volatile terpenoids of *Juniperus scopulorum* (Cupressaceae). Am. J. Botany 60, 1041-1050 (1973).

195. Radwan, M.D.: Differences in phenols and monoterpene hydrocarbons among Douglas-fir genotypes. Plant Physiol. 49 (Suppl.), 6 (Abstr.)(1972). Plant Breed. Abstr. 43 (2), 1532 (1973).

196. Rao, B.S., Sood, V.K.: "Pine" needle oil from the Himalayan silver fir, *Abies pindrow*. Soap Perfum. Cosmet. 35, 253-254 (1962).

197. Rife Lamprecht, M.P., Carroscosa Barron, M.T.: The composition of Spanish turpentine. Ann. Inst. Forest Insec. Exp. 37, 113-148 (1965). For. Abstr. (Oxford) 28, 4904 (1967).

198. Riffer, R., Anderson, A.B., Cabotage, E.V., Ynalvez, A.: Composition of turpentine and resin acid in Benguet pine stump heartwood (*Pinus insularis* Endl.). TAPPI 49, 415-417 (1966).

199. Roberts, D.R.: Effect of wounding on the composition of slash pine oleoresin: a preliminary report. Assoc. Southeast. Biol. Bull. 15, 53 (Abstr.)(1968a).

200. Roberts, D.R.: Use of the solid sample technique for pine oleoresin gas chromatographic analysis. J. Gas Chromatogr. 6, 126-127 (1968b).

201. Roberts, D.R.: Within-tree variation of monoterpene hydrocarbon composition of slash pine oleoresin. Phytochemistry 9, 809-815 (1970).

202. Rockwood, D.L.: Aspects of monoterpene composition in loblolly pine. Ph.D. Thesis, N.C. State Univ., Sch. of Forest Resour., 129 pp. (1972).

203. Rockwood, D.L.: Variation in the monoterpene composition of two oleoresin systems of loblolly pine. Forest Sci. 19, 147-153 (1973a).

204. Rockwood, D.L.: Monoterpene-fusiform rust relationships in loblolly pine. Phytopathology 63, 551-553 (1973b).

205. Rockwood, D.L.: Cortical monoterpene and fusiform rust resistance relationships in slash pine. Phytopathology 64, 976-979 (1974).

206. Rottink, B.A., Hanover, J.W.: Identification of blue spruce cultivars by analysis of cortical oleoresin monoterpenes. Phytochemistry 11, 3255-3257 (1972).

207. Rudakov, G.A., Poltavchenko, Y.A.: Genesis of monoterpenes derived from the essential oil of the Scotch pine. Mezhdunar, 4th Kongr. E firnym Maslam, (Mater.) 1968, 1, 289-296 (1971). Chem. Abstr. 78, 140347d (1973).

208. Rudakov, G.A., Poltavchenko, Y.A., Ivanova, L.S.: Biogenesis and interconversions of monoterpenes in the essential oils of some conifers. 7th Intern. Symp. Chem. Nat. Prod., Riga, pp. 439-440 (1970).

209. Rudinsky, J.A., Novak, V., Svihra, P.: Terpenes and pheromones as attractants for spruce bark beetle (*Ips typographus* L.). Lesson 16 (12), 1051-1062 (1970).

210. Rudloff, von, E.: Gas-liquid chromatography of terpenes. Part IV. The analysis of the volatile oils of the leaves of eastern white cedar. Can. J. Chem. 39, 1200-1206 (1961).
211. Rudloff, von, E.: Gas-liquid chromatography of terpenes. VI. The volatile oil of *Thuja plicata* Donn. Phytochemistry 1, 195-202 (1962a).
212. Rudloff, von, E.: Gas-liquid chromatography of terpenes. V. The volatile oils of the leaves of black, white and Colorado spruce. TAPPI 45, 181-184 (1962b).
213. Rudloff, von, E.: Gas-liquid chromatography of terpenes. IX. The volatile oil of the leaves of *Juniperus sabina* L. Can. J. Chem. 41, 2876-2881 (1963).
214. Rudloff, von, E.: Gas-liquid chromatography of terpenes. Part X. The volatile oils of the leaves of Sitka and Engelmann spruce. Can. J. Chem. 42, 1057-1062 (1964).
215. Rudloff, von, E.: Gas chromatographic analysis of the volatile oil from a single conifer needle. J. Gas Chromatogr. 3, 390-391 (1965).
216. Rudloff, von, E.: Gas-liquid chromatography of terpenes. XIV. The chemical composition of the volatile oil of the leaves of *Picea rubens* Sarg. and chemo-taxonomic correlations with other North American spruce species. Phytochemistry 5, 331-341 (1966).
217. Rudloff, von, E.: Chemosystematic studies in the genus Picea (Pinaceae). I. Introduction. Can. J. Botany 45, 891-901 (1967a).
218. Rudloff, von, E.: Chemosystematic studies of the genus Picea (Pinaceae). II. The leaf oil of *Picea glauca* and *P. marianna*. Can. J. Botany 45, 1703-1714 (1967b).
219. Rudloff, von, E.: Gas-liquid chromatography of terpenes. Part XVI. The volatile oil of the leaves of *Juniperus ashei* Buchholz. Can. J. Chem. 46, 679-683 (1968).
220. Rudloff, von, E.: Scope and limitations of gas chromatography of terpenes in chemosystematic studies. In: Recent Advances in Phytochemistry, Vol. 2 (ed. M.K. Seikel, V.C. Runeckles), pp. 127-159. New York: Appleton-Century-Crofts 1969.
221. Rudloff, von, E.: Chemosystematic studies in the genus Pseudotsuga. I. Leaf oils analysis of the coastal and Rocky Mountain varieties of Douglas-fir. Can. J. Botany 50, 1025-1040 (1972a).
222. Rudloff, von, E.: Seasonal variation in the composition of the volatile oil of the leaves, buds, and twigs of white spruce (*Picea glauca*). Can. J. Botany 50, 1595-1630 (1972b).
223. Rudloff, von, E.: Geographic variation in the terpene composition of the leaf oil of Douglas fir. Pure Appl. Chem. 34, 401-410 (1973a).
224. Rudloff, von, E.: Chemosystematic studies in the genus Pseudotsuga. (3). Population differences in British Columbia as determined by volatile leaf oil analysis. Can. J. Forest Res. 3 (3), 443-452 (1973b).
225. Rudloff, von, E.: Gas-liquid chromatography of terpenes. Adv. Chromatogr. 10, 173-230 (1974).
226. Rudloff, von, E., Couchman, F.M.: Gas-liquid chromatography of terpenes. Part XI. The volatile oil of the leaves of *Juniperus scopulorum* Sarg. Can. J. Chem. 42, 1890-1895 (1964).
227. Rudloff, von, E., Holst, M.J.: Chemosystematic studies in the genus Picea (Pinaceae). III. The leaf oil of a *Picea glauca* x *mariana* hybrid (Rosendahl spruce). Can. J. Botany 46, 1-4 (1968).
228. Rudloff, von, E., Irving, R., Turner, B.L.: Reevaluation of allopatric introgression between *Juniperus ashei* and *J. virginiana* using gas chromatography. Am. J. Botany 54 (5, II), 660 (1967).
229. Rudloff, von, E., Sood, V.K.: Gas-liquid chromatography of terpenes. Part XVIII. The volatile oil of the leaves of *Juniperus communis* L. Can. J. Chem. 47, 2081-2086 (1969).
230. Runeberg, J.: The chemistry of the natural order Cupressales. XXVII. Heartwood constituents of *Juniperus utahensis* Lemm. Acta Chem. Scand. 14, 797-804 (1960a).

231. Runeberg, J.: The chemistry of the natural order Cupressales. XXVIII. Constituents of *Juniperus virginiana* L. Acta Chem. Scand. 14, 1288-1294 (1960b).
232. Rybinski, S., Rozmej, Z.: The use of gas chromatography in turpentine analysis. Prace Inst. Tech. Drewna 14, 3-10 (1968). For. Abstr. (Oxford) 30, 3326 (1969).
233. Sakai, T., Maarse, H., Kepner, R.E., Jennings, W.G., Longhurst, W.M.: Volatile components of Douglas fir needles. J. Agr. Food Chem. 15, 1070-1072 (1967).
234. Sakhalov, E., Belova, N.V.: Chemical study of essential oils derived from *Biota orientalis* Endl. and *Cupressus sempervirens* L. cultivated in Azerbaidzhan. Farmatsiya (Moscow) 17, 33-39 (1968). Biol. Abstr. 50, 123138 (1969).
235. Sandermann, W.: Biosynthetic investigations on different Pinus species. Holzforsch. 16, 65-74 (1962). Transl. Commonw. Sci. Ind. Res. Organ. Aust. No. 6793 (1964).
236. Sandermann, W., Bruns, K.: Über die Biogenese des Limonens in *Pinus pinea* L. Naturwissenschaften 49, 258-259 (1962).
237. Sandermann, W., Schweers, W.: Über die Biogenese von α-pinen in *Pinus nigra austriaca*. Tetrahedron Letters 7, 257-258 (1962).
238. Sandermann, W., Schweers, W., Beinhoff, O.: Über die Biogenese von n-Heptan in *Pinus jeffreyi* Murr. Chem. Ber. 93, 2266-2271 (1960).
239. Santamour, F.S., Jr.: Insect-induced crystallization of white pine resins. I. White-pine weevil. USDA Forest Serv. Res. Note NE-38, 8 pp. (1965).
240. Santamour, F.S., Jr., Cunningham, F.E., Peterson, R.J.: Variation in resistance of hard pines to mouse damage. USDA Forest Serv. Res. Note NE-5, 6 pp. (1963).
241. Savory, B.M.: The taxonomy of *Pinus khasya* (Royle) and *Pinus insularis* (Endlicher). Empire Forest Rev. 41, 67-80 (1962).
242. Saylor, L.C., Kang, K.W.: A study of sympatric populations of *Pinus taeda* L. and *Pinus serotina* Michx. J. Elisha Mitchell Sci. Soc. 89, 101-110 (1973).
243. Schantz, von, M.: Über die Zusammensetzung der ätherischen Öle bei verschiedenen Picea-arten. Planta Med. 13, 369-381 (1965).
244. Schantz, von, M., Juvonen, S.: Chemotaxonomische Untersuchungen in der Gattung Picea. Acta Botan. Fenn. 73, 1-51 (1966).
245. Schmidtling, R.C.: Rootstock does not affect cortical monoterpene composition in loblolly pine. Forest Sci. 20, 375-376 (1974).
246. Shaw, A.C.: The essential oil of *Tsuga canadensis* (L.) Carr. J. Am. Chem. Soc. 73, 2859-2861 (1951).
247. Shrimpton, D.M.: Extractives associated with wound response of lodgepole pine attacked by the mountain pine beetle and associated microorganisms. Can. J. Botany 51, 527-534 (1973).
248. Simonsen, J.L., Owen, L.N.: The terpenes. Vol. I. The simpler acyclic and monocyclic terpenes and their derivatives, 2nd ed., 479 pp. Cambridge Univ. Press 1953.
249. Simonsen, J.L., Owen, L.N.: The terpenes. Vol. II. The dicyclic terpenes and their derivatives, 2nd ed., 619 pp. Cambridge Univ. Press 1957.
250. Sindelar, J., Skalska, A.: Variability in the content and certain properties of essential oils in organs of *Larix decidua*. For. Abstr. (Oxford) 28, 141-184 (1967).
251. Singh, A.P.: Oil of turpentine as raw material for chemical industry. ISI Bull. (India) 20, 310-314 (1968).
252. Sinha, G.K., Prakash, S.: Chemical examination of essential oil from the fruits of *Cupressus torulosa* D. Don. Indian Perfum. 15, 11-15 (1971). For. Abstr. (Oxford) 34, 6683 (1973).
253. Sinha, G.K., Prakash, S.: Chemical studies of the plant, *Cupressus torulosa*. Indian Perfum. 17 (2), 31-34 (1973).
254. Sirota, A.Y.: Chemical and enzymic studies on *Pinus radiata* tissues. Anales Fac. Quim. Farm., Univ. Chile 17, 36-40 (1965).

255. Smedman, L.A., Snajberk, K., Zavarin, E., Mon, T.R.: Oxygenated monoterpenoids and sesquiterpenoid hydrocarbons of the cortical turpentine from different Abies species. Phytochemistry 8, 1471-1479 (1969).
256. Smedman, L.A., Zavarin, E., Teranishi, R.: Composition of oxygenated monoterpenoids and sesquiterpenoid hydrocarbons from the cortical oleoresin of *Abies magnifica* A. Murr. Phytochemistry 8, 1457-1470 (1969).
257. Smelyanets, V.P.: Role of terpenoids in (tree) resistance to pests. Anz. Schädlingsk. Pflanzenschutz 42, 33-37 (1969).
258. Smelyanets, V.P., Chursin, L.A.: Effects of terpenoids on the distribution of insect pests in young pine stands. Anz. Schädlingsk. Pflanzenschutz 45, 134-140 (1972).
259. Smith, R.H.: Perennial constancy of the monoterpene synthesis in the wood oleoresin of *Pinus ponderosa*. Nature 202 (4927), 107-108 (1964a
260. Smith, R.H.: Variations in the monoterpene composition of ponderosa pine wood oleoresin. USDA Forest Serv. Res. Pap. PSW-15, 17 pp. (1964b).
261. Smith, R.H.: Variation in the monoterpenes of *Pinus ponderosa* Laws. Science 143, 1337-1338 (1964c).
262. Smith, R.H.: The monoterpenes of lodgepole pine oleoresin. Phytochemistry 3, 259-262 (1964d).
263. Smith, R.H.: Effect of monoterpene vapors on the western pine beetle. J. Econ. Entomol. 58, 509-510 (1965).
264. Smith, R.H.: The monoterpene composition of *Pinus ponderosa* xylem resin and of *Dendroctonus brevicomis* pitch tubes. Forest Sci. 12, 63-68 (1966a).
265. Smith, R.H.: Resin quality as a factor in the resistance of pines to bark beetles. In: Breeding Pest Resistant Trees (eds. H.D. Gerhold, E.J. Schriener, R.E. McDermott, J.A. Winieski), pp. 189-196. New York: Pergamon Press 1966b.
266. Smith, R.H.: Monoterpene composition of pine species and hybrids - some preliminary findings. USDA Forest Serv. Res. Note PSW-135, 14 pp. (1967a).
267. Smith, R.H.: Variations in the monoterpene composition of the wood resin of Jeffrey, Washoe, Coulter and lodgepole pines. Forest Sci. 13, 246-252 (1967b).
268. Smith, R.H.: Intratree measurements of the monoterpene composition of ponderosa pine xylem resin. Forest Sci. 14, 418-419 (1968).
269. Smith, R.H.: Xylem resin as a factor in the resistance of pines to forced attacks by bark beetles. 2nd World Consult. Forest Tree Breed. FO-FTB-69-5/6, 13 pp. (1969).
270. Smith, R.H.: Xylem monoterpenes of *Pinus ponderosa*, *P. washoensis*, and *P. jeffreyi* in the Warner Mountains of California. Madrőno 21, 26-32 (1971).
271. Smith, R.H.: Monoterpenes of ponderosa pine xylem resin in western United States. USDA Tech. Bull. (1976).
272. Smith, R.H., Peloquin, R.L., Passof, P.C.: Local and regional variation in the monoterpenes of ponderosa pine wood oleoresin. USDA Forest Serv. Res. Pap. PSW-56, 10 pp. (1969).
273. Snajberk, K., Lee, C.J., Zavarin, E.: Chemical composition of volatile from cortical oleoresin of *Pseudotsuga menziesii*. Phytochemistry 13 185-188 (1974).
274. Snedecor, G.W., Cochran, W.G.: Statistical methods, 593 pp., 6th ed. Ames: Iowa State Univ. Press 1967.
275. Squillace, A.E.: Inheritance of monoterpene composition in cortical oleoresin of slash pine. Forest Sci. 17, 381-387 (1971).
276. Squillace, A.E., Fisher, G.S.: Evidences of the inheritance of turpentine composition in slash pine. USDA Forest Serv. Res. Pap. NC-6, 53-60 (1966).
277. Stairs, G.R.: Monoterpene composition in Larix. Silvae Genetica 17, 182-186 (1968).

278. Stankovic, S.C.: Glasnik Ŝumarskog Fak. Univ. Beogradu 24, 82 (1961).
279. Stepanov, E.V.: Physiological characteristics of the secretion and composition of volatile substances of Siberian fir resin. Fiziol. Biokhim. Osn. Vzaimodeistviya Rast. Fitotsenozakh 3, 96-98 (1972).
280. Stepanov, E.V., Dubovenko, Z.V.: Study of volatile organic substances by direct gas chromatography of plant material. I. Monoterpene hydrocarbons. Izv. Sib. Otd. Akad. Nauk SSSR, Ser. Biol. Nauk 3, 152-158 (1970).
281. Stepanov, E.V., Dubovenko, Z.V.: Direct gas chromatography of plant material for studying volatile organic substances. 2. Mono- and sesquiterpenoids. Izv. Sib. Otd. Akad. Nauk SSSR, Ser. Biol. Nauk 2, 84-88 (1971).
282. Takashima, T., Nii, T., Kodama, S.: Exceptionally rotatory polarization of turpentine obtained from a pine tree. Nippon Mokuzai Gakkaishi 10, 214-216 (1964).
283. Tatro, V.E., Scora, R.W., Vasek, F.C., Kumamoto, J.: Variation in the leaf oils of three species of Junipers. Am. J. Botany 60, 236-241 (1973).
284. Tchertches, H.A., Bardyshev, I.I., Rekunova, E.A.: Composition of coniferous essential oil of *Pinus sylvestris*. Zh. Prikl. Khim. 1, 209-212 (1962).
285. Thor, E., Barnett, P.E.: Taxonomy of Abies in the southern Appalachians: variation in balsam monoterpenes and wood properties. Forest Sci. 20, 32-40 (1974).
286. Thorin, J., Nommik, H.: Monoterpene composition of cortical oleoresin from different clones of *Pinus sylvestris*. Phytochemistry 13, 1879-1881 (1974).
287. Tobolski, J.J.: Variations in monoterpenes in Scotch pine (*Pinus sylvestris* L.). Diss. Abstr. 29, Order No. 69-5974:3567B (1969). Plant Breed. Abstr. 60, 8620 (1970).
288. Tobolski, J.J., Hanover, J.W.: Genetic variation in the monoterpenes of Scotch pine. Forest Sci. 17, 293-299 (1971).
289. Townsend, A.M., Hanover, J.W., Barnes, B.V.: Altitudinal variation in photosynthesis, growth and monoterpene composition of western white pine (*Pinus monticola* Dougl.) seedlings. Silvae Genetica 21, 133-139 (1972).
290. Tsankova, E.: Composition of essential oils from *Pinus peuce*. Riechst., Aromen, Körperpflegem. 19, 104, 107 (1969).
291. Turner, B.L., Flake, R.H.: Volatile constituents, especially terpenes and their utility and potential as taxonomic characters. In: Chemistry in Botanical Classification (ed. G. Bendz, J. Santesson), 320 pp. New York: Academic Press 1974.
292. Uvarova, N.I., Morozova, O.V., Ivanova, P.I.: The composition of turpentine from *P. jezoensis* resin. Z. Prik. Him. 30, 1274-1276 (1957). For. Abstr. (Oxford) 20, 2613 (1959).
293. Valente, H.M., Montes, A.L., Vonesch, E.E.: Chemical composition of oil of turpentine from *Pinus elliottii* grown in Argentina. Rev. Forest Argent. 12, 111-114 (1968). For. Abstr. (Oxford) 30, 4995 (1969).
294. Valente, H.M., Montes, A.L., Vonesch, E.E.: Composition of the turpentine oils of various species of cultivated pine grown in Argentina. Cienc. Invest. 25, 67-71 (1969).
295. Valenzuela, P., Cori, O., Yudelevich, A.: Occurrence of monoterpenes in *Pinus radiata* and utilization of labeled CO_2 and mevalonic acid. Phytochemistry 5, 1005-1011 (1966).
296. Vasek, F.C., Scora, R.W.: Analysis of the oils of western North American junipers by gas-liquid chromatography. Am. J. Botany 54, 781-789 (1967).
297. Vedeneev, K.P.: Determination of the composition of turpentines by gas chromatography. Sb. Tr. Tsent. Nauch.-Issled. Proekt. Inst. Lesokhim. Prom. 17, 40-48 (1966).
298. Verghese, J.: Indian turpentine oil from *Pinus longifolia*. Perfum. Essent. Oil Rec. 57, 767-770 (1966).

299. Vico, A.F.: Study on the terpenical composition of *Pinus pinaster* (seedling, plant, and adult pine). Influence of macronutrients and other ecological factors. Philogenetical implications. Ph.D. Thesis, Granada, Spain, 215 pp. (1968).
300. Vinutha, A.R., Rudloff, von, E.: Gas-liquid chromatography of terpenes. Part XVII. The volatile oil of the leaves of *Juniperus virginiana* L. Can. J. Chem. 46, 3743-3750 (1968).
301. Vlahov, R., Tsankova, E.: Bulgarian essential oil from *Pinus silvestris* leaves. Compt. Rend. Acad. Bulgare Sci. 17, 483-486 (1964).
302. Volskii, L.N., Schmidt, E.N.: Chemical composition of oleoresin of the genus Picea. (2) Monoterpenes of Siberian spruce and Yeso spruce. Izv. Sibir. Otd. Akad. Nauk. SSSR (Ser. Khim. Nauk.) 2, 132-134 (1970). For. Abstr. (Oxford) 33, 7192 (1972).
303. Vrkoc, J., Lukes, V., Skuhravy, V.: Content of monoterpene hydrocarbons in *Pinus mugo* subsp. *mughus* in relation to infestation by *Thecodiplosis brachyntera*. Z. Angew. Entomol. 74, 199-206 (1973). For. Abstr. (Oxford) 35, 3700 (1974).
304. Weiner, J., Pollock, V. (eds.): Turpentine. Bibliographic Ser. 260, 93 pp. Appleton, Wisc.: Inst. Pap. Chem. 1974.
305. Weissmann, G.: Die Zusammensetzung der Nadelöle einiger Tannenarten. Qual. Plant. Mat. Veg. 14, 337-344 (1967).
306. Weissmann, G.: The oleoresin of the whitebark pine (*Pinus heldreichii* Christ.). Holz als Roh- und Werkstoff 31, 341-343 (1973).
307. Weissmann, G., Bruns, K.: Terpenes in the resin of *Araucaria imbricata* (*A. araucana*). Naturwissenschaften 52, 185-186 (1965). For. Abstr. (Oxford) 27, 1512 (1966).
308. Weissmann, G., Charernsri, S., Aranyapooti, S.: Determination of the extractives of *Pinus merkusii* Jungh and *Pinus khasya* Royle (from Thai forests). Vanasarn 29, 262-271 (1971).
309. Weissmann, G., Charernsri, S., Aranyapooti, S.: Determination of extractives of son-songbai (*Pinus merkusii*) and son-sambai (*Pinus kesiya*). Thai J. Agr. Sci. 6, 115-126 (1973).
310. Weissmann, G., Dietrichs, H.H., Sootichanchai, W.: Oleoresin and colophony of Thai Pine species (*Pinus khasya* and *Pinus merkusii*). Holz als Roh- und Werkstoff 27, 189-191 (1969).
311. Westfall, R.D.: Developmental and genetic variation in the cortical terpenes of species of Pinus and Picea. Ph.D. Thesis, Mich. State Univ., East Lansing, Mich., 102 pp. (1972).
312. Westfelt, L.: Terpenes from *Pinus silvestris*. Sven. Kem. Tidskr. 79, 3-23 (1967).
313. Wilkinson, R.C.: Chemical analysis of species relationships in northeast American spruces. Ph.D. Thesis, Mich. State Univ., East Lansing, Mich., 124 pp. (1970).
314. Wilkinson, R.C.: Monoterpenes as possible attractants to the white pine weevil *Pissodes strobi*. Abstr. 2nd North Amer. Forest Biol. Workshop, p. 23 (1972).
315. Wilkinson, R.C., Hanover, J.W.: Geographical variation in the monoterpene composition of red spruce. Phytochemistry 11, 2007-2010 (1972).
316. Wilkinson, R.C., Hanover, J.W., Wright, J.W., Flake, R.H.: Genetic variation in the monoterpene composition of white spruce. Forest Sci. 17, 83-90 (1971).
317. Williams, A.L., Bannister, M.H.: Composition of gum turpentines from twenty-two species of pines grown in New Zealand. J. Pharm. Sci. 51, 970-975 (1962).
318. Yoshida, T., et al.: Chemical constituents in the essential oil of the leaves of *Chamaecyparis taiwanensis* Masaru. J. Pharm. Soc. Japan 87, 434-439 (1967). For. Abstr. (Oxford) 29, 1649 (1968).
319. Zavarin, E.: Chemotaxonomy of the genus Abies. II. Within tree variation of the terpenes in cortical oleoresin. Phytochemistry 7, 99-107 (1968a).

320. Zavarin, E.: Monoterpenoids of coniferales. Bull. Intern. Assoc. Wood Anat. 1988/1, 3-12 (1968b).
321. Zavarin, E.: Qualitative and quantitative co-occurrence of terpenoids as a tool for elucidation of their biosynthesis. Phytochemistry 9, 1049-1063 (1970).
322. Zavarin, E., Cobb, F.E., Jr.: Oleoresin variability in *Pinus ponderosa*. Phytochemistry 9, 2509-2519 (1970).
323. Zavarin, E., Cobb, F.E., Jr., Bergot, J., Barber, H.W.: Variation of the *Pinus ponderosa* needle oil with season and needle age. Phytochemistry 10, 3107-3114 (1971).
324. Zavarin, E., Critchfield, W.B., Snajberk, K.: Turpentine composition of *Pinus contorta* x *Pinus banksiana* hybrids and their derivatives. Can. J. Botany 47, 1443-1453 (1969).
325. Zavarin, E., Critchfield, W.B., Snajberk, K.: Composition of the cortical and phloem monoterpenes of *Abies lasiocarpa*. Phytochemistry 10, 3229-3237 (1971).
326. Zavarin, E., Hathaway, W., Reichert, T., Linhart, Y.B.: Chemotaxonomic study of *Pinus torreyana* Parry turpentine. Phytochemistry 6, 1019-1023 (1967).
327. Zavarin, E., Lawrence, L., Thomas, M.C.: Compositional variations of leaf monoterpenes in *Cupressus macrocarpa*, *C. pygmaea*, *C. goveniana*, *C. abramsiana* and *C. sargentii*. Phytochemistry 10, 379-393 (1971).
328. Zavarin, E., Mirov, N.T., Cooling, E.N., Snajberk, K., Costello, K.: Chemical composition of turpentine from *Pinus khasya*. Forest Sci. 14, 55-61 (1968).
329. Zavarin, E., Mirov, N.T., Snajberk, K.: Turpentine chemistry and taxonomy of three pines of southeastern Asia. Phytochemistry 5, 91-96 (1966).
330. Zavarin, E., Snajberk, K.: Chemotaxonomy of the genus Abies. I. Survey of the terpenes present in the Abies balsams. Phytochemistry 4, 141-148 (1965).
331. Zavarin, E., Snajberk, K.: Geographical variability of monoterpenes from *Abies balsamea* and *A. fraseri*. Phytochemistry 11, 1407-1421 (1972).
332. Zavarin, E., Snajberk, K.: Geographical variability of monoterpenes from cortex of *Pseudotsuga menziesii*. Pure Appl. Chem. 34, 411-433 (1973).
333. Zavarin, E., Snajberk, K., Critchfield, W.B.: Monoterpene variability of *Abies amabilis* cortical oleoresin. Biochem. Syst. 1, 87-93 (1973).
334. Zavarin, E., Snajberk, K., Lee, C.J., Henley, M., Mirov, N.T.: The composition of terpenoid hydrocarbons from *Pinus monophylla* wood oleoresin. Phytochemistry 10, 1857-1862 (1971).
335. Zavarin, E., Snajberk, K., Reichert, T., Tsien, E.: On the geographic variability of the monoterpenes from the cortical blister oleoresin of *Abies lasiocarpa*. Phytochemistry 9, 377-395 (1970).
336. Zubyk, W.J., Conner, A.Z.: Analysis of terpene hydrocarbons and related compounds by gas chromatography. Anal. Chem. 32, 912-917 (1960).

CHAPTER 7

Isolation and Analysis of Plant Phenolics from Foliage in Relation to Species Characterization and to Resistance Against Insects and Pathogens

J. LUNDERSTÄDT

Introduction

Phenolic compounds have been shown to play key roles in the resistance of plants to disease-causing microorganisms (Farkas and Kiraly, 1962; Cruickshank, 1963; Tomiyama, 1963; Rohringer and Samborski, 1967; Kuc, 1972). In addition, phenolics may also serve as protectants against leaf-eating insects as, for example, they may influence the availability or digestibility of proteins which insects obtain from leaves (Feeny, 1970). For these reasons, tree breeders and forest pathologists should be well acquainted with plant phenolics and the methods for their analysis.

The phenolics that comprise resistance systems against disease organisms and insects are divided into three groups based upon the relation of their formation to pest attack (Rohringer and Samborski, 1967):

1. Phenols normally present in healthy plants (preformed resistance factors).

2. Phenols formed from plant metabolites as a result of pest attack (host-parasite interaction).

3. Phenols formed as host-specific defense products (phytoalexins).

Trees are rich in phenolic compounds, especially those lignified portions resistant to decay (Scheffer and Cowling, 1966). Some of the information about phenolics in trees has been obtained in relation to: 1. seasonal changes in leaves (Feeny and Bostock, 1968; Thielges, 1972a), 2. chemotaxonomy (Erdtman et al., 1966; Hillis, 1967), 3. speciation (Wellendorf et al., 1971), and clone identification (Sauer et al., 1973), 4. study of introgressive hybridization (Hanover and Wilkinson, 1970), and 5. study of interspecific (Thielges, 1972a) and intraspecific (Thielges, 1972b) hybrids.

Because of the importance of phenolics in the protection of plants from pests, and because of our scant knowledge of the phenolics important to trees in these situations, I have endeavored to detail the methods used in my laboratory to study phenolics.

Methods

Phenolic compounds are a complex group of organic chemicals. Because of their heterogeneity, a host of different extraction and purification procedures have been developed. Phenolic compounds cannot be separated completely from other extractives on the basis of solubility alone (Browning, 1967), and, therefore, every extraction procedure yields phenolics somewhat contaminated with nonphenolic extractives, unless further purification is attempted, especially for simple ether-soluble phenols.

Phenolics are classified into three main groups based upon their solubility: 1. simple phenols (ether-soluble), 2. polyphenols (ether-insoluble but n-butanol soluble), and 3. tannins (water soluble). In addition to the soluble phenolics above, bound phenols occur in some plants at high concentrations (Lunderstädt, 1972) which can be extracted only after alkali hydrolysis. These phenolic classes may be obtained regardless of previous sample treatment, such as fresh (Feeny and Bostock, 1968); frozen (Wellendorf et al., 1971); oven-dried (Thielges, 1972a, b); and freeze-dried (Lunderstädt, 1972; Sauer et al., 1973) material.

Following extraction, a number of different analytical techniques may be used. Among these methods are paper or thin layer chromatography, column chromatography, and gas chromatography. Selection of the analytical technique depends largely upon the class of phenolics to be assayed. Polyphenols have been used most for the studies of hybridization or species identification, although gas chromatography of simple phenols (Pierce, 1968) should give useful results.

For rapid screening of plant material to separate clones of plant material, methods must allow quick analysis of small samples with simple inexpensive equipment. The method I use in my laboratory is to compare the fluorescent pattern of phenolic compounds in extracts of fresh plant material (Börtitz, 1963), or in extracts from small samples of freeze-dried material. The water content of the plant and manufacture of a press capable of expressing plant sap are two factors that limit use of fresh sap.

Phenolic Extraction from Leaves, Especially Conifer Needles

Free Phenolics (Free Phenols and Phenolic Acids)

Procedure 1. (Thielges, 1972b): the dried leaves precut into 1 cm long pieces are homogenized in ethyl ether (1/12.5, w/v) in a blender for 4 min and extracted for 48 h on a shaker. Then at 12-h intervals, the liquid is decanted and replaced with 50 ml of fresh ethyl ether. The insoluble material recovered by vacuum filtration is then reextracted with n-butanol and with cold distilled water, each for 48 h.

Procedure 2. (Hanover and Wilkinson, 1970; Wellendorf et al., 1971; Sauer et al., 1973): 5 g fresh leaves or 100-400 mg freeze-dried leaves are extracted for 3 h with 20 or 150 ml hot water, respectively. The filtered hot water extract is extracted sequentially with 5-50 ml ethyl ether and 5-50 ml n-butanol.

Procedure 3. (Lunderstädt and Claus, 1972): 2 g of freeze-dried needle powder are extracted sequentially in a Soxhlet extraction apparatus with ethyl ether/pentane (40°C), benzene/ethanol (85°C), and methanol/water (100°C). Between extraction steps, the insoluble material is air-dried. Each extraction is continued until the solvent does not extract more colored material. The ethyl ether/pentane and benzene/ethanol fractions are combined, adsorbed onto Celite, and reextracted with ethyl ether. Ether-insoluble compounds are extracted with water and combined with the methanol/water fraction for the analysis of polyphenols. To detect phenols bound as glycosides of the soluble fraction, the respective sample is hydrolyzed for 30 min with 2 m HCl, cooled, and extracted with ether.

Tannins (Feeny and Bostock, 1968): 50 g of oak leaves are placed in 250 ml acetone/water (70/30) under a N_2 atmosphere and kept in darkness for 7 to 10 days, then homogenized in a blender and vacuum filtered. Addition of NaCl to the filtrate causes the solvent to separate into acetone and NaCl-saturated water phases. The acetone (upper) phase is recovered, then mixed with two volumes of ethanol and 5-10 volumes of ethyl ether, which causes tannin and NaCl to precipitate. The precipitate is recovered by centrifugation (2500 rpm for 10 min), washed with ethyl ether, and dried with N_2. The tannin-NaCl mixture is dissolved in a small volume of water and dialyzed against water until no positive reaction with $AgNO_3$ and $FeCl_3$ occurs.

Bound Phenolics (Kurth and Smith, 1954): The method given here was developed for use with Douglas-fir bark, but can be used with the insoluble residue from freeze-dried needles (Lunderstädt and Claus, 1972). 500 mg of oven dry (5 h at 105°C) insoluble residue of spruce needles from one of the above procedures is extracted in 200 ml of 1% NaOH (w/v) for 1 h at 90°C. The soluble portion is recovered by filtration and acidified to pH 3.7 with conc. HCl, which causes phenolic acids to precipitate. The precipitate is recovered by filtration, washed with 5% NaCl, dissolved in 150 ml dioxane, and dehydrated over anhydrous Na_2SO_4. The water-free solution is filtered, concentrated with N_2 to 25 ml, and added to 250 ml ethyl ether and stirred to precipitate the phenolic acid. The phenolic acid precipitate is recovered on a fritted glass filter, washed with 100 ml ethyl ether, then oven dried at 105°C.

Characterization of Phenols

Qualitative Methods

1. Paper and Thin Layer Chromatography. All samples must be concentrated to about 1 ml. 10 to 120 µl of each concentrated extract are applied to Whatman 3MM paper (Hanover and Wilkinson, 1970; Thielges, 1972b), Schleicher and Schüll No. 2043b paper (Sauer et al., 1973) or thin layer plates of Kieselgur (Harborne, 1973; Sauer et al., 1973). A number of solvents are available (Block et al., 1958; Seikel, 1964; Smith, 1969; Harborne, 1973) to develop the chromatograms, but n-butanol/acetic acid/water at 60/15/25 (v/v/v) (Feeny, 1970) or 4/1/5 (v/v/v) (Feeny and Bostock, 1968; Wellendorf et al., 1971; Thielges, 1972b) in one direction followed by an acidic solvent such as 2% acetic acid (Feeny, 1970) or aqueous sodium formate (Thielges, 1972a) at right angles are good choices. Many detection reagents also are available (Seikel, 1964; Browning, 1967; Smith, 1969; Harborne, 1973), but detection by UV light is best because the detected compounds can be recovered unaltered.

2. Spectrophotometric Methods. Small quantities of dried phenolics are dissolved in an appropriate solvent, such as n-butanol, benzene, or dioxane, and diluted until an extinction value of about 1.0 absorbance unit at the absorption maximum is reached. A spectral scan in the range of 200-400 nm is then obtained. Spectral data for identification purposes are given by Harborne (1973).

Quantitative Methods

1. Photometric Methods. To quantify total phenolics the modified Folin-Denis method is used (Seikel, 1964; Thielges, 1972b). The same method may be used to quantify individual spots eluted from

paper chromatograms if reference substances are available (Seikel, 1964). Large amounts of phenols separated on thick sheets of Whatman 3 or 3MM paper are eluated with 70% ethanol which is dispensed from a reservoir bottle by capillary action (Harborne, 1967). If great care is taken in the sample extraction procedure, relative differences in phenolic content of separate samples may be estimated by the intensity of fluorescence at 366 nm of spots on paper and thin layer chromatograms (Sauer et al., 1973).

2. Gas Chromatography. The method has been elaborated for heartwood extractives of *Pinus resinosa* AIT (Sato and von Rudloff, 1964) but works also with leaf phenols: the ether extracts or n-butanol extracts containing phenols are concentrated to dryness and redissolved in dry pyridine. 0.1 ml of hexamethyl disilazane is added to 0.1 ml of this pyridine solution followed by 0.1 ml of trimethyl chlorosilane. The mixture is shaken for a few minutes, allowed to stand for 5 min and aliquots of the supernatant liquid are injected into the GLC-apparatus. The column (3 ft. x 1/4 in.O.C., 15% SE-52 silicone polymer on Chromosorb W, 60-80 mesh) is maintained at 160°C, 192°C and 242°C to determine the presence of low and high boiling phenolsililethers. Reference compounds are similarly converted to the TMS-derivatives and the relative retention times are determined.

Furuya (1965) separated TMS-derivatives of flavones, flavonols, flavanones, flavanonols, chalcones, leucoanthocyanins, isoflavones (2.25-m-long stainless steel column, 4 mm internal diameter, column packing Chromosorb W 60-80 mesh coated with 1.5% of silicone rubber SE-30, N_2 flow rate 110.5 ml/min, column temp. 240°C, detector temp. 240°C, flash heater temp. 305°C). Furuya and Kojima (1967) analyzed coumarins as acetone solutions or TMS-derivatives either on SE-30 or HI-EFF-iPB columns. (1.5% SE-30 Chromosorb W 60-80 mesh), column temp. 180°C, detector temp. 230°C, flash heater temp. 250°C, N_2 flow rate 70.4 ml/min; 12% HI-EFF-iPB on Gaschrom P (80-100 mesh) column temp. 210°C, detector temp. 230°C, flash heater temp. 250°C, N_2 flow rate 69.1 ml/min.

Micromethods with Fresh Plant Sap and Freeze-Dried Plant Tissue

Isolation of Plant Sap from Fresh Leaves. With a homemade press (1 cm piston compressing tissue into a cylinder with a bottom disk perforated with 0.5 mm diam. holes), approximately 250 µl of plant sap can be expressed from 2.5 g of spruce needles. A single drop of sap can be obtained from about 500 mg of fresh needles.

Isolation of Phenols from Freeze-Dried Plant Tissue. 10 mg of freeze-dried spruce needles (about 25 mg fresh weight) pulverized in liquid N_2 before drying are mixed with 0.5 ml of 2-butanol in a test tube and heated for 3 h at 80°C. A marble placed on top of the tube is used to slow evaporation of butanol while heating the mixture.

Separation of the Phenolics. The plant sap and the 2-butanol extract are separately centrifuged for 20 min at 30,000 g at 15°C. 5 µl portions of the supernatants are separated on paper or thin layer chromatograms developed in 2-butanol/water (4/1, v/v) (van Sumere et al., 1961) and detected by UV light or by anisaldehyde/H_2SO_4 reagent (Stahl and Kaltenbach, 1961). Results are recorded either by measuring the fluorescence of spots in UV light or by photographing the spots under standard conditions (e.g., quartzlamp NN 8130 L 001645 G3 by Hanau Quarzlampen GmbH; film Agfa ortho 25, shutter 5.6; exposure time 1-6 min depending upon fluorescent intensity; printed on Agfa-Gaevert BH 1 (or BN 1) paper (Fig. 1a and b).

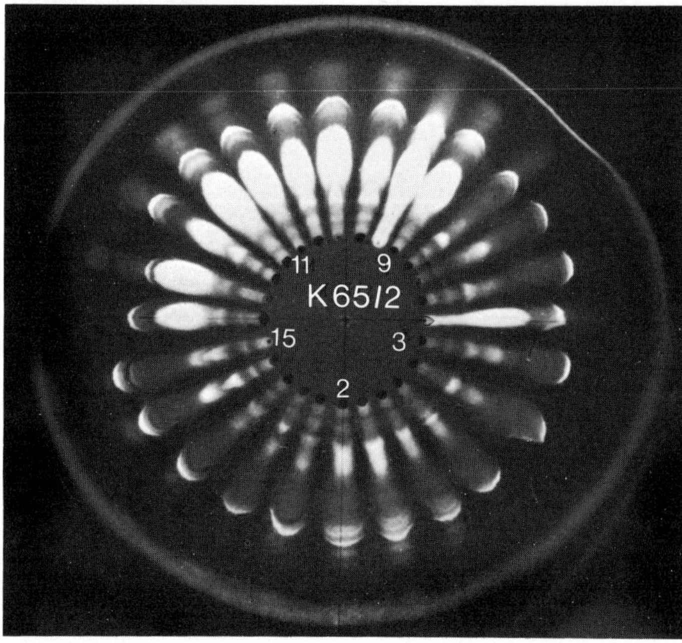

a

Fig. 1a and b. Separation of phenolic compounds from spruce needles of 5 different clones each on 5 different stands by paper chromatography and detection with UV light. Phenols are extracted as press sap from fresh material harvested in June 1973 (a) or with 2-butanol from freeze-dried matter harvested in October 1973 (b), respectively

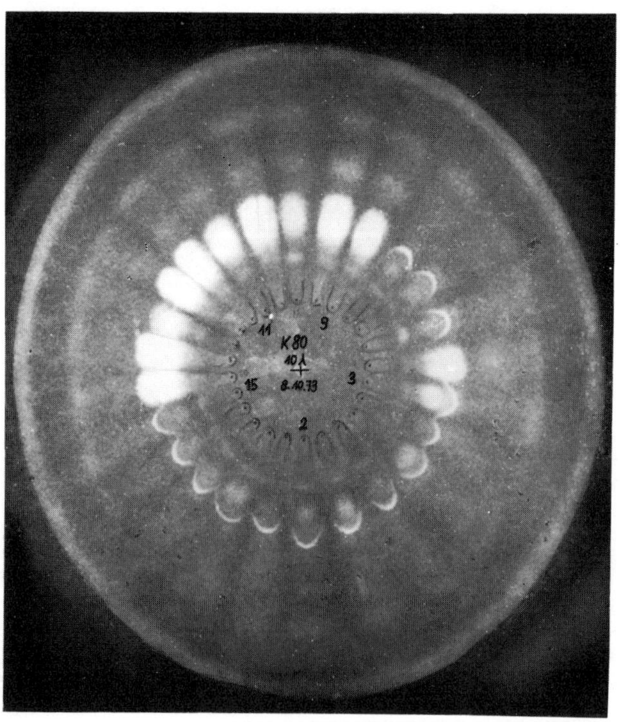

b

Discussion

The methods given here are simplistic but useful for tree breeders and forest pathologists because they obtain information on phenolic presence and relative concentration with a minimum of expensive equipment or over-complex procedures. While the absence of a specific phenolic in a tree susceptible to insect or disease attach does not necessarily mean that the phenolic is part of the resistance mechanism, such information is a reasonable point to begin a search for resistance mechanisms. Later, if such correlations appear to be factual, more precise methods of extraction and quantification may be used, such as by gas chromatography (Burchfield and Storrs, 1962; Harborne, 1967; Pierce, 1968), column chromatography (Harborne, 1967), or high pressure liquid chromatography.

References

Block, R.J., Durrum, E.L., Zweig, G.: A Manual of Paper Chromatography and Paper Electrophoresis, 2nd edition. New York: Academic Press 1958.

Börtitz, S.: Ermittlung von *Picea abies*-Varianten durch Fluoreszenzmuster. Flora 153, 320-325 (1963).

Brown, B.R., Love, C.W.: Protein-Fixing Constituents of Plants. Report on Forest Research, 90-92, London, H.M.S.O. (1961).

Browning, B.L.: Methods of Wood Chemistry, Vol. 1. New York, London, Sydney: Interscience Publishers 1967.

Burchfield, H.P., Storrs, E.E.: Biochemical Applications of Gas Chromatography, p. 408-410. New York, London: Academic Press 1962.

Cruickshank, I.A.M.: Phytoalexins. Ann. Rev. Phytopath. 1, 351-374 (1963).

El-Basyouni, S.Z., Neish, A.C., Towers, G.H.N.: The phenolic acids in wheat. III. Insoluble derivatives of phenolic cinnamic acids as natural intermediates in lignin biosynthesis. Phytochemistry 3, 627-639 (1964).

Erdtman, H., Kimland, B., Norin, T.: Pine phenolics and pine classification. Bot. Mag. Tokyo 79, 499-505 (1966).

Farkas, G.L., Kiraly, Z.: Role of phenolic compounds in the physiology of plant diseases and disease resistance. Phytopathol. Z. 44, 105-150 (1962).

Feeny, P.: Seasonal changes in oak leaf tannins and nutrients as a cause of spring feeding by winter moth caterpillars. Ecology 51, 565-581 (1970).

Feeny, P., Bostock, H.: Seasonal changes in the tannin content of oak leaves. Phytochemistry 7, 871-880 (1968).

Furuya, T.: Gas-liquid chromatography of trimethylsilylethers of some flavonoids and related compounds. J. Chrom. 19, 607-610 (1965).

Furuya, T., Kojima, H.: Gas-liquid chromatography of coumarins. J. Chrom. 29, 382-388 (1967).

Hanover, J.W., Wilkinson, R.C.: Chemical evidence for introgressive hybridization in *Picea*. Silvae Genetica 19, 17-22 (1970).

Harborne, J.B.: Chromatography of phenolic compounds. In: Chromatography (ed. E. Heftmann), pp. 677-698, 2nd edition. New York: Reinhold Publ. Corp. 1967.

Harborne, J.B.: Phenolic compounds. In: Phytochemical Methods, pp. 33-88. London: Chapman and Hall 1973.

Hillis, W.E.: Polyphenols in the leaves of Eucalyptus: A chemotaxonomic survey. II. The sections Renantheroideae and Renantherae. Phytochemistry 6, 259-274 (1967).

Kuć, J.: Phytoalexins. Ann. Rev. Phytopath. 10, 207-232 (1972).
Kurth, E.F., Smith, J.E.: The chemical nature of the lignin of Douglas-fir bark. Pulp and Paper Mag. Can. 55, 125-132 (1954).
Lunderstädt, J., Claus, G.: Zur Nahrungsqualität von Fichtennadeln für forstliche Schadinsekten. 2. Vergleichende biochemische Analysen von Fichtennadeln und Kot der Larven von *Gilpinia hercyniae* Htg. Z. angew. Entomol. 70, 386-403 (1972).
Pierce, A.E.: Silylation of Organic Compounds. Rockford, Ill.: Pierce Chemical Co. 1968.
Rohringer, R., Samborski, D.J.: Aromatic compounds in the host-parasite interaction. Ann. Rev. Phytopath. 5, 77-86 (1967).
Sato, A., von Rudloff, E.: The heartwood extractives of *Pinus resinosa* AIT. Can. J. Chem. 42, 635-640 (1964).
Sauer, A., Kleinschmit, J., Lunderstädt, J.: Charakterisierung von Fichten-Klonen (*Picea abies* Karst.) mit Hilfe morphologischer, physiologischer und biochemischer Methoden. Silvae Genetica 22, 173-182 (1973).
Scheffer, T.C., Cowling, E.B.: Natural resistance of wood to microbial deterioration. Ann. Rev. Phytopath. 4, 147-170 (1966).
Seikel, M.K.: Isolation and identification of phenolic compounds in biological materials. In: Biochemistry of Phenolic Compounds (ed. J.B. Harborne), pp. 33-76. London, New York: Academic Press 1964.
Smith, I.: Chromatographic and electrophoretic techniques. Vol. I, Chromatography. London: William Heinemann, Medical Books Ltd. 1969.
Stahl, E., Kaltenbach, U.: Dünnschichtchromatographie: VI. Mitteilung Spurenanalyse von Zuckergemischen auf Kieselgur G-Schichten. J. Chromatog. 5, 351-355 (1961).
Thielges, B.A.: A chromatographic study of foliage polyphenols in pine hybrids (Subsection sylvestres). Silvae Genetica 21, 109-114 (1972a).
Thielges, B.A.: Intraspecific variation in foliage polyphenols of *Pinus* (Subsection sylvestres). Silvae Genetica 21, 114-119 (1972b).
Tomiyama, K.: Physiology and biochemistry of disease resistance of plants. Ann. Rev. Phytopath. 1, 295-324 (1963).
van Sumere, C.F., Teuchy, H., Parmentier, F.: Quantitative paper chromatographic determinations. 1. Coumarins and phenolic acids, especially Esculetin, Daphnetin and Ferulic acid. J. Chromatog. 6, 481-483 (1961).
Wellendorf, H., Kaufmann, U., Hansen, M.: Thin layer chromatography of fluorescent phenolic compounds in needles. Forest Tree Improvement 2, 19-39 (1971).

CHAPTER 8

Mineral Analyses

F. H. EVERS and W. BÜCKING

Introduction

Foliar analysis is widely applied as a tool in the diagnosis of the nutrient status of forest trees. The method has been developed by Mitchell and Chandler (1939), Ulrich (1943), Lundegardh (1945), Goodall and Gregory (1947), Heiberg and White (1951), Tamm (1954), Leyton (1957), Wittich (1957), Wehrmann (1959), Kenworthy (1961), Smith (1962) and others. Many recent and detailed investigations have defined standard conditions for performance of analyses and exact methods for interpreting the results (Raupach, 1967; Ulrich and Hills, 1967; Armson, 1973). Several comprehensive reviews have summarized the present status of foliar analysis in forestry (Kral, 1965; Mustanoja and Leaf, 1965; Leaf, 1972; van den Driessche, 1973). Forest geneticists have made use of mineral analyses of leaves in clone identification, in the selection of genotypes for survival on difficult sites under adverse environmental conditions (lack of nutrients, emissions, heavy attack of parasites), or to show different responses to fertilization (Steinbeck, 1966; Pritchett and Goddard, 1967; Rediske et al., 1968; Zobel and Rober, 1970; Knight, 1972). Differences in mineral metabolism among genotypes under identical conditions are useful in tree breeding programs as mineral analysis techniques facilitate the early selection of best adapted provenances and provide one possible avenue for the detection of genetically superior trees from young selections.

Diagnosis of nutrient status is based on the assumption that individuals of a population of a given species grown under the same conditions will exhibit nearly identical mineral metabolism. Considerable differences may exist between different provenances as was shown by Gerhold (1959) for *Pinus silvestris*, by May et al. (1962) for provenances of *Pinus elliottii*, by Forrest and Ovington (1971) for progenies of *Pinus radiata*, and by Evers (1973) for rooted cuttings of different *Picea abies* clones (see Table 1).

Growth rate, as well as resistance to parasite attack (Shigo, 1973), depends on the ability of the trees to take up soil nutrients in adequate amounts. In plant breeding, concentrations of certain nutrient elements may serve as a basis for early selection. Kral (1965) found a close correlation between growth rate and potassium concentration in *Pseudotsuga menziesii*. Needles from fast growing provenances showed higher potassium concentrations than needles from slow growing seed sources, corresponding to higher biomass production. Steinbeck (1966) established similar relationships between Mg-concentration in needles and growth rate for *P. silvestris*. Unbalanced nutrient status can frequently be corrected by means of fertilization; however, it would be better to select genotypes that are adapted to difficult site conditions. These selections would be able to absorb nutrients even in impoverished soils or in soils with strong nutrient fixation. For instance, Knight (1972) found healthy trees in boron deficient stands of *P. radiata* which evidently were capable of exploiting the soil stock

Table 1. Nitrogen concentrations in 1971 and 1972 in needles of different clones of rooted spruce cuttings on the same site

Clone	N%	
no.	1971	1972
7002	1.77	1.59
7003	1.39	1.33
7008	1.24	0.98
7009	1.28	1.13
7010	0.97	1.17
7014	2.05	1.81
7015	1.45	1.18
7016	1.34	1.16
7017	1.31	1.24
7018	1.48	1.33

of boron. Jahromi (1971) showed that significant increases in the growth rate of *P. elliottii* can be obtained by the selection of genotypes according to special soil conditions. This effect can be increased even more if selected clones respond well to fertilization. Differences in nutrient content within a given species may be associated with high growth rates (Teinberg, 1966; Forrest and Ovington, 1971; Bevege and Simson, 1972). Pritchett and Goddard (1967) defined the problem by stating, "With careful screening, one may encounter the rare tree that grows well on less productive sites and at the same time responds well to site improvements". Selection of fast growing tree varieties that also retain desirable wood qualities even with fertilization (Borsdorf, 1965) is another objective.

Floral initiation (flowering) is frequently stimulated by fertilization (Kleinschmit, 1961; Barnes and Bengtson, 1968; Barnes, 1969). Clones with a higher capacity for assimilating nutrients are thought to flower more abundantly (Goddard and Strickland, 1966). It seems, therefore, that nutrition uptake is correlated with the flowering process.

As indicated by the few examples above, forest geneticists may have to employ mineral analysis techniques to reach their research objectives. Therefore, we wish to outline some of the main prerequisites for plant analysis, describe some commonly applied methods that can be used in genetic studies, and report on previous experience in plant nutrition research.

Methods

Sampling Time

Conifer needles can be collected throughout the year with the knowledge, however, that the concentrations of nutrient elements may vary considerably from one season to another. Nutrient concentrations in leaves collected during dormancy cannot be compared with those collected during the growing season (Gerhold, 1959; Wehrmann, 1961; Tamm, 1968; Hunger, 1974). Results from systematic investigations have shown that analysis from one collection does not necessarily reflect the true nutritional status of a stand. The mineral concentrations may vary considerably from year to year even when winter samples are compared (Wehrmann, 1963; Tamm, 1968; Höhne and Fiedler, 1970; Evers, 1972, 1973). Therefore, a single foliar analysis may lead to serious misinterpretations and the analyses should be repeated over several consecutive years in order to obtain reliable results. Since concentration variations are generally smallest at dormancy, samples for nutrient analysis should be taken at this time. In order to analyze and compare samples from different clones, provenances or stands, only samples collected at known phenologically and physiologically comparable times should be used for analysis. The most suitable time for sampling deciduous trees is difficult to determine. Usually the period between completion of foliar differentiation and beginning of the autumn color change is chosen. Another view is that nutrient concentrations would be more closely related to physiological conditions in the plant and thus to growth if leaves are sampled just after full expansion or when they are most active physiologically (Hunger, 1974).

Analytical values, however, are not absolute. As already mentioned, concentrations fluctuate with season and year. Finally it must be kept in mind that needles of trees not of the same age show different nutrient levels even if the plants have been cultivated under exactly the same conditions and the needles have been taken from comparable parts of the crown. In needles of three year-old slash pine (*P. elliottii*), for instance, Pritchett (1968) found higher P concentrations than in five to eight year-old trees.

However, if analytical data from provenances of the same age, tested under the same conditions, are compared, few difficulties are encountered. Under unusual circumstances, such as water stress or unknown nutrient deficiencies, deviations in nutrient status of leaves are nevertheless possible. Phosphorus (P) concentration, for example, may be considerably increased and lose its correlation to growth rate (Pritchett, 1968).

Sample Position

Results from leaf analyses are highly dependent upon the sample position on the tree. When the sampling position has not been exactly defined, data are not comparable either with further results of the same researcher or with those of other investigators. In coniferous species, however, the side branches of the uppermost whorl are usually analyzed. The terminal shoot is not used.

In order to circumvent problems inherent in single-time sampling, the procedure of sampling older and younger needles simultaneously has been investigated. Needles adjacent to the stem are collected from the top down to, and including, the fifth whorl (Fig. 1). Thus, the collection represents a five year sequence of needles. In our investigations, needles from the seventh whorl, ranging in age from one to five years (Fig. 2) are collected for analysis (Reemtsma, 1964). Deficiencies in Mg and Ca content, otherwise difficult to discern, are easily apparent by this technique. Pritchett (1968) utilizes a different sampling method and van Goor et al. (1971) and van den Driessche (1974) review sampling procedures.

Deciduous tree sampling is difficult and systematic investigations of sampling procedures should be investigated. No uniform sampling method exists, but as a rule only those leaves fully exposed to the light from the uppermost part of a crown are collected for analytical purposes. In the case of young plants, leaves from the entire shoot system of the current year may be used.

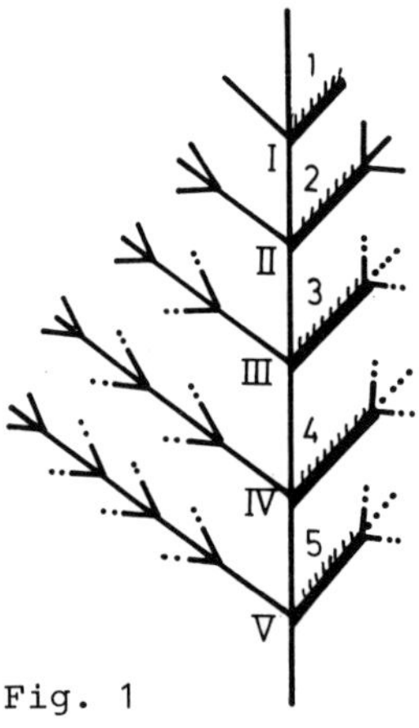

Fig. 1

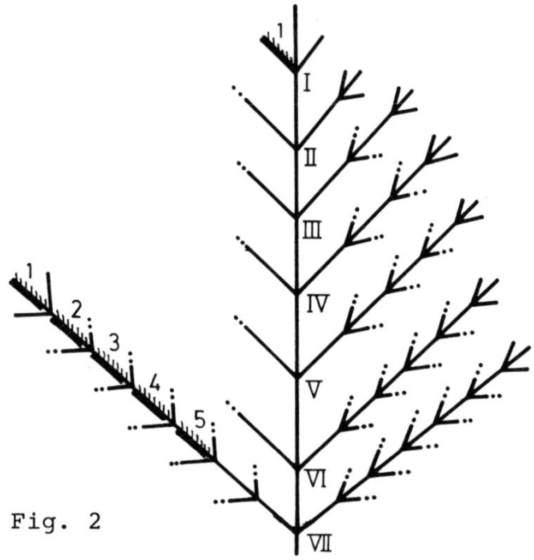

Fig. 2

Fig. 1. Sampling of five age classes of needles (1-5) from first to fifth whorl

Fig. 2. Sampling of five age classes of needles (1-5) from seventh whorl

Preparation for Sample Analysis

Fresh foliar material should be processed as quickly as possible. The laminae are removed immediately wherever petioles are not included in analysis in order to avoid movement of nutrient elements back into the stem or branches of the detached leaves. Needles are usually not separated from branches before drying. In most cases, leaves should be washed after collection and before drying to remove surface dust and salt contamination, as described by Arkley et al. (1960).

Leaves and needles should be dried immediately. The same drying method should be employed for all samples. Oven drying at 65°C to a constant weight is most commonly used. Dried material should be ground as soon as possible and may be stored for analysis at a later time. Storage in tightly closed glass jars protected from daylight is advisable.

The material should be ground as finely as possible (40 to 60 mesh) according to Müller and Fiedler (1969) in order to avoid a differential separation of parenchymatous, sclerenchymatous and phloem tissues. Otherwise the reproducibility of the results may be questionable. The weight difference between the ground material dried at 65°C and a subsample dried for two h at 105°C is determined. Analytical results are later computed on the basis of dry weight at 105°C.

Interpretation of Analytical Data

Nutrient concentration per unit of dry matter is the usual method of presentation and is adequate most of the time. For certain purposes, however, analytical data must be related to foliar area or needle mass (e.g., the weight of 1,000 needles). Thus, more comprehensive information can be obtained on nutrient assimilation, relations between mineral concentrations, and production of organic matter. To relate analytical data to foliar area, small disks of defined area are stamped out of the laminae. The weight of the disks is determined and a weight per m^2 foliar area can be computed.

The degree to which percent concentration may differ from content per unit area is shown for chlorotic and green poplar leaves (Table 2). Chlorotic leaves show distinctly higher element and percent ash concentration than the green leaves. However, there is no difference between the chlorotic and green leaves in the nutrient content per unit of foliar area. Interpretation, therefore, of the results from foliar analysis is sometimes not possible without data on total amounts.

Various nutrient ratios are useful in diagnosis of nutrient status in plants. Kral (1965) found that Ca/K ratio is more closely correlated with growth rate than percent dry weight of K. In the interpretation of analytical data from needles, additional ratios such as N/P, N/Mg, and Ca/Mg have proven equally important. Results are summarized in Table 3 which indicate the order of magnitude of the nutrient element concentrations found in the foliar organs of some tree species. Stone (1968) has reviewed trace element concentrations, and Leaf (1968) gives corresponding data for K, Mg and S.

Analytical Methods

The analytical procedures described below are alternative methods for determining major and some important trace elements in ashed plant material, and have been used for a long time in our Institute. Procedures are therefore immediately applicable and suitable for routine use. New working groups may adopt such methods quickly and without complications. Furthermore, we refer to some important methodological development which might be considered for special investigations. It is, however, not always possible or reasonable to integrate all such methods in a normal laboratory routine. More complete summaries of methods including all trace nutrients have been published, for instance, by Humphries, 1956; Müller and Fiedler, 1969; Müller, 1970. International comparison of methods and working procedures organized by Iufro - Section 21 - revealed important variations of results between research groups even when the same methods were applied (van Goor et al., 1971; Müller, 1973). Exact analytical work and continuity in methods during one research project will therefore be premises for comparable results. In order to simplify analytical descriptions, some details applying to the subsequent sections are summarized:

1. All chemicals and filter papers should be of analytical quality. The chemicals required are commercially available (e.g., from most chemical supply houses). For flame photometry and atomic absorption analysis, purified gases of technical quality have proven adequate.

2. Water (H_2O) means distilled water or ion-exchanged water of high purity.

3. The following abbreviations are generally used:

C correction term
DM dry matter (g)
Ex extinction
f proportionality coefficient
P weight of precipitate (mg)
Pb weight of blank test precipitate (mg)

Table 2. Nutrient element, ash concentrations, and content of nutrient elements per unit leaf area in healthy and chlorotic two year-old cuttings of Populus x euramericana cv. Androscoggin

			N%	P%	K%	Ca%	Mg%	Ash %
Percent dry weight	Green leaves		3.27	0.323	1.15	0.750	0.237	5.05
	Chlorotic leaves		5.73	0.693	2.51	1.240	0.382	10.45
	% of wt/m^2	1 m^2 foliar area g	N mg/m^2	P mg/m^2	K mg/m^2	Ca mg/m^2	Mg mg/m^2	Ash mg/m^2
Content per unit leaf area	Green leaves	52.8	1.730	171	609	396	125	2,670
	Chlorotic leaves	30.0	1.730	208	752	372	129	3,140

Table 3. Mean ranges of % dry weight composition of the most important nutrients in needles and leaves

	N%	P%	K%	Ca%	Mg%
Coniferous trees	0.7-3.0	0.06-0.4	0.3-1.6	0.3-1.3	0.04-0.3
Deciduous trees	0.9-4.0	0.10-0.6	0.5-3.1	0.5-2.5	0.05-0.5

Ps weight of sample precipitate (mg)
T titration value (ml)
Tb titration value of blank test (ml)
Ts titration value of sample (ml)
V volume (ml)
Va volume of aliquot (ml)
Vt total volume (ml)
W weight of subsample used for analysis (g)

4. Temperatures are given in Centigrade (°C). All other physico-chemical units are metric.

General Plan of Analysis

Total nutrient analysis includes 1 wet method of digestion and 2 ashing procedures (see Table 4):

I. Wet digestion in concentrated sulfuric acid for total nitrogen.

II. Dry ashing in a muffle furnace for total ash weight and P-, K-, Ca-, Na-, Mg-, Fe-, Mn-concentrations.

III. Dry ashing with addition of alkali in order to fix volatile sulfur and chloride compounds. Two separate digestions, identical in principle, have to be performed for sulfur and chloride.

Table 4 also shows the elemental subsample weights commonly used. Concentration ranges are limited for numerous methodological reasons; possibilities for shifting these ranges are given later.

Analytical Protocols

Total Nitrogen

1. Theoretical Background. The Kjeldahl procedure for the determination of nitrogen (Kjeldahl, 1883; Bradstreet, 1965) is adapted to mass production methods where many analyses must be made as quickly as possible, especially in the micro and semi-micro Kjeldahl dimensions using small samples and small volumes of digestion flasks. In this process a sample of organic material is heated with concentrated sulfuric acid and a catalyst-mixture to convert the organic nitrogen to ammonium-ion. After dilution with water, the solution is made alkaline with sodium hydroxide and the liberated ammonia distilled into a standardized solution of acid. The volume of acid solution neutralized by the ammonia is a quantitative measure of the nitrogen in the foliar sample. Nitrates and nitrites can be lost as nitrous gases during , digestion process; Kjeldahl procedure is nevertheless believed to be a total nitrogen detecting method as oxidized inorganic nitrogen fractions amount to less than 1% of total nitrogen.

2. Reagents. a) Concentrated sulfuric acid (~ 98%, d ≈ 1.84); b) 0.1 N sulfuric acid; c) 0.1 N sodium hydroxide; d) selen reaction mixture as described by Wieninger (commercial product from Merck, FGR; composition Fiedler, 1965: 8 g Se/8 g $CuSO_4$ (dehydrated)/0.5 g Na_2SO_4); e) sodium hydroxide 50% (dissolve 500 g NaOH in 500 ml H_2O); f) indicator (methyl red; 0.2% alcoholic solution; Fiedler, 1965) or preferably mixed indicators for ammonia titrations (dissolve 0.099 g bromcresol green and 0.66 g methyl red in 100 ml absolute alcohol (Bremner,

Table 4. Plan of analytical procedure and measuring ranges of elements

Procedure Subsample weights	Elements analyzed									
	N	S	Cl	K	Na	Mg	Ca	P	Fe	Mn
I. Wet digestion (H_2SO_4 conc.) 0.600 g	0.05- 4.67%									
II. Dry ashing 2.000 g				0.005- 0.700%	0.003- 0.350%	0.005- 0.300%	0.005- 1.000%	0.006- 0.500%	12- 180 ppm	50- 2,500 ppm
III. Alkali ashing										
0.600 g		0.02- 5.30%								
1.000 g			0.035- 3.540%		0.005- 0.700%					

1960) or use commercial products such as "Mischindikator 5" from Merck, FGR).

Equipment. Micro Kjeldahl heater; steam distillation apparatus (Parnas and Wagner, 1921; Bremner, 1960; Müller and Fiedler, 1969; electric distillation apparatus was recently developed by Büchi Ltd., Switzerland); Kjeldahl flasks (50 ml); volumetric flasks (100 ml); Erlenmeyer flasks (200 ml); graduated cylinders (100 ml); burettes (10 ml); funnels (Ø 2 cm, 8 cm); pipettes (50 ml).

3. Working Procedure. Weigh 0.600 g dry matter into a Kjeldahl flask, mix with an amount of selen reaction mixture (d) equal to the tip of a small spatula (~ 0.5 g), add 25 ml concentrated sulfuric acid (a) using a graduated cylinder, cover with 2 cm funnel as drop collector and heat at first cautiously (be careful of sudden foaming!) until white SO_2-fumes evolve, then boil more vigorously, keeping the sulfuric acid at a gentle boil until the reactant solution is clear. Continue boiling for at least 1/2 h. Transfer total digest to 100 ml volumetric flask and make to volume with water. Distill a 50 ml aliquot into exactly 10 ml 0.1 N H_2SO_4 (b; burette) contained in a 200 ml Erlenmeyer flask. Immerse the condenser of the distillation apparatus before starting distillation. Alkalize the aliquot with about 50 ml of 50% NaOH (e), distill 6-8 min to a 50 ml distillate volume. Keep distillation time and, above all, volume constant. Lower Erlenmeyer flask and clean the inside of condenser by continued distillation (3 min), the outside by water. Add 3 drops of indicator and titrate the excess sulfuric acid by 0.1 N NaOH (c).

Calculation:

$$1 \text{ ml excess } 0.1 \text{ N } H_2SO_4 \triangleq 1.4008 \text{ mg N}$$

$$\frac{(V_{H_2SO_4} - Ts) \cdot C \cdot Vt \cdot 1.4008}{W \cdot Va \cdot 10} = \% \text{ N by weight}$$

C = correction term for relation of normalities H_2SO_4/NaOH

C = a/b; a - volume of sulfuric acid (ml)
b - volume of NaOH used for neutralization of a (ml)

Computation formula simplified for standard conditions:

$$\frac{(10 - Ts) \cdot C \cdot 100 \cdot 1.4008}{0.600 \cdot 50 \cdot 10} = (10 - Ts) \cdot C \cdot 0.467 \triangleq \% \text{ N by weight}$$

Measuring range:

~ 0.14 - 14 mg N per aliquot $\triangleq$ ~ 0.05 - 4.67 % N by weight

4. Literature. Humphries (1956), Müller and Fiedler (1969) and others recommend boric acid as a trapping solution facilitating titration procedure by the use of only one standard solution (a stable acid titration solution). To obtain analytical range different from standard procedure: change sample amount weighed in, distillation aliquot, trapping acid volume, normality of trapping solution and titration acid (down to 0.01 N NaOH and 0.01 N H_2SO_4). Ammonia concentrations in Kjeldahl digestions can be measured by specific electrodes (Cammann, 1966; Orion, 1972; with additional references). Nitrates can be included by Fe-reduction method, Peterson and Chesters (1964); by salicylic acid method, Humphries (1956). Detection of nitrates only, Balks and Reekers (1954); Scharrer and Seibel (1956); Bremner and

Keeney (1965); Wellenstein et al. (1975). Investigations on more rapid or more effective digestion procedures are numerous (Bremner, 1960; Kunze et al., 1964; Bornmann, 1972; Concon and Soltess, 1973; Batey et al., 1974; and others). Besides the classical Kjeldahl method, automated analytical equipments employ colorimetric assays (Kessen, 1966a; Holz, 1971) or the dry combustion method after Dumas (Kreutzer and Merz, 1973; Merz, 1973). The Dumas procedure will yield in some cases higher N-values than the Kjeldahl digestion which cannot degrade all of the N-containing compounds or is subject to gaseous nitrogen losses (Bornmann, 1972). In normal biological material, however, comparable values were found with both methods (Hübsch and Nehring, 1960).

Dry Ashing, Total Ash and Ash Extract

1. Theoretical Background. An organic sample is dry ashed without admixtures in a muffle furnace at a controlled temperature. The difference between dry ash weight and dry weight corresponds to the organic material. Elements to be analyzed are then dissolved in hydrochloric acid.

2. Reagents. a) Hydrochloric acid 1:1 volume to volume (500 ml HCl conc., d ≈ 1.19, : 500 ml H_2O); b) hydrochloric acid 5% (114 ml HCl conc., d ≈ 1.19, and 890 ml H_2O); c) hydrochloric acid 10% (232 ml HCl conc., d ≈ 1.19 and 774 ml H_2O).

Equipment. Muffle furnace with temperature regulation, platinum or quartz crucibles for ashing; volumetric flasks (100 ml); funnels; filter papers (quick or medium filtering quality).

3. Working Procedure. Weigh 2.000 g plant material into a crucible, ash 30 min at 150°C, then 3-4 h at 550°C (do not exceed this recommended temperature as K losses may occur) until a homogenous white to light gray colored ash is obtained. If necessary, mix ash and incinerate a second time. For ash-content incinerate to a constant weight. Place crucibles on a heater and add some drops of water, boil down twice with 3 ml HCl 1:1 (a), being careful of splashing, heat with 10 ml HCl 5% (b) for a short period and filter through paper filter into a 100 ml volumetric flask. Clean the crucible carefully. Add 45 ml HCl 10% (c) and fill with water. Final HCl-concentration is about 5%.

Determination of Phosphates

1. Theoretical Background. Colorimetric determination of phosphate by means of vanadate and molybdate. Vanadate and molybdate, anions of heavy-metal acids, react with ortho-phosphate ions in nitric acid, forming a yellow acid-soluble complex compound (heteropolyacid). Pyrophosphates condensed during ashing must be rehydrolyzed and the hydrochloric acid extract must be transferred to a nitric acid medium.

2. Reagents. a) Ammonium-mono-vanadate 0.25% (dissolve 2.5 g NH_4VO_3 in about 500 ml boiling water, then fill volumetric flask to 1,000 ml); b) ammonium-hepta-molybdate 5% (dissolve 50 g $(NH_4)_6Mo_7O_{24}\cdot 4H_2O$ in about 800 ml water at 50°C and fill 1,000 ml volumetric flask); c) nitric acid 1:2 volume to volume, ~ 28%, d ≈ 1.16 (330 ml HNO_3 conc., ~ 67%, d ≈ 1.40, and 660 ml H_2O); d) phosphorus standard solution for standard curve (dissolve 0.2195 g KH_2PO_4 (predried for 6 h at 40°C) in about 400 ml H_2O and fill to 1,000 ml; 1 ml ≙ 50 µg P).

Equipment. Beakers (100 ml), volumetric flasks (100 ml, 1,000 ml), pipettes (10 ml, 20 ml), photometer (filter 436 nm), cuvets (2 cm absorbing layer), heater.

3. Working Procedure. Prepare standard curve with 0-20 ml of P-standard solution (d), ≙ 0-1,000 μg P, in 100 ml volumetric flasks. Add to respective volume 10 ml HNO_3 (c), 10 ml vanadate solution (a), 10 ml molybdate solution (b), fill and shake well. Measure extinction after 30 min at 436 nm in 2 cm cuvets against blank solution. For ashed samples carefully boil down 10 ml of hydrochloric acid ash extract on heater (danger of splashing), reboil for a short time with 10 ml HNO_3 (c), and dilute with 10 ml H_2O; transfer to 100 ml volumetric flask, add 10 ml (a) and (b), dilute to volume and mix. Measure the extinction as done for calibration values. Evaluation: the standard curve is highly reproducible and, once established, it can be used without further repetition for a long period. Since the standard curve is linear in the 0-0.740 extinction range, computation of results by an empirical conversion coefficient "f" (determine from the standard curve) is possible. "f" for the conditions of our laboratory is 1.346 (for conversion from extinction to amount of phosphorus in mg).

Computation formula:

$$\frac{Ex \cdot f \cdot Vt}{W \cdot Va \cdot 10} = \% \text{ P by weight}$$

Simplified for standard conditions:

$$\frac{Ex \cdot 1.346 \cdot 100}{2 \cdot 10 \cdot 10} = Ex \cdot 0.673 = \% \text{ P by weight}$$

Measuring range:

~ 0.006 to 0.500 % P by weight.

4. Literature. Gericke and Kurmies (1952). Interference due to color of the ash extract can be avoided by comparing the extinction of a prepared sample to the color of the respective extract. Silicone dioxide, though it interferes in small concentrations, can be neglected in practice since it is not soluble in hydrochloric acid. In addition to the molybdate-vanadate method, more sensitive detection methods based on the formation of molybdenum blue are used (Humphries, 1956; Müller, 1970). Wet digestion by $HClO_4$ is also applied instead of ashing (Allen, 1940). Method for automatic analyzer apparatus is presented by Kessen (1966a).

Flame Spectroscopy and Atomic Absorption Analysis

Precise methods for determining alkaline (Na, K) and alkaline earth metals (Ca, Mg) are flame spectroscopy and atomic absorption. In flame spectroscopy, i.e., flame emission photometry, light intensities are measured at defined wavelengths that are specific for atomic emissions of the respective metals. Emissions are excited by a suitable flame. Light intensities are in direct proportion to atom concentrations. In atomic absorption flame photometry ("atomic absorption analysis") energy absorptions, i.e., "extinctions", are recorded when a flame-excited atom absorbs additional energy at a defined wavelength, again specific for the element in question. This additional energy is irradiated by a hollow cathode lamp. As in colorimetric techniques, the extinction (defined as the log of the ratio of light intensity before and after absorption by sample) is proportional to element concentration.

Both of the above procedures are strongly influenced by properties of the detection apparatus and the working conditions, so that samples must be compared with standards under the same conditions. To meet the conditions of standardization some general advice should be considered:

1. Choice of fuel gases, by means of which flame temperature and flame excitation energy is fixed. As a rule of thumb, alkali metals are best excited in propane/air mixture ~ 1:10 (~ 1,925°C), alkaline earth metals in acetylene/air mixture ~ Λ:5 (~ 2,325°C). (Hermann and Alkemade, 1960, 1963; Netheler and Hinz, 1956). Correct mixture is adjusted according to flame characteristics of individual apparatus (see subsequent sections).

2. Excitation of analyzed element is strongly dependent on the accompanying solution components, so that blank solutions, calibration standards, and sample solutions, must be as similar as possible. In actual sample preparation, hydrochloric acid concentration of ash extract and of calibration solutions should be exactly 5.0%.

3. Examine the linearity of the calibration curve in the detection range. If necessary, a standard curve has to be established for the exact computation of analytical data and for learning equipment characteristics. Atomic absorption technique will demand a calibration curve for each run (see Magnesium).

4. Analytical Technique. After each sample measurement the blank solution is inserted and the zero point is corrected. Thus, contamination by preceding samples can be avoided. Change blank solutions frequently since pollution will occur, especially after samples with high element concentrations; determinations of subsequent samples with low concentrations might then be highly erroneous.

Potassium

1. Theoretical background of flame photometry has been presented above.

2. Reagents. a) Hydrochloric acid 5% (see "Dry ashing technique"); b) potassium standard solution (dissolve 1.9069 g KCl (predried at 105°C) in 1,000 ml 5% HCl (volumetric flask); the solution contains 1.000 g/l K); c) potassium calibration solution (dilute K-standard solution (b) 1/9 volume to volume with 5% HCl. [K] = 100 mg/l); d) potassium standard curve (dilute 0-20 ml K standard solution (b), ≙ 0-200 mg/l K, with 5% HCl in volumetric flasks of 100 ml); e) propane (C_3H_8).

Equipment. Flame photometer with K-filter (maximum of transmissibility at 767 nm), volumetric flasks (1,000 ml, 100 ml), pipettes (10 ml).

3. Working Procedure. Adjust air-propane mixture for correct flame (clear blue without yellow borders). Best adjustment is achieved by reducing fuel gas supply by about 10% from slightly yellow flame (Netheler and Hinz, 1956). Standard curve: space the scale in such a way that the calibration solution (c) with 100 mg/l K corresponds to 50 scale divisions and the blank solution to the zero point. Then measure the

other calibration solutions. Plot scale divisions versus K concentration. The resulting standard curve, though not completely linear, will in most cases be of sufficient proportionality for practical use in the 0-70 scale division range. Between 70 and 100 scale units, the curve deviation from a straight line increases to about 10%, yielding highly erroneous results.

Samples. Compare the hydrochloric acid extract of samples directly with the blank solution and the calibration solution (c) adjusted to 50 scale divisions. Thus, for the standard procedure, direct reading of the K concentration in the sample is possible, since the scale reading divided by 100 corresponds to % K by weight. Because of the reasons mentioned above, include only readings up to 70 scale units and dilute more highly concentrated samples 1:9 with 5% HCl. Diluted sample extracts are computed by dividing scale reading by 10. Refer to the calibration curve for more exact analyses.

Measuring range:

$\sim$ 1 - 140 mg/l K $\hat{=}$ $\sim$ 0.005 - 0.700 % K by weight.

4. Literature. Humphries (1956); Herrmann and Alkemade (1960, 1963). Method for automatic analysis of alkaline metals is given by Kessen (1966a). *Chemical method:* Humphries (1956).

Sodium

1. Theoretical Background. Sodium determination by flame spectroscopy in hydrochloric acid extract or in alkali ash extract.

2. Reagents. a) Hydrochloric acid 5% (see "Dry ashing technique") or nitric acid 3%, d = 1.015 (32 ml conc. HNO_3 ($\sim$ 68%, d $\approx$ 1.405) and 970 ml H_2)); b) sodium standard solution (dissolve 1.270 g NaCl (predried at 105°C) or 1.850 g $NaNO_3$ (keep dry over silica gel) in 5% HCl or 3% HNO_3, respectively, in 1,000 ml volumetric flask). The solution contains 500 mg/l Na; c) sodium calibration solution (dilute Na-standard solution (b) 1:9 volume to volume with the respective acid. [Na] = 50 mg/l); d) sodium standard curve (dilute 0-20 ml Na standard solution (b), $\hat{=}$ 0-100 mg/l Na, in 100 ml volumetric flasks with 5% HCl or 3% HNO_3, respectively; e) propane (C_3H_8).

Equipment. As in potassium method, but Na filter (maximum of transmissibility at 589 nm).

3. Working Procedure. Follow the same procedure as described under Potassium determination. Small flame is recommended (Zeiss, 1968). Adjust calibration solution with 50 mg/l Na (c) to 50 scale divisions. As Na-concentration in the calibration solution amounts to half the K-concentration, divide scale readings by 200 for hydrochloric acid extract and by 100 for alkali ash extract. Pay special attention to uniformity of acid concentration in blank, calibration, and sample solutions. For nonlinear standard curve errors and dilutions, recommendations as given under potassium determination should be followed.

Measuring range:

$\sim$ 1-70 mg/l Na $\approx$ 0.003-0.350 % Na by weight (HCl-extract)

$\approx$ 0.005-0.700 % Na by weight (alkali ash extract)

4. Literature. References cited under potassium analysis are applicable.

Calcium

1. Theoretical Background. Calcium determination by flame spectroscopy.

2. Reagents. a) Hydrochloric acid 5% (see "Dry ashing technique"); b) calcium standard solution (dissolve 2.4975 g $CaCO_3$ (predried at 105°C) in 1,000 ml 5% HCl (volumetric flask)). Solution contains 1.000 g/l Ca; c) calcium calibration solution (dilute Ca-standard solution (b) 1:9 volume to volume with 5% HCl, [Ca] = 100 mg/l); d) calcium standard curve (dilute 0-20 ml of Ca-standard solution (b), $\triangleq$ 0-200 mg/l Ca, with 5% HCl in 100 ml volumetric flasks); e) acetylene (C_2H_2).

Equipment. Same as used in potassium and sodium analysis, but calcium filter (maximum of transmissibility at 622 nm).

3. Working Procedure. Follow same procedures as given under potassium analysis, but use acetylene/air flame. In contrast to K and Na, however, the standard curve is sufficiently linear throughout the test range, so that the whole scale can be used without correction factor.

Measuring range:

~ 1-200 mg/l Ca ≈ 0.005-1.000 % Ca by weight.

4. Literature. See potassium analysis section. Atomic absorption analysis is of special interest for low Ca-content samples (see manuals of producers of atomic absorption photometers, e.g., Zeiss, 1968). Recently, a photometric method for automatic analyzers has been developed (Holz, 1973).

Magnesium

1. Theoretical Background. Magnesium determination by atomic absorption analysis. The addition of strontium buffer solution eliminates an alkaline-induced increase of extinction and phosphorus or sulfur-induced reduction of extinction.

2. Reagents. a) Hydrochloric acid 5% (see "Dry ashing technique"); b) strontium-chloride-buffer solution (dissolve 152.59 g $SrCl_2 \cdot 6H_2O$ in 1,000 ml H_2O; Sr $\hat{=}$ 50 g/l); c) magnesium standard solution (dissolve 0.835 g $MgCl_2 \cdot 6H_2O$ in 100 ml H_2O (volumetric flask)). Solution contains 1.000 g/l Mg. d) Magnesium calibration curve (prepare dilution 1:99 volume to volume of Mg-standard solution (c) with H_2O; 1 ml $\triangleq$ 10 µg Mg. Pipette 0-30 ml diluted Mg-solution into 100 ml volumetric flasks, add 5 ml $SrCl_2$ solution and 5 ml HCl 5% (a) and fill up with H_2O. Final concentrations of Mg 0-3 mg/l, of HCl 0.25%, of Sr 0.25%; e) diluted ash solution (pipette 5 ml ash extract and 5 ml $SrCl_2$ solution into 100 ml volumetric flask and make to volume with H_2O); f) acetylene (C_2H_2).

Equipment. Atomic absorption photometer, hollow cathode lamp for Mg (emission peak 285.2 nm), volumetric flasks (100 ml, 1,000 ml), pipettes (10-30 ml).

3. Working Procedure. Adjust acetylene/air mixture for a yellow reducing flame (Herrmann, 1965; Zeiss, 1968). The determination of magnesium in any sample series demands calibration curve consisting of an extinction vs. concentration plot. The standard curve is not completely linear. Moreover, the system parameters (slit width, amplification, cathode lamp characteristics, supply and purity of gases) cannot be kept constant over long periods, and frequent control is necessary by means of standard solutions. As the sensitivity of the method is very high and discontinuities from flame and apparatus make direct readings unstable, recorder registration during determination period (about 1 min/sample) is recommended. Registration is averaged graphically. Computation of results by means of standard curve: divide readings in ppm Mg (= mg/l Mg) by 10.

Measuring range:

~ 0.05-3.00 mg/l Mg $\hat{=}$ ~ 0.005-0.300 % Mg by weight.

4. Literature. Zeiss (1968); Welz (1972); Elwell and Gidley (1966); Hoffmann (1967); Müller (1970). Some chemical procedures are described but they are not sufficiently sensitive in the relevant concentration range, e.g., the method with 8-oxyquinoline (Scharrer and Mengel, 1957; Müller, 1970, 1973); the technique with Titan Yellow (Humphries, 1956). Automatic analyzer method (Holz, 1973). Flame spectroscopic determination has narrow sensitivity with possible high interference (Gärtel, 1959).

Iron

1. Theoretical Background. The colorimetric determination of iron is based on the formation of a red stain complex "Ferroin" established by the combination of Fe^{++}-ions and o-Phenanthroline. As iron is oxidized after ashing it must be reduced to Fe^{++}-oxidation stage by means of hydroquinone.

2. Reagents. a) Hydrochloric acid 5% (see "Dry ashing technique"); b) hydrochlorid acid 10% (see "Dry ashing technique"); c) brom-phenol-blue 0.04% (dissolve 40 mg brom-phenol-blue in 100 ml H_2O); d) sodium acetate solution 25% (dissolve 250 g Na-acetate in 750 ml H_2O); f) o-Phenanthroline solution 0.5% (dissolve 500 mg o-Phenanthroline (Synonyms: 1,10-Phenanthroline, 4,5-Phenanthroline) in 100 ml H_2O); g) iron standard solution (dissolve 143.0 mg Fe_2O_3 in 500 ml 10% HCl and dilute to 1,000 ml with water (volumetric flask). Solution contains 100 mg/l Fe; h) iron calibration solution (dilute Fe-standard solution (g) 1:9 ([Fe] = 10 μg/ml) and 1:99 ([Fe] = 1 μg/ml), volume to volume, with 5% HCl).

Equipment. Photometer, filter 492 nm, cuvette (2 cm absorbing layer), volumetric flasks (50 ml, 100 ml, 1,000 ml), pipettes (5 ml, 10 ml), burette (50 ml).

3. Working Procedure. First determine exact volume of Na-acetate solution (d) necessary for neutralization of 10 ml 5% HCl (volume is about 12-13 ml). Before titration add brom-phenol-blue indicator (c).

Standard Curve. Pipette 0-10 ml Fe dilution 1:99 ($\hat{=}$ 0-10 μg Fe) and 1.5-4 ml Fe dilution 1:9 ($\hat{=}$ 15-40 μg Fe) into 50 ml volumetric flasks and add 5% HCl to a volume of 10 ml. For sample analysis, pipette 10 ml ash extract into flasks of the same size. Then add to calibration solutions, as well as to sample solutions, 4 ml hydroquinone solution (e), 4 ml o-Phenanthroline solution (f) and the exact volume Na-acetate solution as determined previously. Fill up with H_2O and compare extinctions at 492 nm 1 h later. Standard curve is linear in 0.020-0.300 extinction range and yields an empirical coefficient "f" of about 120.6 for conversion from extinction to amount of Fe as µg Fe.

Computation formula:

$$\frac{Ex \cdot f \cdot Vt}{W \cdot Va} = \text{ppm Fe by weight}$$

Simplified for standard conditions:

$$\frac{Ex \cdot 120.6 \cdot 100}{2 \cdot 10} = Ex \cdot 603.0 \text{ ppm Fe by weight}$$

Measuring range:

~ 12-180 ppm Fe by weight

4. Literature. Humphries (1956); automatic analyzer apparatus, Holz (1969).

Manganese

1. Theoretical Background. Colorimetric determination of manganese as permanganate. Manganese, present as Mn^{++}, is oxidized to violet permanganate ion by sodium periodate:

$$2\ \overset{+2}{Mn}{}^{++} + 5\ \overset{+7}{IO_4}{}^{-} + 3\ H_2O \rightarrow 2\ \overset{+7}{MnO_4}{}^{-} + 5\ \overset{+5}{IO_3}{}^{-} + 6\ H^{+}.$$

Reducing substances have to be removed before periodate oxidation by boiling down the hydrochloric acid ash extract and reboiling with HNO_3. Interfering Fe^{+++}-ions are masked by added phosphoric acid.

2. Reagents. a) Sulfuric acid 5%, d ≈ 1.033 (29 ml H_2SO_4 conc., ~ 98%, d ≈ 1.84, and 979 ml H_2O); b) diluted nitric acid, 3%, d ≈ 1.015 (see sodium method); c) phosphoric acid (85%, d ≈ 1.71), commercial product; d) sodium sulfide (Na_2S); e) sodium-meta-periodate ($NaIO_4$); f) manganese standard solution (dissolve 143.8 mg $KMnO_4$ in small volume of H_2O (1,000 ml volumetric flask), acidify with 4 ml H_2SO_4 5% (a), reduce to Mn^{++} by means of the amount of Na_2S held on the tip of a spatula, ~ 0.3 g, (reaction:

$$8\ \overset{+7}{MnO_4} + 5\ \overset{-2}{S}{}^{--} + 8\ H_2O \rightarrow 8\ \overset{+2}{Mn}{}^{++} + 5\ \overset{+6}{SO_4}{}^{--} + 16\ H^{+})$$

and fill to volume with H_2O. 1 ml ≙ 50 µg Mn); g) diluted Mn-standard solution (dilute reduced Mn-standard solution 1:9, volume to volume, with H_2O. 1 ml ≙ 5 µg Mn).

Equipment. Photometer, filter 546 nm, cuvettes (2 cm absorbing layer), beakers (100 ml), volumetric flasks (50 ml, 100 ml, 1,000 ml).

3. Working Procedure. For standard curve, pipette 0-10 ml diluted Mn solution (g; ≙ 0-50 µg Mn) and 2-10 ml Mn standard solution (f; ≙ 100-500 µg Mn) into 50 ml volumetric flasks and add about 10 ml diluted nitric acid (b).

Samples. Boil down cautiously 10 ml of hydrochloric acid ash solution, take up the residue with 5-10 ml diluted HNO_3 (b) and reboil for a short time, transfer quantitatively to 50 ml volumetric flask, rinsing beaker well with diluted HNO_3. Pipette 5 ml H_3PO_4 (c) to standard and sample solutions, make almost to volume with diluted HNO_3, add one spatula (0.5 g) of $NaIO_4$ (3) and shake cautiously. Boil for 2 h in a water bath. Fill exactly with diluted HNO_3, shake and measure at 546 nm. The standard curve is linear from 0.015-0.900 extinction units with a coefficient of about 571.2 for conversion from extinction to amount of manganese as µg Mn. A special calibration curve is recommended for low extinctions.

Computation formula:

$$\frac{Ex \cdot f \cdot Vt}{W \cdot Va} = \text{ppm Mn by weight}$$

Simplified for standard conditions:

$$\frac{Ex \cdot 571.2 \cdot 100}{2 \cdot 10} = Ex \cdot 2{,}856 = \text{ppm Mn by weight}$$

Measuring range:

~ 50-2500 ppm Mn by weight

4. Literature. Sandell (1950); improvement of test range, Kreutzer (1972); automatic analyzer apparatus, Kessen (1966b); principles of manganese photometry, Gottschalk (1965).

Ashing with Alkali

Total Sulfur

1. Theoretical Background. Gravimetric sulfur determination as based on barium-sulfate precipitate formation. Volatility of sulfur and chloride components at higher temperatures is the reason why ashing must be done with added alkali. Less volatile alkaline and alkaline earth sulfates and chlorides are formed. Sulfur is determined by gravimetry after precipitation as heavy soluble barium-sulfate.

2. Reagents. a) Eschkas mixture (2 parts MgO and 1 part Na_2CO_3 by weight; store in dessiccator); b) hydrogen peroxide 30% (H_2O_2), commercial product; c) hydrochloric acid 1:1 volume to volume (add 500 ml HCl conc., ~ 38%, d ≈ 1.190 to 500 ml H_2O); d) ammonium acetate; e) 0.2 N barium chloride (dissolve 20.825 g $BaCl_2$ or 24.428 g $BaCl_2 \cdot 2\,H_2O$ in H_2O to 1,000 ml final volume); f) litmus paper.

Equipment. Muffle furnace; platinum or quartz crucibles for ashing; beakers (250 ml, 600 ml); funnels; filter paper (slow and quick filtering qualities); heater or Bunsen burner.

3. Working Procedure. Mix thoroughly 0.600 g dry matter with about 3 g Eschkas mixture (a) in a crucible and cover mixture with another 2 g Eschkas mixture. Blank sample contains only 5 g Eschkas mixture. First heat samples up for 45 min starting with a cold muffle furnace, than ash for 75 min at about 750°C. Transfer the ashed material from the crucible to a 250 ml beaker and rinse well with about 100 ml H_2O. Add 2 ml H_2O_2 (b) and keep boiling for 30 min on a heater or Bunsen burner. Filter into a 600 ml beaker through a quick filtering filter

paper and rinse the filter and the precipitates well (final volume in beaker about 500 ml). Boil down to about half the volume, neutralize with HCl 1:1 (c)(color change of litmus paper chip), continue boiling for another 5 min and add one small spatula (0.5 g) of NH_4-acetate (d). Add drop by drop about 10 ml 0.2 N $BaCl_2$ (e) from a pipette in order to precipitate sulfate in the hot solution. Keep standing overnight and filter afterwards through a slow filtering filter paper. Ash filter in a crucible to weight constancy. Weight of $BaSO_4$-precipitation in mg (subtract blank value) is converted to sulfur content by the coefficient 0.1374.

Computation formula:

$$\frac{(Ps - Pb) \cdot 0.1374}{W \cdot 10} = \% \text{ S by weight}$$

Simplified for standard conditions:

$$\frac{(Ps - Pb) \cdot 0.1374}{0.6 \cdot 10} = (Ps - Pb) \cdot 0.0229 = \% \text{ S by weight}$$

Measuring range:

~ 0.02-5.3 % S by weight

4. Literature. Humphries (1956). Instead of alkaline dry ashing, wet digestion can be employed (Blanchar et al., 1965). Differentiation between organically and inorganically bound sulfur is possible by determining the difference between total sulfur content (method described) and the sulfur fraction extracted by 20% HCl (Dijkshoorn et al., 1960). Instead of $BaSO_4$ gravimetri-$BaSO_4$-turbidimetric determination may also be used (Bardsley and Lancaster, 1965; Blanchar et al., 1965), which seems to yield slightly lower values (Boukhris and Lossaint, 1970).

Total Chloride

1. Theoretical Background. Titrimetric chloride determination as AgCl (Volhard's "rest" method). Alkaline ashing as described above. After dissolving the Eschkas mixture with excess nitric acid, an exact volume of silver nitrate solution is added. The resulting AgCl-precipitate is filtered and the non-precipitated Ag^+ titrated with ammonium rhodanide. Difference to total Ag^+ added is proportional to Cl^- in sample.

2. Reagents. a) 0.1 N silver nitrate solution (dissolve 16.987 g $AgNO_3$ in H_2O to a total volume of 1,000 ml; store solution in brown glass bottle); b) 0.02 N ammonium-thiocyanate (= ammonium-rhodanide) solution (dissolve 1.5224 g NH_4SCN in H_2O to a total volume of 1,000 ml); c) saturated Fe^{3+}-indicator-solution (take $Fe(SO_4)_3$ or $Fe(NH_4)(SO_4)_2 \cdot 12\ H_2O$); d) Eschkas mixture (see sulfur method); e) nitric acid 10%, d ≈ 1.055 (add 110 ml HNO_3 conc. (~ 65%, d ≈ 1.400) to 900 ml H_2O).

Equipment. Platinum or quartz crucibles, muffle furnace, beakers (400 ml), funnels, filter paper (slow filtering quality), burettes (10 ml), volumetric flasks (100 ml, 250 ml, 1,000 ml).

3. Working Procedure. Incinerate 1.000 g dry substance as described under total sulfur analysis. For the blank sample, ash Eschkas mixture only. Dissolve ashed samples in 400 ml beakers with HNO_3 (e) added in excess. About 25 ml of 10% HNO_3 is necessary per 1 g Eschkas

mixture. (There is a risk of foaming when acid is added). Add 5.00 ml of 0.1 N $AgNO_3$-solution (a; burette) to the sample dissolved in nitric acid, filter into a 250 ml volumetric flask and wash filter thoroughly, filling the flask to volume. Take a 50 ml aliquot, add 5 ml Fe^{3+}-indicator-solution (c) and titrate with 0.02 n NH_4SCN-solution (b) to the stable yellowish-brown coloration. The correction term "C" for the normality ratio $AgNO_3/SCN$: titrate 2.00 ml $AgNO_3$-solution (a) that has been complemented by 5 ml 10% HNO_3 (e), 10 ml H_2O and 2 ml Fe^{3+}-solution (c).

Computation formula:

$$1 \text{ ml } (Tb - Ts) \hat{=} 1 \text{ ml } 0.02 \text{ N } NH_4SCN \hat{=} 0.709 \text{ mg } Cl^-$$

$$\frac{(Tb - Ta) \cdot C \cdot 0.709 \cdot Vt}{W \cdot Va \cdot 10} = \% \ Cl^- \text{ by weight}$$

$$C = \frac{a \cdot 5}{b}$$ a, volume of $AgNO_3$-solution
b, volume of NH_4SCN used for titration of $AgNO_3$

Simplified for standard conditions:

$$\frac{(Tb - Ta) \cdot C \cdot 0.709 \cdot 250}{1 \cdot 50 \cdot 10} = (Tb - Ta) \cdot C \cdot 0.354$$
$$= \% \ Cl^- \text{ by weight}$$

Measuring range:

0.035-3.54 % Cl^- by weight

4. Literature. Humphries (1956). Determination methods using ion selective electrodes are pertinent (Durst, 1969; Bazzelle, 1971; see also the documentation of numerous electrode producers) and the even more highly recommended potentiometric determination and separation of halogenides (Scharrer and Jung, 1957; Kjellerup, 1970). Iodometric procedure is applied to pressed juices of fresh plant material (Prikladowizky and Apollonow, 1928). Instead of ashing, acid extraction of plant material is frequently sufficient in itself for detection of total chlorides (Glatzel, 1974).

Acknowledgement. We gratefully acknowledge the help of Dr. Richard E. Dickson, Plant Physiologist, North Central Forest Experiment Station, Rhinelander, Wisconsin, in the preparation of this manuscript.

References

Allen, R.: The estimation of phosphorus. Biochem. J. 34, 858 (1940).

Arkley, T., Munns, D.N., Johnson, C.M.: Preparation of plant tissue for micronutrient analysis. Removal of dust and spray contaminants. Agr. Food Chem. 8, 318-321 (1960).

Armson, K.A.: Soil and plant techniques as diagnostic criteria for evaluating fertilizer needs and treatment response. Forest Fertilization Symposium Proc. USDA For. Serv. NE-3, 155-166 (1973).

Balks, R., Reekers, I.: Zur Methode der Nitratbestimmung in Pflanzensubstanz. Landwirtsch. Forschung 6, 121-126 (1954).

Bardsley, C.E., Lancaster, J.D.: Sulfur. In: Methods of Soil Analysis (ed. C.A. Black), pp. 1102-1116. Wisconsin: 1965.

Barnes, B.V.: Effects of thinning and fertilizing on production of western white pine seed. USDA, For. Serv. Res. Pap. Int-58. (1969).

Barnes, R.L., Bengtson, G.W.: Some aspects of nitrogen and metabolism in relation to fertilizer response in southern pines. In: Forest Fertilization, Theory and Practice, pp. 58-63. Knoxville, Tenn.: TVA 1968.

Batey, T., Cresser, M.S., Willet, I.R.: Sulphuric-perchloric acid digestion of plant material for nitrogen determination. Anal. Chimica Acta 69, 484-487 (1974).

Bazzelle, W.E.: Application of ion selective electrodes. Anal. Chimica Acta 54, 29-39 (1971).

Bevege, D.I., Simpson, J.A.: Statement on current activities in forest nutrition. Queensland Dept. Forestry, 25-28 (1973).

Blanchar, R.W., Rehm, G., Xadwell, A.C.: Sulfur in plant material by digestion with nitric and perchloric acid. Soil Sci. Soc., Am. Proc. 29, 71-72 (1965).

Bornmann, P.: Stickstoffbestimmungsmethoden in Abhängigkeit vom Molekülaufbau. Unpublished. Ed. by Süddt. Kalkstickstoffwerke Trostberg (FRG), 13 p., 42 Tables (1972).

Borsdorf, W.: Über den Einfluß mineralischer Düngung auf Wuchsleistung und Darrdichte junger Pappelpflanzen. Arch. Forstwesen 14, 61-78 (1965).

Boukhris, M., Lossaint, P.: Sur la teneur en soufre de quelques plantes gypsophiles de Tunisie. Oecol. Plant. 5, 345-354 (1970).

Bradstreet, R.B.: The Kjeldahl Method for Organic Nitrogen. New York: Academic Press 1965.

Bremner, J.M.: Determination of nitrogen in soil by the Kjeldahl method. J. Agric. Sci. 55, 11-33 (1960).

Bremner, J.M., Keeney, D.R.: Steam distillation methods for determination of ammonium, nitrate, and nitrite. Anal. Chimica Acta 32, 485-495 (1965).

Cammann, K.: Bestimmung von Ammoniumionen mit einer ionenspezifischen Glaselektrode. Beckmann-Report 2, 13-14 (1966).

Concon, J.M., Soltess, D.: Rapid micro Kjeldahl digestion of cereal grains and other biological material. Analytical Biochemistry 53, 35-41 (1973).

Dijkshoorn, W., Lampe, J.E.M., van Burg, P.F.J.: A method of diagnosing the sulphur nutrition status of herbage. Plant Soil 13, 227-241 (1960).

Durst, R.A. (ed.): Ion Selective Electrodes. Special Publication 314, 474 pp. Washington, D.C.: National Bureau of Standards 1969.

Elwell, W.T., Gidley, J.A.F.: Atomic absorption spectrophotometry. New York: Pergamon Press 1966.

Evers, F.H.: Die jahrweise Fluktuation der Nährelementkonzentrationen in Fichtennadeln und ihre Bedeutung für die Interpretation nadelanalytischer Befunde. Allgem. Forst- und Jagdzeitg. 143, 68-74 (1972).

Evers, F.H.: Genetische Unterschiede im Mineralstoffgehalt der Nadeln junger Fichten (*Picea abies* (L.) Karst.). Mitteilungen des Vereins f. Forstl. Standortskunde und Forstpflanzenzüchtung 23, 67-71 (1973)

Fiedler, H.J.: Die Untersuchung von Böden. Vol. 2, 256 pp. Dresden und Leipzig: Verlag Theodor Steinkopf 1965.

Fiedler, H.J.: Neuere Entwicklungen der flammenphotometrischen Analysenverfahren. Zeitschr. Anal. Chemie 212, 1-18 (1965).

Forrest, W.G., Ovington, J.D.: Variation in dry weight and mineral nutrient content of *Pinus radiata* progeny. Silvae Genetica 20, 174-179 (1971).

Gärtel, W.: Flammenspektrometrische Magnesium-Bestimmung in Bodenauszügen und Pflanzenaschen mit dem Flammenzusatz zum Zeiss'schen Spektralphotometer PMQ II. Landwirtsch. Forsch. 13, Sonderheft, 97-98 (1959).

Gerhold, H.D.: Seasonal variation of chloroplast pigments and nutrient elements in the needles of geographic areas of Scotch pine. Silvae Genetica 8, 113-123 (1959).
Gericke, S., Kurmies, B.: Die kolorimetrische Phosphorsäure-Bestimmung mit Ammonium-Vanadat-Molybdat und ihre Anwendung in der Pflanzenanalyse. Zeitschr. Pflanzenernährung, Düngung und Bodenkunde 59, 235-247 (1952).
Glatzel, G.: Analytische Methoden zum Nachweis der Schädigung von Pflanzen durch Auftausalze. Europ. J. Forest Path. 4, 52-53 (1974).
Goddard, R.E., Strickland, R.K.: Eighth Progress Report. Cooperative Forest Genetics Research Program, Univ. Fla., School of Forestry Research, Rep. No. 17, 17 pp. (1966).
Goodall, D.W., Gregory, F.G.: Chemical composition of plants as an index of their nutritional status. Commonw. Bur. of Hort. Plant. Crops, Techn. Comm. 17 (1947).
Gottschalk, G.: Grundlagen photometrischer Analysenverfahren. V. Manganbestimmung als Permanganat. Zeitschr. anal. Chemie 212, 303-317 (1965).
Heiberg, S.O., White, D.P.: Potassium deficiency of reforested pine and spruce stands in northern New York. Soil Sci. Soc., Am. Proc. 15, 369-376 (1951).
Hermann, R., Alkemade, C.Th.J.: Flammenphotometrie. Berlin-Göttingen-Heidelberg: Springer 1960.
Hermann, R., Alkemade, C.Th.J.: Flame photometry. New York, London, Sydney: Interscience Publishers 1963.
Hoffmann, F.: Untersuchungen zur atomabsorptionsflammenphotometrischen Magnesium-Bestimmung in Lösungen von Pflanzenaschen und Doppellaktat- sowie Ammoniumlaktatessigsäure-Extrakten aus dem Boden. Thaer-Archiv 11, 405-414 (1967).
Höhne, H., Fiedler, H.J.: Beiträge zur Stickstoffdüngung mittelalter Kiefernbestände (IV). Arch. Forstwesen 19, 877-898 (1970).
Hollis, C.A., Smith, W.H., Schmidt, R.A., Pritchett, W.L.: Some effects of fertilizers and soil properties on the incidence of fusiform rust. Agron. Abst. p. 139 (1972).
Holz, F.: Die automatische Analyse der Lösungen von Pflanzenaschen mit dem Auto-Analyzer. 4. Mitt.: Die automatische Bestimmung von Eisen. Landwirtsch. Forschung 22, 123-129 (1969).
Holz, F.: Die automatische Bestimmung des Stickstoffes als Indophenolgrün in Böden und Pflanzen. Landwirtsch. Forschung 26, Sonderheft, 177-191 (1971).
Holz, F.: Die automatische Analyse der Lösungen von Pflanzenaschen mit dem Auto-Analyzer. 5. Mitt.: Die photometrische Simultanbestimmung von Calcium und Magnesium. Landwirtsch. Forschung 26, 56-80 (1973).
Hübsch, H., Nehring, K.: Vergleichende Untersuchungen über die Bestimmung des Stickstoffgehaltes nach den Methoden von Kjeldahl und Dumas. Zeitschr. Anal. Chemie 173, 278-284 (1960).
Humphries, E.C.: Mineral components and ash analysis. In: Modern Methods of Plant Analysis (eds. K. Peach, M.V. Tracey), Vol. 1, pp. 468-502. Berlin-Göttingen-Heidelberg: Springer 1956.
Hunger, W.: Über den Einfluß der Lufttrocknung auf den Nährelementgehalt der Fichtennadel. Flora 163, 265-275 (1974a).
Hunger, W.: Untersuchungen über die jahreszeitliche und jahrweise Fluktuation der Nährelementkonzentrationen in den Nadeln eines jüngeren Fichtenbestandes. Flora 163, 422-442 (1974b).
Jahromi, S.T.: Genetic variation in nutrient absorption in slash pine. Ph.D. Dissertation, Univ. of Florida, Gainesville. 133 p. (1971).
Kenworthy, A.L.: Interpreting the balance of nutrient-elements in leaves of fruit trees. In: Plant Analysis and Fertilizer Problems (ed. W. Reuther). Amer. Inst. of Biol. Sci., Pub. No. 8, pp. 28-43.

Kessen, G.: Die automatische Analyse der Lösungen von Pflanzenaschen mit dem Auto-Analyzer. 1. Mitt.: Die automatische Ammoniakbestimmung in Kjeldahlaufschlüssen und die automatische Bestimmung von Phosphor, Kalium und Natrium in Pflanzenaschen. Landwirtsch. Forschung 19, 108-116 (1966a).
Kessen, G.: Die automatische Bestimmung der Lösungen von Pflanzenaschen mit dem Auto-Analyzer. 2. Mitt.: Die automatische Bestimmung von Mangan. Landwirtsch. Forschung 19, 227-230 (1966b).
Kjeldahl, J.: Neue Methode zur Bestimmung des Stickstoffes in organischen Körpern. Zeitschr. Analyt. Chemie 22, 366-382 (1883).
Kjellerup, V.: Comparison of electrometric and colorimetric determination of chloride in crops. Tidskrift for Planteavl 74, 367-371 (1970).
Kleinschmit, R.: Stickstoffdüngungsversuche in einer Samenplantage. 2. Mitt. Forst- und Holzwirt. 16, 403-404 (1961).
Knight, P.J.: Interclonal variations in foliar nutrient concentrations of *P. radiata*. Ann. Rep. New Zealand Forest Res. Inst., Wellington, N.Z. (1972).
Kral, F.: Physiologische Frühtestversuche an Herkünften der grünen Douglasie. Centralblatt Ges. Forstw. 82, 129-149 (1965).
Kreutzer, H., Merz, W.: Stickstoffschnellbestimmung nach Merz in der Futter- und Düngemittelanalyse. Landwirtsch. Forschung 28, Sonderheft, 120-129 (1973).
Kreutzer, K.: Die Wirkung des Manganmangels auf die Farbe, die Pigmente und den Gaswechsel von Fichtennadeln (*Picea abies* Karst.). Forstwiss. Centralblatt 91, 80-98 (1972).
Kunze, A., Ebeling, R., Vielemeyer, H.P., Bergmann, W.: Einige apparative Verbesserungen beim Halbmikroaufschluß nach der Kjeldahl-Methode in Verbindung mit verschiedenen Vergleichsuntersuchungen. Z. f. Landwirtsch. Versuchs. Untersuchungswesen 10, 55-63 (1964).
Leaf, A.L.: K, Mg and S deficiencies in forest trees. Forest fertilization/theory and practice. Symposium 1967 Gainesville, Fla., Publ. 1968 by TVA. Muscle Shoals, Ala., 88-122 (1968).
Leaf, A.L.: Plant analysis as an aid in fertilizing forests. In: Soil Testing and Plant Analysis, 2nd ed. Madison, Wisc.: Soil Sci. Soc. Amer. Inc. 1972.
Leyton, L.: The relationship between the growth and mineral nutrition of conifers. In: The Physiology of Forest Trees (ed. K.V. Thimann), pp. 323-345. New York: The Ronald Press Co. 1958.
Lundegard, H.H.: Die Blattanalyse. Jena: G. Fischer Verlag 1945.
May, J.T., Johnson, H.H., Gilmore, A.R.: Chemical composition of southern pine seedlings. Georgia Forest Res. Pap. 10, 11 (1962).
Merz, W.: Ein neuer Automat zur Bestimmung von Stickstoff in Dünge- und Futtermitteln. Landwirtsch. Forschung 26, 311-319 (1973).
Mitchell, H.L., Chandler, R.F.: The nitrogen nutrition and growth of certain deciduous trees of North Eastern United States. Black Rock Forestry Bull. No. 11 (1939).
Müller, W.: Verfahren zur Bestimmung der Makro- und Mikronährstoffe in Koniferennadeln. Arch. Forstwesen 19, 861-876 (1970).
Müller, W.: Zur Bewertung analytischer Methoden auf dem Gebiet der Blattanalyse. Wiss. Zeitschr. Techn. Univ. Dresden 22, 368-369 (1973).
Müller, W., Fiedler, H.J.: Zur Bestimmung des Gesamtstickstoffes in Boden- und Pflanzenmaterial nach der Kjeldahl-Methode. Arch. Forstwesen 18, 945-956 (1969).
Mustanoja, K.J., Leaf, A.L.: Forest fertilization research. Botan. Review 31, 151-246 (1965).
Netheler, Hinz, GmbH (ed.): Flammenphotometer Eppendorf (unpublished manual), 85 pp. Hamburg 1956.
Orion Research Inc. (ed.): Orion Research Newsletter 4, 17-20 (1972).

Parnas, J.K., Wagner, R.: Über die Ausführung von Bestimmungen kleiner Stickstoffmengen nach Kjeldahl. Biochem. Z. 125, 253-256 (1921).

Peach, K., Tracey, M.: Moderne Methoden der Pflanzenanalyse. Vol. 1, 542 pp. Berlin-Göttingen-Heidelberg: Springer 1956.

Peterson, I.A., Chesters, G.: A reliable total nitrogen determination on plant tissue accumulating nitrate nitrogen. Agro. J. 56, 89-90 (1964).

Piper, C.S.: Soil and plant analysis. New York, London, Sydney: 1950.

Prikladowizky, S., Apollonow, A.: Jodometrische Methode zur Bestimmung der Chloride. Biochem. Z. 200, 135 (1928).

Pritchett, W.L.: Progress in the development of techniques and standards for soil and foliar diagnosis of phosphorus deficiency in slash pine. Forest fertilization/theory and practice. Symposium 1967 Gainesville, Fla., Publ. 1968 by TVA, Muscle Shoals, Ala., 81-87 (1968).

Pritchett, W.L., Goddard, R.E.: Differential response of slash pine progeny lines to some cultural practices. Soil Sci. Soc. Am. Proc. 31, 280-284 (1967).

Raupach, M.: Soil and fertilizer requirements for forests of *Pinus radiata*. Adv. Agron. 19, 307-353 (1968).

Rediscke, J.H., Gauditz, I., Campbell, R.K.: Family response to levels of fertilizer. Ann. Meeting West. Forest Genet. 13, Corvallis, Oregon (1968).

Reemtsma, J.B.: Untersuchungen an Fichte und anderen Nadelbaumarten über den Nährstoffgehalt der lebenden Nadeljahrgänge und der Streu, 139 p. Dissertation Hann.-Münden (1964).

Sandell, E.G.: Colorimetric determination of traces of metals. New York: Interscience Publishers 1950.

Scharrer, K., Jung, J.: Die Bestimmung des Chloridgehaltes pflanzlicher und tierischer Substanzen auf potentiometrischem Wege. Zeitschr. f. Tierernährung u. Futtermittelk. 12, 14-19 (1957).

Scharrer, K., Mengel, K.: Die Bestimmung kleinster Mengen Magnesiums für Zwecke der agrikulturchemischen Analyse. Zeitschr. Pflanzenernährung, Düngung u. Bodenk. 77, 18-36 (1957).

Scharrer, K., Seibel, W.: Vergleichende Überprüfung verschiedener Methoden zur Nitratbestimmung in Futterpflanzen. Zeitschr. Tierernährung u. Futtermittelk. 11, 145-156 (1956).

Shigo, A.L.: Insect and disease control: forest fertilization relations. In: Forest Fertilization Symposium Proceedings, USDA Forest Serv. Gen. Tech. Rep. NE-3, 117-121 (1973).

Smith, H.W., Goddard, R.E.: Effects of the genotype on the response to fertilizer. FAO/IUFRO Symposium on Forest Fertilization, Paris 1973, Solicited Paper (1973).

Smith, P.F.: Mineral analysis of plant tissues. Ann. Rev. Plant Physiol. 13, 81-108 (1962).

Steinbeck, K.: Site, height, and mineral nutrient control relations of Scotch pine provenances. Silvae Genetica 15, 42-50 (1966).

Stone, E.L.: Microelement nutrition of forest trees; a review. Forest fertilization/theory and practice. Symposium 1967 Gainesville, Fla., Publ. 1968 by TVA, Muscle Shoals, Ala., 132-175 (1968).

Tamm, C.O.: The effects of nitrogen fertilization on tree growth and foliage composition in a forest stand. Plant analysis and fertilizer problems. Inst. Rech. Huiles et Oléagineux, p. 150-157, Paris (1956).

Tamm, C.O.: An attempt to assess the optimum nitrogen level in Norway spruce under field conditions. Studia Forestalia Suecica 61, 1-65 (1968).

Ulrich, A.: Plant analysis as a diagnostic procedure. Soil Science 55, 101-112 (1943).

Ulrich, A., Hills, F.J.: Principles and practices of plant analysis. In: Soil Testing and Plant Analysis. Part II. Plant Analysis, p. 11-24. Soil Science Soc. Am. (1967).

van den Driessche, R.: Foliar nutrient concentration differences between provenances of Douglas fir and their significance to foliar analysis interpretation. Can. J. Forest Res. 3, 323-327 (1973a).

van den Driessche, R.: Prediction of mineral nutrient status of trees by foliar analysis. Botan. Rev. 40, 347-394 (1973b).

van Goor, C.P., de Wit, M., van den Burg, J.: International comparison of methods for soil and plant analysis. Report on activities in 1970, IUFRO Sect. 21. WG, 38 p. Wageningen, Netherlands: Forest Res. St. 1971.

Wehrmann, J.: Methodische Untersuchungen zur Durchführung von Nadelanalysen in Kiefern-Beständen. Forstwiss. Centralblatt 78, 77-97 (1959).

Wehrmann, J.: Möglichkeiten und Grenzen der Blattanalyse in der Forstwirtschaft. Landwirtsch. Forschung 16, 130-145 (1963).

Wellenstein, G., Eichner, M., Scholl, W.: Das Rückstandsproblem nach Anwendung von Wuchsstoff-Herbiziden in der Forstwirtschaft. Allg. Forst- u. Jagdzeitung 146, 63-72 (1975).

Welz, B.: Atom-Absorptions-Spektroskopie, 216 p. Weinheim: Verlag Chemie 1972.

Wittich, W.: Auswertung eines forstlichen Düngungsversuches auf einem Standort mit für weite Gebiete Deutschlands typischem Nährstoffhaushalt, p. 1-48. Bochum: Ruhr-Stickstoff 1958.

Zeiss, C. (ed.): Analytische Methoden für Atomabsorption und Flammenemission. (Unpublished manual by Zeiss, Oberkochen, FGR)(1968).

Zobel, B.J., Roberds, J.H.: Differential genetic response to fertilizers within tree species. First North Amer. For. Biol. Workshop, Michigan State Univ., East Lansing (Abstr.) 1 (1970).

CHAPTER 9

Pollution Responses

K. F. JENSEN, L. S. DOCHINGER, B. R. ROBERTS, and A. M. TOWNSEND

Introduction

Air pollution has been with man since he started his first fire. The problem has intensified rapidly in the last 50 years; and air pollutants are reaching concentrations that are harmful to plants and animals.

Many compounds classified as air pollutants are produced by natural processes and are indigenous components of the atmosphere. The main cause of the air-pollution problem has been man's increasing demand for energy, goods, and services. The production of energy - mainly by the burning of fossil fuels - and the use of this energy to smelt metals and produce an endless array of products have increased the concentration of air pollutants in the atmosphere.

Air pollution was first noted to be harmful to forests around ore smelters that emitted large amounts of sulfur dioxide (SO_2). Trees in the vicinity of smelters at Anaconda, Montana, were investigated in 1910 and 1911; and those near the smelters at Trails, British Columbia, were investigated around 1930 (Scheffer and Hedgcock, 1955). In the more recent past, the trees near the Sudbury, Ontario, smelting complex have shown injury due to SO_2 and have been extensively studied (Linzon, 1971).

Injury noted on trees in California in about 1950 was also found to result from air pollution. The pollutant responsible, ozone (O_3), is generated in an atmosphere containing large amounts of hydrocarbons and oxides of nitrogen. This pollutant has injured many acres of conifer forest in the mountains around Los Angeles (Miller and McBride, 1973).

Economic losses caused by air pollutants have been evaluated in only a few cases. Linzon (1971) attempted to evaluate the loss in white pine around the smelters at Sudbury, Ontario, and concluded that the loss was about \$ 117,000 annually. In Czechoslovakia, Lux (1971) estimated that SO_2 emissions were causing about \$ 90 million worth of damage to forests yearly. The total loss in forests and nurseries in the United States due to air pollution in 1971 was estimated at \$ 19 million (Benedict et al., 1971).

The economic loss due to air pollution will probably increase in the near future. World population is expected to continue its rapid rate of increase; and it is only realistic to expect that energy consumption and industrial production and pollutant emissions will also continue to increase.

Material and Methods

Methods of Investigating Air-Pollution Effects

Many approaches have been used to investigate the effects of air pollutants on trees. They include fumigation studies, air monitoring, field surveying, vegetation sampling, and biological monitoring (Donaubauer, 1971).

Fumigation Studies. Fumigation studies are usually carried out in some type of chamber. The chambers used range from specially constructed growth chambers with elaborate control systems to polyethylene-covered boxes with minimal controls.

Hill (1967) designed an environmental chamber for fumigating plants. Temperature, light, relative humidity, wind velocity, and carbon dioxide and pollutant concentration could be controlled. A second chamber in which these factors, except for wind velocity, could be controlled was constructed from a commercially available environmental chamber (Wood et al., 1973). The chamber was modified by coating the interior surfaces with epoxy-based paints and by using electrical wiring with Teflon insulation inside the chamber.

Small fumigation chambers have also been placed inside environmental chambers for the control of environmental factors (Heck et al., 1968). This type of system is less expensive to construct than modified environmental chambers and allows for fumigation of several treatments at once.

Two general types of chambers have been designed for field studies. Heagle et al. (1972) described an enclosed chamber about 2.4 x 2.4 x 2.1 m with a sloping top constructed with an aluminum frame and covered with Teflon film (closed-top chamber). The second type of field chamber is the open-top chamber (Heagle et al., 1973). The chambers are cylinders 2.4 m high and 3.0 m in diameter covered with polyethylene plastic. This type of chamber was developed to provide an environment more closely resembling ambient conditions than the conditions found in closed-top chambers.

Fumigation chambers used in studies with trees have been described by Berry (1970), Spierings (1967), and Dochinger and Seliskar (1970). Berry described a lean-to type of greenhouse for growing trees up to 2 m tall. The system developed by Spierings enclosed a branch of a tree, while Dochinger and Seliskar enclosed the top part of a small tree.

Each chamber or chamber system must be equipped with pumps or blowers to move air through the chambers. They have been designed with both negative and positive air-flow systems, depending on how the chamber was to be used and where it was to be located. Each system must also have air filters to purify the air before it enters the chamber and some method for adding or generating pollutants within the chamber.

Air Monitoring. Several methods are available for measuring each of the major pollutants. The method selected will depend on the sensitivity of the instrument, the type of data needed, the cost, and the maintenance requirements of the equipment.

Sulfur dioxide is best determined by means of the flame photometry or coulometry (Thomas, 1974). Instruments utilizing either method can measure ambient concentrations of SO_2, but flame photometry is more sensitive than coulometry. The flame photometric instruments are total sulfur instruments and measure SO_2, hydrogen sulfide, and organic sulfur compounds. For analysis of only one pollutant, scrubbers must be placed in line to remove other contaminants. Hydrogen sulfide and atmospheric oxidizing and reducing agents interfere with the coulometric methods. To measure SO_2 in areas without electricity or where a mean concentration for several days is sufficient, lead sulfation plates can be used (Huey, 1968).

Ozone may be determined with chemiluminescence photometry or instruments using an iodine-based reaction (Thomas, 1974). The chemiluminescence instruments utilize the reaction of ozone with ethylene to measure ozone. These instruments usually have low maintenance requirements. The iodine-based instruments are subject to error. However, if they are properly maintained and calibrated, and equipped with appropriate scrubbers, they will provide satisfactory data (Costomis and Linzon, 1971).

No automatic methods are presently available for analyzing ambient concentrations of fluorides. The fluoride gases or particles must first be trapped in a filtering device, dissolved in a liquid, and analyzed for fluoride in a conventional method or with a specific ion electrode (Thompson et al., 1971).

Particles are collected in several different ways. Dust-fall buckets may be used to collect particles that settle out of the air. High-volume samplers or tape samples collect particles on filters that can be used to measure the suspended particles in the air. The particles may also be removed from the filters, and chemical analysis may be made to determine their composition. Automatic analyzers are also available for measuring particle size.

For a recent discussion on air monitoring see "Methods of Air Sampling and Analysis" by the American Public Health Service (1974).

Field Surveying. Field surveying is a very important means of judging the status of vegetation in areas where air pollution may be causing injury (Lacasse and Moroz, 1969; Weinstein and McCune, 1970). First, the important pollutants in an area must be determined, and then a list of plants native to the area and sensitive to the pollutants is compiled. The survey is then conducted according to standard procedures.

The main problem associated with field surveys is to determine whether the observed injury is caused by a pollutant or by some other biotic or abiotic factors. Many symptoms caused by disease organisms, insects, senescence, or environmental stresses are similar to symptoms caused by pollutants. The personnel conducting the survey must be well-trained and experienced to identify the different causal agents.

Vegetation Sampling. Vegetative sampling can be used to measure the dispersion area of a pollutant or to study the absorption of pollutants by vegetation. The vegetation should be sampled according to a statistical design and in such a manner that the sample is representative of the vegetation in the area (Weinstein and McCune, 1970).

The two major gaseous pollutants for which tissue analysis has been used most frequently are sulfur and fluorides. Sulfur may be determined by conventional gravimetric techniques or by the much faster combustion-titration procedure (Bremanis et al., 1967). Fluorides may also be determined by conventional quantitative techniques or with the recently developed fluoride ion electrode (Carlson and Keeney, 1971; Jones and MacLaughlin, 1971). The details for most of the conventional methods can be found in quantitative chemistry books or in books on plant analysis (Association of Official Agricultural Chemists, 1960).

Tissue analysis has also been used to evaluate the effect of heavy metal pollutants on vegetation. Analyses for these pollutants are often performed with atomic absorption techniques (see contribution by Evers and Bücking, this volume).

Biological Monitoring. Biological monitors are plants known to be sensitive to a certain pollutant; for example, ponderosa pine seedlings are sensitive to fluorides. These monitors may be used as part of a field survey for supplying additional information, or they may be used in fumigation investigation for determining the effect of environmental factors on known symptom patterns. Weinstein and McCune (1970) list the guidelines that should be considered in using biological monitors.

Vegetation Responses to Air Pollution

Air pollutants affect leaf tissue and the inherent related physiological processes in leaves. Hence, assaying foliar injury is the primary means of determining the impact of air pollutants on vegetation. Other variables that have been measured are plant growth and biomass production. General reviews of the effects of air pollution on vegetation have been published by Scurfield (1960), Thomas (1961), Brandt and Heck (1968), Treshow (1970), Heggestad and Heck (1971), and others. Knabe (1972) edited a series of papers on effects of air pollution on forest trees.

Foliar Symptoms. Leaf symptoms induced by air pollutants consist of acute and chronic patterns of injury. Acute injury is initiated by a toxic concentration for a short time. Symptoms may appear within 24 to 48 h after exposure. Expressions of injury are necrotic lesions that may extend to premature defoliation followed by plant mortality.

Chronic injury results from a sublethal dosage of pollutant after intermittent or continuous exposure. Vegetation exposed to low pollution levels usually exhibits a chlorosis or a similar disruptive pigmentation in leaf tissues. Occasionally, premature senescence of foliage is evident, which in time can curtail tree growth. Unlike acute injury, which is generally characterized by necrosis, chronic symptoms do not immediately destroy the physiological functions of leaves. In fact, plants continue to photosynthesize, though at a reduced level. Both acute and chronic injury can be viewed simultaneously on affected trees.

Although characteristic symptoms are induced by specific air pollutants, there can also be a range in the plant response. Numerous investigators, including Thomas (1951), Pelz (1956), Dochinger et al. (1972), and Townsend and Dochinger (1974), have indicated that genetic variation can influence the sensitivity of plants to air pollutants. Reviews by Katz (1949), Zahn (1963), Heck (1968), and Wentzel (1968) further clarify the variability in plant response to air pollutants as conditioned by maturity of tissues, stomatal movements, soil influences, levels and fluctuations in ambient concentrations of air pollutants, pollutant composition, duration, and interaction with other atmospheric constituents. Responses are also influences by relative humidity, temperature, and related environmental factors.

Innumerable air pollutants are responsible for toxic effects on vegetation. Although their relative importance as phytotoxicants differs from location to location, international estimates of the economic damage in forest and shade tree-plantings indicate that the primary offenders are sulfur dioxide, ozone, and fluorides.

Sulfur Dioxide

Sulfur dioxide originates from the combustion of organic fossil fuels. Combustion of sulfur-containing compounds releases vast amounts of sulfur dioxide, sulfur trioxide, hydrogen sulfides, and mercaptans into the atmosphere.

Sulfur dioxide effluents from industrial complexes are considered to be a major cause of pollution injury to forest plantings throughout the world. In the last 100 years, more investigations have been conducted on the effects of this gaseous pollutant upon vegetation than on the effects of any other pollutant.

Markings indicative of acute exposure to SO_2 in conifers begin near the apex of a needle and progress downward. In trees exposed to high concentrations of SO_2, needles develop distinctive bands, their terminals appear to be water-soaked, and tissues turn reddish-brown. With repeated exposures, premature needle drop may occur. Sensitive conifers grow slowly and may die prematurely.

In conifers, symptoms of chronic exposure to low levels of SO_2 develop over an extended period. Needles are chlorotic, often discolored over their entire length. Pigmented necrotic lesions are occasionally scattered along the yellow-flecked needles. Senescence of older needles on sensitive trees causes early needle loss, and these trees appear sparse and unthrifty (Dochinger, 1968).

Symptoms of acute injury in broadleaved trees appear after exposure to high concentrations of SO_2 over a short period. Immediately after exposure, tissues become dull and water-soaked, after which they collapse, usually assuming a gray-green color. Affected areas later dry and bleach to a light tan or ivory in many species. The final pattern of acute injury is either marginal or intercostal discoloration, with minimum effect near the veins of leaves unless injury is severe.

Broadleaf foliage chronically exposed to low levels of SO_2 are mottled. Injured leaves are colored with multicolored lesions - yellow, silver, bronze, and black.

Oxidants

Sources of injurious oxidants include activities of man, the upper atmosphere, and electrical storms. Photolytic reactions in polluted atmospheres are probably the major source of ozone in the lower atmosphere. Internal-combustion engines and industries waft tons of hydrocarbons and oxides of nitrogen into the atmosphere, which are then subsequently energized by the rays of sunlight in the presence of oxygen to form oxidant-type pollutants.

Ozone and peroxyacetylnitrate (PAN) are the two main oxidants, but O_3 is recognized as the most serious phytotoxicant in densely populated areas.

In conifers, maturing needles are most susceptible to O_3 during the midsummer months, whereas young foliage is fairly tolerant to O_3. Acute injury causes a chlorotic flecking and tipburn, which is associated with the stomatal region. These flecks develop into lesions that can band the needles. Cellular collapse follows, and an orange-red necrosis permeates the needle tip of one to all needles in a fascicle. On genetically sensitive trees, chronic exposure of 6-week-old foliage effects a chlorosis and a mottled condition of the leaves. Older foliage prematurely abscisses, and conifers become stunted in growth.

Hill et al. (1970) have classified visible symptoms of O_3 injury to plants into four general types: pigmented lesions, surface bleaching, bifacial necrosis, and chlorosis. Usually a single type is present, but more than one form of marking may be observed on a single leaf or on various species within a specific planting location. Pigmented lesions and surface bleaching are common types of injury observed on the upper surfaces of the foliage of deciduous trees exposed to low levels of O_3. Individual lesions normally appear as small discrete spots of pigmented, chlorotic, or necrotic tissues when viewed with a hand lens. Bleached lesions range from a few to an infinite number. When the lesions combine, an overall bronzed, silvery, or purple cast is imparted to the leaf surface. Some leaves display only a slight loss in green color, but bifacial necrosis on the foliage of other species may appear during acute episodes.

Fluorides

In the past 30 years, plant injury from fluorine compounds has increased because of the expansion of industries involved with aluminum production, mining of phosphate deposits, and fertilizer manufacture. Phytotoxic pollutants emitted from these commercial processes are the hydrogen fluorides and the silicon tetrafluorides. At present, particulate fluorides are not a major problem to trees. Generally, injury from gaseous fluorides results from a gradual accumulation of fluorides in leaf tissues.

Older needles of conifers are more tolerant to sublethal doses than newly emerging ones. Injury is first expressed as a yellowing and finally as a reddish-brown necrosis of needle tips or a complete collapse of all needles in a fascicle.

Fluoride compounds generally accumulate in broadleaf trees at the leaf apices and margins and cause a reddening of these areas. Treshow and Pack (1970) observed that, under continued exposure, numerous bands of maroon lines develop, which tend to impart a wavy, zonate appearance to the leaves. Fumigation with high concentrations of fluorides for short periods can cause acute injury in the form of interveinal and marginal necrosis similar to that induced by SO_2 (Thomas et al., 1952). Treshow (1969) concluded that fluorides may induce necrosis in very sensitive plants, growth suppression in less sensitive plants, and possibly growth increases in tolerant plants exposed to the same pollutant dose.

Miscellaneous Air Pollutants

Oxides of nitrogen, hydrogen chloride and chlorine, ammonia, and ethylene are other aerial phytotoxicants; but they are of minor importance in causing tree injury. They may become more destructive if they become more widely distributed.

Oxides of nitrogen (NO_x) originate from the combustion of fuels and organic wastes, petroleum refining, and internal-combustion engines. At ambient concentrations, no symptoms are reported to occur on vegetation. However, in conifers, growth reduction, needle mottling, and premature needle drop of older foliage may result from combinations of NO_x and other gaseous pollutants (Skelly et al., 1972).

Chlorine and hydrogen chloride fumes result from accidental spills, incineration of polyvinyl chlorides, and leaks in water-purification systems. Chlorine injury consists of an acute necrosis, although some chronic leaf injury and shedding are observed on trees close to sublethal sources of chlorine. On conifers, Treshow (1970) has reported foliar symptoms of stipple, chlorosis, reddish discoloration, and necrosis. When vegetation is near sources of high concentrations, acute injury and death can occur.

Ammonia injury to trees results primarily from seepage in industrial operations, accidental spills from tank and truck ruptures, and leakage of liquid ammonia fertilizers. When concentrations are above 400 ppm, tissue blackening and breakdown similar to symptoms reported for plants near accidental spillage are seen in laboratory studies (Brennan et al., 1962).

Ethylene is emitted during the combustion of organic wastes and is present in the fumes from internal-combustion engines. Symptoms of injury due to this pollutant are manifested in growth aberrations and in leaf yellowing, leaf loss, and tissue necrosis. Ethylene is also a natural constituent of plants and is classified as an inhibitory hormone. It has the natural function of inducing or promoting senescence.

Tolerance of Tree Species

The sensitivity ratings of tree species presented in Tables 1 and 2 are based on the observations of numerous researchers. Discrepancies result from differences in methods of diagnosis, in genetic makeup of plants and tree vigor, and in modifying influences in the environment. Until more definitive research information becomes available, species sensitivity ratings are of limited value in assessing relative susceptibility.

A tolerant (T) designation indicates that the tree species is the least sensitive to the air pollutant. Trees of intermediate (I) sensitivity do show some response. Sensitive (S) trees are susceptible and display various air-pollution injury symptoms.

Table 1. Tolerance of broadleaved plant species to air pollution

Broadleaved plant	Ozone	Hydrogen fluoride	Sulfur dioxide
Acer campestre (Hedge maple)	-	I	T
Acer negundo (Boxelder)	S	S	-
Acer platanoides (Norway maple)	T	-	I
Acer rubrum (Red maple)	T	-	T
Acer saccharinum (Silver maple)	S	I	-
Acer saccharum (Sugar maple)	-	T	-
Alnus Sp. (Alder)	S	-	-
Alnus glutinosa (European black alder)	-	T	-
Amelanchier sp. (Serviceberry)	-	-	S
Amelanchier canadensis (Serviceberry)	-	I	-
Betula sp. (Birch)	-	-	S
Betula pendula 'Gracilis' (Cutleaf European birch)	-	T	-
Betula verrucosa (European white birch)	I	-	-
Catalpa speciosa (Catalpa)	S	-	S
Cornus florida (White dogwood)	T	-	T
Cornus mas (Cornelian cherry)	-	T	-
Fraxinus americana (White ash)	S	-	-
Fraxinus excelsior (European ash)	-	I	-
Fraxinus pennsylvanica (Green ash)	S	I	T
Fraxinus velutina 'Modesto' (Modesto ash)	-	T	-
Gleditsia triacanthos (Honeylocust)	S	-	-
Hamamelis sp. (Witch hazel)	-	-	S
Ilex aquifolium (English holly)	-	I	T
Juglans nigra (Black walnut)	T	I	-
Juglans regia (English walnut)	-	I	-
Kalmia latifolia (Mountain laurel)	-	-	S
Liquidambar styraciflua (Sweetgum)	S	-	-
Liqustrum vulgare (Common privet)	S	-	-
Liriodendron tulipifera (Tulip tree)	S	-	T
Malus sp. (Apple)	-	-	S
Malus baccata (Siberian Crab)	S	-	-
Morus microphylla (Texas mulberry)	-	-	S
Nyssa sylvatica (Black gum)	-	-	T
Oxydendron arboreum (Sourwood)	-	-	T
Physocarpus capitatus (Ninebark)	-	-	S
Platanus sp. (Planetree)	-	T	-
Platanus occidentalis (American planetree)	S	-	T
Platanus orientalis (Oriental planetree)	-	-	T
Populus balsamifera (Balsam poplar)	-	-	I
Populus canadensis eugenei (Eugene poplar)	-	I	-
Populus deltoides (Eastern cottonwood)	-	-	T
Populus grandidentata (Bigtooth aspen)	-	-	I
Populus nigra 'Italica' (Lombardy poplar)	-	I	S
Populus tremuloides (Quaking aspen)	S	I	S
Prunus americana (Flowering apricot)	-	S	-
Prunus cerasifera (Flowering plum)	-	T	-
Prunus domestica 'Bradshaw' (Bradshaw plum)	-	S	-
Pyracantha sp. (Firethorn)	-	T	-
Pyrus communis (Pear)	-	-	S
Prunus serrulata (Oriental cherry)	-	T	-
Quercus alba (White oak)	S	-	-
Quercus coccinea (Scarlet oak)	S	-	-
Quercus gambelii (Gambel oak)	S	-	-
Quercus imbricaria (Shingle oak)	T	-	-
Quercus palustris (Pin oak)	S	-	T

Table 1 (continued)

Broadleaved plant	Ozone	Hydrogen fluoride	Sulfur dioxide
Quercus robur (English oak)	T	-	T
Quercus rubra (Red oak)	T	-	T
Rododendron sp. (Rhododendron)	-	I	-
Rhus glabra (Smooth sumac)	-	I	-
Ribes sp. (Currant)	-	T	-
Robinia pseudoacacia (Black locust)	S	I	T
Rosa sp. (Rose)	-	I	-
Salix sp. (Willow)	-	T	S
Salix babylonica (Weeping willow)	S	-	-
Sorbus aucuparia (European mountain ash)	-	T	-
Spirea prunifolia (Bridal wreath spirea)	S	T	-
Symphoricarpos albus (Snowberry)	S	-	-
Syringa vulgaris (Lilac)	S	I	-
Tilia americana (American linden)	-	T	-
Tilia cordata (Little leaf linden)	T	-	-
Tilia europea (European linden)	-	I	-
Ulmus americana (American elm)	S	T	S
Vaccinium sp. (Blueberry)	-	S	-

T = Tolerant; S = Sensitive; I = Intermediate sensitivity.

Table 2. Tolerance of conifer plant species to air pollution

Conifers	Ozone	Hydrogen fluoride	Sulfur dioxide
Abies balsamea (Balsam fir)	T	-	I
Abies concolor (White fir)	T	-	-
Juniperus sp. (Juniper)	-	T	T
Larix sp. (Larch)	-	-	S
Larix decidua (European larch)	S	-	-
Larix leptolepis (Japanese larch)	S	-	-
Larix occidentalis (Western larch)	-	S	-
Picea abies (Norway spruce)	T	-	-
Picea glauca (White spruce)	T	T	T
Picea glauca 'Densata' (Black Hills spruce)	T	-	-
Picea mariana (Black spruce)	-	-	S
Picea pungens (Colorado spruce)	T	S	-
Pinus banksiana (Jack pine)	S	-	S
Pinus nigra (Austrian pine)	S	T	T
Pinus resinosa (Red pine)	T	-	-
Pinus rigida (Pitch pine)	S	-	-
Pinus strobus (Eastern white pine)	S	S	S
Pinus sylvestris (Scotch pine)	S	S	I
Pinus taeda (Loblolly pine)	-	S	-
Pinus virginiana (Virginia pine)	S	-	-
Pseudotsuga menziesii (Douglas fir)	T	S	I
Taxus cuspidata (Japanese yew)	-	I	-
Thuja sp. (Arborvitae)	T	I	-
Thuja occidentalis (White cedar)	-	-	T
Tsuga canadensis (Eastern hemlock)	S	-	S

T = Tolerant; S = Sensitive; I = Intermediate sensitivity.

Physiological Responses

Growth

Air pollutants may reduce plant growth and even cause death of some trees. Linzon (1971) found suppression of growth of white pine trees up to 25 miles from the pollution source. Katz (1939) observed reduction in growth of Douglas fir, yellow pine, and lodgepole pine trees for 12 to 18 miles from a smelter in British Columbia. In Poland, Greszta et al. (1969) reported that air pollution reduced the growth of Scotch pine stands.

The reduction in growth depends on many factors, including the kind of pollutant and dose. With SO_2, some data suggest that foliar injury must occur before growth is reduced. Katz and McCallum (1952) found that SO_2 had no effect on the growth of trees with no leaf symptoms or on trees whose symptoms affected less than 5% of the leaf area. Scheffer and Hedgcock (1955) reported similar results in alpine fir, Douglas fir, lodgepole pine, and ponderosa pine.

Dochinger et al. (1972) and Dochinger and Jensen (1975) investigated the growth response of hybrid poplar cuttings to SO_2 fumigations. Fumigations with 13,100 $\mu g/m^3$ (5 ppm) for 1 to 12 h and with 655 $\mu g/m^3$ (0.25 ppm) for 6 weeks reduced height growth. Height-growth reduction was always associated with a large amount of foliar injury.

Reduction in diameter growth is also correlated with foliar injury. Greszta (1972) reported that a 35 to 50% loss in the number of needles resulted in a 46 to 50% reduction in diameter growth. This observation supported the work of Greszta et al. (1969). As the distance from the pollutant source increased, injury decreased, and volume and timber quality improved.

Ozone will also cause a reduction in growth of trees. Jensen (1973b) fumigated nine tree species with 588 $\mu g/m^3$ (0.3 ppm) for 5 months. Growth of sycamore, sugar maple, and silver maple was significantly reduced. In a study of hybrid poplar cuttings, Jensen and Dochinger (1974) reported that a fumigation with 490 $\mu g/m^3$ (0.25 ppm) O_3 for 6 weeks reduced height growth, while 1,960 $\mu g/m^3$ (1 ppm) for 2 to 8 h had no significant effect.

Fluorides also affect tree growth. Brewer (1960) reported that tree height and crown volume of many trees were reduced by fluoride fumigation. However, the reduction in growth was not always accompanied by foliar injury (Treshow, 1970). Treshow et al. (1967) reported that needle length of Douglas-fir trees exposed to fluorides was positively correlated with fluoride content of the needles.

Stone and Skelly (1974) investigated the growth of yellow-poplar and eastern white pine that were periodically fumigated with nitrogen oxides and SO_2. An inverse relationship was found to exist between the periods of fumigation and the annual growth increment of both species.

Metabolic Responses

When air pollutants cause foliar injury, photosynthesis is adversely affected. However, when no visible injury is noted, the effects on plant metabolism are still unclear.

Sulfur dioxide can reduce the rate of photosynthesis of trees without causing visible plant injury. Jensen and Kozlowski (1974) found that fumigation with 2,620 $\mu g/m^3$ (1 ppm) SO_2 for 2 or 4 h reduced the apparent photosynthetic rates of white ash and trembling aspen, but no foliar injury was observed. In these cases and also with seedlings fumigated with 10,480 $\mu g/m^3$ (4 ppm) where no injury occurred, the photosynthetic rate returned to normal after 24 h.

Botkin et al. (1971) studied the effect of ozone on white pine by exposing branches to consecutive doses. The initial dose suppressed the net photosynthetic rate, but the rate later returned to normal. The second exposure reduced the net photosynthetic rate by about 80%. Their data suggest that the reduction in net photosynthesis occurred before visible symptom expression, but they did not indicate whether suppression of photosynthesis could occur without visible evidence of injury.

In another study, Botkin et al. (1972) concluded that visible symptom expression was not a good index of the timing or the severity of O_3-induced photosynthetic suppression. Barnes (1972b) fumigated seedlings of slash pine, pond pine, white pine, and loblolly pine with 98 $\mu g/m^3$ (0.05 ppm) or 294 $\mu g/m^3$ (0.15 ppm) O_3 for up to 18 weeks. Photosynthesis was usually depressed by the O_3 fumigation. These data agree with those of Miller et al. (1969) - that photosynthesis of conifers can be impaired at O_3 concentrations too low to cause visible foliar injury.

The influence of pollutants on photosynthesis may be due in part to their effect on chlorophyll. Jensen (1975) measured the chlorophyll content of hybrid poplar cuttings fumigated with SO_2, but did not find any reduction in chlorophyll even though visible injury was found. He concluded that this may have been due to harvesting the seedlings immediately at the end of the fumigation. Williams et al. (1971) found that the chlorophyll content of the leaves of Durmast oak in a polluted environment decreased 19% as the sulfur content increased 50%. They suggested that hydrolysis of chlorophyll could be due to the free H^+ ion associated with bisulfate production.

Miller et al. (1963) found that the chlorophyll content of ponderosa pine exposed to O_3 was generally less than that of the control treatments. This observation was supported by Kadota and Yamada (1970), who concluded that O_3 may have caused a considerable decrease in chlorophyll content by the time that visual injury was noted.

Many other physiological processes are affected by air pollutants. Sulfur dioxide has been found to depress catalase, peroxidase, and polyphenoloxidase activity in maple and birch leaves (Nikolaevskiy, 1970). This may be due to the accumulation of sulfate in plant tissue (Brennan et al., 1964; Lihnell, 1969), which may affect the buffering capacity of the cells. Grill and Haertel (1972) reported that homogenates of leaves exposed to SO_2 were less buffered than unexposed ones.

Activity of acid phosphatase increased in cells during O_3 exposure (Evans and Miller, 1972) before visual symptoms were evident. Barnes (1972a) found a significant increase in total soluble carbohydrates, reducing sugar, and ascorbic acid in pine seedlings exposed to O_3. Miller et al. (1969) reported similar results. Fluorides also affect enzyme activity (Keller and Schwager, 1971).

Two other important processes in herbaceous plants influenced by air pollutants are stomatal activity (Rich and Turner, 1972) and membrane permeability (Evans and Ting, 1973).

Reproduction

Reproduction may be reduced by air pollution (Krutikow, 1969). Air pollution can damage the germinating seedling, the seed, or the reproductive organs of the parent. Mamajev and Shkarlet (1972) found that SO_2 weakened trees in a Scotch pine stand and that both the number of seeds per cone and the size of the germinative organs were reduced. Hedgecock (1912) found that cones of conifers were extremely abortive in areas near a smelter producing SO_2, and that natural regeneration was limited. Germination, however, was not suppressed by high SO_2 fumigation.

Sorption and Cleansing of Atmospheric Contaminants by Trees

The literature suggests that trees can function as biological filters. The subject of vegetal sorption and cleansing, however, is not without controversy. Published reports by Lampadius (1963), Flemming (1967), and Weidensaul (1973) suggest that the significance of trees in purifying the air is questionable.

For a tree to be an efficient air purifier, it must, by absorbing and detoxifying or metabolizing a pollutant, reduce the concentration of a given pollutant from an undesirable to an acceptable level without suffering significant harmful effects to its vegetative or reproductive stages of development (Weidensaul, 1974). Trees or woody vegetation not only have an active influence in filtering contaminants from the air, but individuals or groups of trees may also have a very important passive effect on the cleansing of the atmosphere. The passive result is caused by the tree acting as an obstacle in the path of the pollutant and thereby causing changes in the velocity, direction, and pressure of air currents that carry the pollutant (Lampadius, 1963; Blum, 1965; Pryakhin, 1969). Depending on the terrain and the meteorological conditions, this influence could have an appreciable effect on the sedimentation of aerial contaminants.

Methods for Studying the Sorption Process. Techniques employed to study the sorption process in plants are fumigation studies, field investigations, and vegetative sampling.

Chambers may be used in several ways to study pollutant sorption. Air may be recycled continuously through the fumigation chamber, the pollutant being added to the chamber at the same rate that it is removed to maintain a constant concentration (Hill, 1967). The rate of removal is a measure of the rate of sorption. Townsend (1974) used a closed chamber and measured the time required for the O_3 concentration to decline to some predetermined level inside a mylar chamber. Foliar sorption was calculated as the difference in O_3 removal between an intact plant and by the same plant after defoliation.

Thorne and Hanson (1972) and Roberts (1974) used single-pass fumigation chambers. Thorne and Hanson measured sorption by comparing the effluent pollutant concentration from two identical fumigation chambers connected in parallel, one with plant material and one without. Roberts (1974) used a single chamber and measured the pollutant concentration in the air stream entering and leaving the chamber. Langer

(1965) used a wind tunnel to measure particle deposition and reentrainment on cedar foliage. Gaseous sorption has also been measured in the field (Lampadius, 1968; Martin and Barber, 1971).

Much of the information about the particle-filtering capacity of vegetation originated from field studies. Dust-fall buckets, tape samplers, high-volume samplers, and natural air-flow impactors have been used to measure particle concentration (White and Turner, 1970; Dochinger, 1972). Rohmeder (1960) measured the filtering capacity of seedlings by weighing the above-ground parts of potted seedlings treated with particles.

Visual observations of the effects of trees in filtering particle contaminants (Czaja, 1951; Podgorow, 1967; Pryakhin, 1969) and gaseous contaminants (Closuit, 1966; Knabe, 1973) from the air have been reported.

Vegetative sampling is another method used in measuring the cleansing ability of trees. Analysis of plant tissue to study sorption of gaseous pollutants by trees grown in polluted environments have been reported by De Cormis and Bonte (1968), Guderian (1970), Knabe (1970), Sato (1970), Kadota (1971), Matsuoka et al. (1973), Pyatt (1973), Keller (1974), and Jensen (1975). Plant-tissue analysis has also been used to study particle cleansing (Popov, 1969; Vigorov, 1969) and heavy metal contamination (Smith, 1971, 1972, 1973).

Another method used to assess the ability of trees to reduce atmospheric pollution is radioisotope labeling (Materna and Kohout, 1969; Rich, 1971; Godzik, 1972; Meyer, 1973).

<u>Sorption of Atmospheric Contaminants</u>. The atmosphere is a complex mixture of gaseous and particulate substances that originate from natural and man-made sources. Although air-pollution problems are generally most acute in urban and industrialized areas, aerial contamination can be a problem in localized rural environments as well. Even though some reduction in atmospheric emissions is possible (Vandegrift et al., 1971), a certain base level of air pollution is inevitable. Thus, an understanding of the natural scavenger processes by which aerial wastes are removed from the atmosphere is extremely important. Three of the largest natural sinks are oceans, soil, and vegetation (Jensen, 1973a). Brief reviews of the sorption phenomenon in woody vegetation have been published by Smith (1970, 1974), Roberts (1971a, b), Dochinger (1972), and Heggestad (1972).

Sorption of Gaseous Pollutants

More published information is available on the sorption of SO_2 than on any of the other gaseous pollutants. Roberts (1974) measured the foliar sorption of several woody species. He found appreciably greater uptake of SO_2 by leaves of maple, birch, and sweetgum than by leaves of ash. Jensen and Kozlowski (1975) studied the absorption of SO_2 by maple, aspen, ash, and birch and found the highest uptake rates in aspen and the lowest in maple. In fumigation studies with pear tree seedlings, Matsuoka et al. (1973) reported that leaves exposed to the same concentration of SO_2 for different lengths of time showed decreasing uptake rates for the longer exposure periods. Another Japanese worker (Kadota, 1971) measured the increase in sulfur content of black pine foliage fumigated with SO_2 for about 7 h. He estimated that black pine has a theoretical capacity to absorb 40 mg of sulfur/m^2/day. Martin and Barber (1971) measured the ambient SO_2 levels near a hawthorne hedge and found atmospheric SO_2 levels reduced from 44 μg/m³ (0.017 ppm) to 21 μg/m³ (0.008 ppm) from March through June at a point 150 mm above the hedge. The uptake of SO_2 in this study occurred by gaseous assimilation through the stomata during the day and by adsorption on moist leaf surfaces at night.

Sato (1970) analyzed the foliar sulfur content of Zelkova trees growing in different ambient levels of SO_2 and reported that the water-soluble inorganic sulfur content was consistently greater in trees grown in the higher ambient SO_2 air. Similar observations were reported by Roberts (1975) who used potted seedlings of pin oak and white birch grown in environments with high- and low-ambient SO_2 concentrations. He found that the growth and sensitivity of these two tree species in different SO_2 environments was related to their effectiveness in removing atmospheric SO_2.

Guderian (1970) found a high correlation between sulfur content of plants and SO_2 exposure for a variety of woody plants, including species of oak, beech, walnut, larch, pine, and spruce. He reported that the absorption of SO_2 by plants is relatively small at high pollutant concentrations, but may become more important at low pollutant concentrations over long periods of time.

In field studies with various fruit trees, De Cormis and Bonte (1968) reported an increase in sulfur content during the period from June through October for foliage sampled from trees in a high-ambient SO_2 location. Popov (1969) also observed that pine and spruce needles from smoke-polluted forest park areas in Moscow contain appreciably more sulfur on a dry-weight basis than corresponding needles in smoke-free sections of that city.

Materna and Kohout (1963, 1969) used radiosulfur dioxide to study SO_2 sorption. They reported a substantial uptake of $^{35}SO_2$ by the foliage of 15 to 25 year-old pine and spruce trees and a variety of deciduous trees. In the latter study, they observed differential SO_2 absorption at leaf tips and margins of maple, birch, hornbeam, beech, oak, locust, willow, and elm species. Uptake of SO_2 through lenticels in the bark was also demonstrated for some of the species. In similar studies in which labeled SO_2 was used, Godzik (1972) reported significant differences in the uptake of 524 μg/m³ (0.2 ppm) SO_2 by detached shoots of Scotch pine, pitch pine, Austrian pine, and eastern white pine. In other studies (Godzik, 1968), larger amounts of sulfur were absorbed by leaves of birch, while substantially smaller quantities were absorbed by spruce. Nowak and Czapla (1971) have also used labeled sulfur to demonstrate the absorption of SO_2 by Scotch pine.

Ozone is also sorbed, and studies by Thorne and Hanson (1972) suggest that certain plants have the potential to remove most of the ozone in their microenvironment. The authors found a strong positive correlation between the O_3 absorption rate and transpirational water loss, which reflects the primary importance of stomata in gaseous pollutant exchange between foliage and the surrounding air (Mansfield and Majernik, 1970; Majernik and Mansfield, 1970, 1972; Unsworth et al., 1972). Townsend (1974) reported significant interspecific and intraspecific variation in foliar uptake of O_3 by species of oak, maple, birch, buckeye, sweetgum, and ash. White oak and white birch removed the largest quantities of O_3, while red maple and white ash removed the least. The O_3 sorption data for birch and ash in Townsend's study correspond with SO_2 uptake for the same two species reported by Roberts (1974).

Keller (1974) analyzed the fluoride content of foliage from different forest tree species exposed to fluoride pollutants from an aluminum smelter. He found that foliar samples from trees near the smelter had consistently higher fluoride levels than corresponding control samples. Bredemann and Radeloff (1937) reported about the accumulation of fluoride from waste gases in the bark of numerous tree species during the winter months. These data show that some plants can continue to function as sinks for certain gaseous contaminants throughout the year - not only during the active growing season. Foliar absorption of atmospheric ammonia was investigated by Meyer (1973). He reported that the aerial portions of young pine seedlings absorbed ammonia linearly over a concentration range from 22 $\mu g/m^3$ (0.031 ppm) to 299 $\mu g/m^3$ (0.43 ppm).

Sorption of Particulate Pollutants

Studies indicate that woody plants can trap particulate pollutants from the air. Keller (1971) estimated the capacity of beech and spruce trees to retain dust. He calculated that a single acre of beech forest could remove 27.6 tons of dust as compared to 12.9 tons by a comparable spruce forest. In studies designed to measure the cleansing capacity of trees, Dochinger (1972) assayed dust-fall and suspended-particle pollution at three sampling locations in a high-ambient particulate area. The total accumulation of airborne contaminants over an 8-month period revealed that woody plant canopies can appreciably reduce particulate pollution when compared to areas without trees.

Zinke (1967) measured fire-retardant compounds dropped from aircraft and calculated that the retention of aerial materials ranged from 17 to 57% in pine stands and from 82 to 86% in hardwood forests. In experiments with a wind tunnel, Langer (1965) reported that the efficiency of individual cedar leaves in removing small airborne dust particles was 0.5 and 6.0% for the broadside and edge leaf positions, respectively. White and Turner (1970) estimated deposition velocities of 3.0 to 7.1 cm/sec for airborne particles in a woodland ecosystem.

Ricks and Williams (1974) observed the accumulation of particles on the lower surface of oak leaves subjected to a differential gradient of particulate pollution. They found partial stomatal occlusion caused by particulate accumulation in stomatal pores and suggested that this condition could enhance the uptake of gaseous pollutants such as SO_2. The capacity of various tree species to retain airborne dust has also been reported by Podgorow (1967) and Berindan (1969).

Smith (1971) assayed lead contamination of white pine trees along a major highway and found lead accumulation to be greater in foliage sampled from the side of the tree closest to the road. Smith (1973) also reported on the metal content of leaf and twig samples from six woody species growing in an urban environment. In similar studies in which he used species of oak, maple, ailanthus, linden, hemlock, pine, and yew, Smith (1972) concluded that woody plants may play an important role in heavy metal cycling in urban areas by acting as a short- and long-term repository for these contaminants.

Relative Importance of the Cleansing Process

There is gaseous exchange between a plant and its surrounding atmosphere. It seems reasonable to assume that vegetation can function as a sink for certain aerial contaminants. Although most people would probably agree that the sink effect does occur, the quantity of pollutants extracted from the air remains unresolved.

Air-quality improvement by trees absorbing gaseous pollutants is even less well understood than their function as particle filters. Certain gaseous contaminants like SO_2 are absorbed by plants and are altered chemically (Daines, 1968). The capacity of leaves to perform this chemical transformation may be exceeded at high SO_2 levels; and subsequently, the plant may be injured, thus reducing its ability to absorb additional pollutants.

Theoretical estimates have been made on the potential sink capacity of vegetation (Zinke, 1967; Hill, 1971; Kadota, 1971; Keller, 1971; Hanson and Thorne, 1972), but reduction of high ambient levels of atmospheric contaminants for prolonged periods seems unlikely. However, it does seem that trees can be extremely important in lessening pollution levels in moderately contaminated microenvironments. Thus, the strategic location of greenbelt plantings containing specifically designated sink plants may be an effective way to provide acceptable air-quality levels in restricted environments.

Genetic Aspects of Air Pollution in Trees

Interspecific variability in tolerance and sensitivity to pollutants offers one solution to economic losses due to pollution. Species-to-species variation has been mentioned previously, and much progress can be made by planting tolerant species of trees in polluted areas.

Genetic variation in resistance to pollutants also occurs within tree species. Such intraspecific variation between clones or seed sources has been demonstrated with SO_2 and O_3 with Norway spruce (Rohmeder et al., 1962), eastern white pine (Dochinger and Seliskar, 1970; Houston and Stairs, 1973; Houston, 1974), ponderosa pine (Karpen, 1970), American sycamore (Santamour, 1969), and red maple (Townsend and Dochinger, 1974). In addition, significant interclonal variation in tolerance to SO_2 was demonstrated by hybrid poplars derived from the same parentage (Dochinger et al., 1972). Therefore, the feasibility of breeding and selecting pollutant-tolerant trees within a species seems very promising because of the range of genetically controlled variability within species.

Methods for evaluating and utilizing genetic variability in response to pollutants: A first step in developing pollutant-tolerant trees is to estimate the intraspecific genetic variation of seed sources over the species range. Once the material is collected, several statistical and biological factors become important. Because the response to air contaminants is regulated by the environment (Heggestad and Heck, 1971), such factors as light intensity, photoperiod, temperature, nutrition, humidity, and plant water status must be kept uniform.

To eliminate environmental responses, the preconditioning environment before fumigation must be identical for all clones or progenies. Ramets screened for tolerance should be grown from cuttings taken from about the same position on the tree to avoid topophytic effects.

Any ontogenetic or developmental change that occurs during the life of the tree is another important factor for consideration. Studies must be completed to show that sensitivity does not change between different ontogenetic stages. For example, Townsend and Dochinger (1974) determined the relative susceptibility to O_3 of four groups of red maple seedlings at four growth stages. The objective was to determine whether ontogenetic factors such as size and age of seedlings and leaves would override or mask phylogenetic differences among progenies.

Different biological responses can be used to examine genetic differences. Height growth, as a percentage of growth shown by untreated or control progenies or clones of the same parentage, is a useful parameter (Dochinger et al., 1972; Townsend and Dochinger, 1974) for detecting tolerance or susceptibility. Tolerant lines usually show less growth depression than susceptible lines. Foliar injury, estimated as the percentage of each leaf damaged or as a percentage of all leaves showing symptoms, may also be used. Again, it is important to compare seedlings or ramets of the same size and age (Townsend and Dochinger, 1974). Leaf-growth suppression can also be used as an indicator of sensitivity to a pollutant. Comparisons of leaf-growth suppression among clones or progeny is valid only if the leaves are of the same age and in the same stage of development.

Stomata act as avenues through which atmospheric pollutants enter the leaf, and they may be a good biological parameter to use. Gabelman (1970) found that certain inbred lines of onion are ozone-resistant because stomata close in the presence of O_3. He used leaf disks and found that ozone-tolerance and sensitivity could be detected readily by observing stomatal response. This technique probably could be used to survey pollutant response of trees.

Certain pollutants can cause a suppression of the rate of net photosynthesis (Roberts et al., 1971). This parameter might be useful as a subtle tool for judging genetic variability within species in response to gaseous contaminants.

In selecting and breeding pollutant-tolerant trees, ultimately the tree must be planted in the field. A tree that proves to be tolerant under laboratory and controlled-environment conditions is of no value unless it proves to be superior outdoors in a natural environment under the normally prevailing climatic conditions and pollutant levels.

Field tests are invaluable and essential for testing the real worth of a certain progeny or clone. However, they may be less valuable than laboratory methods for identifying cause-and-effect relationships between tree and pollutant. Differences may be found in growth rates, fruiting, and germination among different genetic lines planted in a replicated design in polluted air. However, it is invalid to conclude that the differences expressed are due solely to variability in response to pollution. It must be kept in mind that genotype-environment interaction occurs and that pollution is only one of many environmental factors, which include wind, photoperiod, soil, temperature, humidity, and precipitation. If field tests are initiated, they should be replicated in both polluted and unpolluted areas, over as wide a range of environments as possible.

Trees may either be planted in polluted areas or may be fumigated with pollutants. Spierings (1967) developed a fumigation technique that does not disturb the prevailing climatic conditions because ambient air and the test gas are mixed and blown onto a branch of a tree. Dochinger and Seliskar (1970) used a similar technique on eastern white pines in the field.

Dochinger and Seliskar (1970) suggested several ways to increase the probability of selecting genetically tolerant white pines. They recommended planting only seedlings showing no off-color foliage or banded or mottled needles. Seedlings that have made vigorous growth should be selected over smaller ones. The authors believe that selection can be undertaken in the nursery. This suggestion, of course, presupposes that the nursery is exposed to polluted air. Such a supposition becomes increasingly valid as time advances and pollutants act as selective pressures (Hepting, 1968; Sinclair, 1969).

Another technique for finding pollutant-tolerant trees is to survey urban trees that have remained healthy in areas of high-ambient pollutant concentration. These trees should be vegetatively propagated, and ramets from selected trees should be compared with ramets from susceptible trees of the same species. This kind of comparison ensures that the pollution tolerance of the ortet is genetically controlled and not environmentally induced by such factors as nutrition or soil type.

After selection, breeding work can be initiated to determine the degree to which pollutant tolerance can be transmitted to the offspring. Seedlings resulting between a variety of crosses, such as between tolerant parents or between sensitive parents, should be transplanted to progeny-test plantations in both contaminated and uncontaminated areas. These progeny tests will sample a range of soils, sites, and climates and will determine how these factors influence pollutant tolerance (Townsend, 1972).

Air Pollution Tolerance

Conifers

Useful levels of tolerance and much variation have been found within several coniferous species. Genetic differences in resistance to SO_2 and fluoride injury have been found in Norway spruce and Scotch pine (Rohmeder et al., 1962). There is "striking tree-to-tree (or phenotypic) variability" in susceptibility of ponderosa pine to smog and fluorides (Hepting, 1968). Differences in chlorotic spotting among 10 geographic races of ponderosa pine were observed in a stand 65 miles south of Seattle. Karpen (1970) hypothesized that an O_3-SO_2 synergism was responsible for the damage.

Gerhold and Palpant (1968) have crossed several races of Scotch pine and screened the progenies for resistance to SO_2. Dochinger and Seliskar (1970) found genetic differences in susceptibility to air pollutants in eastern white pine. Susceptible individuals are dwarfed and chlorotic. They found that O_3 and SO_2, acting independently or synergistically, are the principal pollutants involved (Dochinger et al., 1970). Resistant and susceptible white pine clones have been vegetatively propagated, both for studies and for seed orchards (Berry, 1968).

Angiosperms

Less genetics research has been devoted to deciduous trees and their responses to pollutants. Dochinger et al. (1972) found significant differences among clones of hybrid poplar (*P. deltoides* x *P. trichocarpa*) in their response to SO_2 fumigation. Rooted stem cuttings of four hybrid poplar clones were exposed to 13,100 $\mu g/m^3$ (5 ppm) SO_2 for 0, 3, 6, 9, or 12 h in environmental chambers. The following factors were statistically significant in determining growth response and foliar injury: 1. genotype or clone, 2. duration of treatment, and 3. interaction between genotype and treatment. The results of this study have encouraged selection of clones tolerant to short-term exposures to high levels of SO_2.

Another recent study (Townsend and Dochinger, 1974) demonstrated significant variation in tolerance to O_3 (1,470 $\mu g/m^3$ or 0.75 ppm) among half-sib seedling progenies of red maple from four seed sources. Trees from all sources were fumigated at four growth stages. Differences in seedling response occurred in the extent of injury but not in the type of injury. Seedlings from an Alabama source showed the least average foliar injury and least suppression of leaf growth after exposure to O_3 in all four stages of growth and four stages of leaf development. The Alabama seedlings also exhibited the largest increase in height growth after each of the four fumigations. Foliar injury and suppression of leaf growth was greatest in Pennsylvania and Minnesota seedlings. Fumigation of the Minnesota seedlings in the youngest growth stage caused an almost complete cessation of growth for 3 weeks.

In a related study (Townsend, 1974), significant variation was demonstrated among the same four progenies in the rate of foliar sorption of O_3. The Pennsylvania group of red maple seedlings showed the highest sorption ability of the four groups. As already mentioned, this group showed the greatest foliar injury and suppression of leaf growth. Therefore, the high O_3 sensitivity of the Pennsylvania progeny may be directly related to their superior ability to sorb greater quantities of O_3.

Discussion and Conclusions

Short-term laboratory fumigation experiments may lead to false conclusions if results of genetic comparisons are extrapolated to field conditions. However, several investigators have demonstrated that short-term fumigation of seedlings, grafted plants, and even cut branches in water enables discrimination of tolerance differences that are in good agreement with field assessments (Gerhold and Palpant, 1968).

Significant progress has already been made in determining the degree of genetic control in tolerance to one or more pollutants. Clonal and progeny differences have been shown in several species. These results indicate strong genetic control in tolerance.

Townsend and Dochinger (1974), in working with red maple, found a consistent exhibition of tolerance of the same progeny in every one of four size classes of leaves and in four plant-growth stages. For red maple, at least, such a repetitive response signifies quantitative genetic control over factors controlling O_3 sensitivity. Therefore, it is probable that high levels of O_3 resistance may be developed by breeding and selection. Also, the stability of O_3 tolerance or susceptibility over several plant and leaf ontogenetic stages should facilitate the selection of ozone-tolerant trees. This correlation among different-aged seedlings from the same seed source would enable one to select plants at any growth stage. Selection for tolerance to foliar injury could even be based on observations made at one leaf development stage.

The above statements refer to results obtained in using only one uniform environment, including a constant concentration of a single pollutant. Unfortunately, such conditions do not exist in areas where pollutant-tolerant trees would be planted. Such areas usually are characterized by fluctuating levels of several air contaminants, temperature, wind, moisture, and pest attacks. We must concentrate our future research efforts on screening and testing progenies and clones under natural conditions, such as near streets, freeways, industrial complexes, and other sources of pollution. If controlled fumigation studies are conducted in the laboratory, promising ramets and progenies should be exposed to several pollutants simultaneously under a variety of environmental conditions, including stress factors such as extremes of temperature, moisture, salt, and wind. Future research for pollutant-tolerant trees must be directed to relating laboratory results to field results. If laboratory methods are used to screen and select pollutant-tolerant trees, environmental parameters must be manipulated and replicated to simulate natural conditions.

Glossary of Air-Pollution Terms

Acid rain	Atmospheric precipitation that is composed of the hydrolyzed byproducts from oxidized halogen, nitrogen, and sulfur substances.
Acute injury	Injury, usually involving necrosis, which develops within several hours to a few days after a short-term exposure to a pollutant, and is expressed as a fleck, scorch, bifacial necrosis, etc.
Additive effects	The combined toxic effects of several air pollutants, acting together or independently, to cause a total response equal to the sum of the separate reactions.
Aerosol	Solid or liquid particles, usually less than 1 micron in diameter, suspended in a gaseous state.
Air pollution	An accumulation of aerial wastes beyond the concentrations that the normal properties of the atmosphere can absorb and, in turn, can cause injury to the human environment.
Ambient air	The outside air surrounding man's environment.
Carbon monoxide	A colorless, odorless gaseous pollutant evolving primarily from the incomplete combustion of carbon-containing compounds.
Chemiluminescent ozone monitoring	An electric instrument for quantifying ozone levels in the atmosphere based on the light reactions between ozone and ethylene.
Chronic injury	Injury that develops only after long-term or repeated exposure to an air pollutant and is expressed as chlorosis, bronzing, premature senescence, reduced growth etc.; can include necrosis.
Damage	A measure of the decrease in economic or aesthetic value resulting from plant injury by pollutants. (Term considered by some to be synonymous with injury.)

Dose	A specific concentration of an air pollutant for a specific period of time to which an object is exposed.
Fluorides	Gaseous and particulate pollutants containing fluorine matter.
Fumigation	The controlled or natural exposure of vegetation to phytotoxic pollutants.
Microgram per cubic meter ($\mu g/m^3$)	A measure of the concentration of a pollutant in micrograms per cubic meter of air at standard temperature and pressure.
Monitoring	The use of sampling instruments for measuring concentrations of air pollutants.
NO_x	A generalized formula for all oxides of nitrogen.
Oxidant	The ability of some oxygen-containing compounds to oxidize other compounds. Specific examples include ozone, peroxyacetyl nitrates, and nitrogen dioxide.
Ozone	The triatomic allotrope of oxygen; a colorless to bluish gas produced by photochemical reactions with hydrocarbons and oxides of nitrogen.
PAN	The acronym for peroxyacetyl nitrate (CH_3C^O-O-NO_2); the principal constituent in a homologous series of compounds referred to as peroxyacyl nitrates or PAN's. Formed as a product of photochemical reactions involving nitrogen dioxide and hydrocarbons. A cause of plant injury resulting in PAN-type symptoms.
Particle	Dispersed matter, solid or liquid, in a range from 2 x 10^{-4} to 5 x 10^2 microns in diameter.
Particulate	Particles of soil or liquid matter; e.g., smoke, dust, fly ash, aerosol pollutants.
Peroxyacetyl nitrate (PAN)	The primary compound in a homologous series of photochemical reactions between hydrocarbons and nitrogen dioxide.
Photochemical process	The chemical alterations caused by radiant energy in the form of light energy acting on various gases in the atmosphere.
Phytotoxic	Any chemical agent injurious to vegetation.
Pollutant (air)	Any gas, liquid, or solid air contaminant that causes undesirable effects on living organisms or materials.
pphm	Parts per hundred million
ppm	Parts per million.
Primary air pollutants	Emissions directly evolved from an identifiable source of pollution; e.g., fluorides from smoke stacks.

Secondary pollutants	Those pollutants that result from the chemical reactions involving primary pollutants or related atmospheric contaminants; e.g., oxidants from photochemical activity.
Sensitivity	A physiological condition of susceptible plants, or particular plant tissues, whereby they are prone to injury by pollutants.
Smog	A combination of primary pollutants (fossil fuels combustion), particulates, and photochemical oxidants. Often, a mixture of fog and smoke may be responsible for smog.
Sorption	Intake of one substance by another's molecules through the processes of absorption and adsorption.
Sulfur oxides	The combustion of fossil fuels that yields a pungent, colorless toxicant gas.
Synergism	The combined toxic effects of several air pollutants, acting collectively or independently, to cause a total response greater than the sum of the separate reactions.
Tolerance	The capacity of plants to withstand the effects of pollutants without significant reduction in quality or quantity of yield.

References

American Public Health Service: Methods of air sampling and analysis. Washington, D.C.: Am. Public Health Assoc. 1974.

Association of Official Agricultural Chemists: Official methods of analysis of the association of official agricultural chemists. Washington, D.C.: Assoc. Off. Agric. Chem. 1960.

Barnes, R.L.: Effects of chronic exposure to ozone on soluble sugar and ascorbic acid contents of pine seedlings. Can. J. Botany 50, 215-219 (1972a).

Barnes, R.L.: Effects of chronic exposure to ozone on photosynthesis and respiration of pines. Environ. Pollut. 3, 133-138 (1972b).

Benedict, H.M., Miller, C.J., Olson, R.E.: Economic impact of air pollutants on plants in the United States. Menlo Park, California: Stanford Res. Inst. Rep. 1971.

Berindan, C.: Interrelation pollution atmospherique et espaces verts en tant que principe protecteur des villes industrielles. Pollut. Atmos. (Paris) 11, 143-153 (1969).

Berry, C.R.: Eastern white pine, a bioindicator of air pollution. Proc. Ann. Meet. Soc. Am. Foresters Ann. Meet. Proc. 26 (1968).

Berry, C.R.: A plant fumigation chamber suitable for forestry studies. Phytopathology 60, 1613-1615 (1970).

Blum, W.: Luftverunreinigung und Filterwirksamkeit des Waldes. Forst u. Holzwirtsch. 20, 211-212, 214-215 (1965).

Botkin, D.B., Smith, W.H., Carlson, R.W.: Ozone suppression of white pine net photosynthesis. J. Air Pollut. Control Assoc. 21, 778-780 (1971).

Botkin, D.B., Smith, W.H., Carlson, R.W., Smith, T.L.: Effects of ozone on white pine saplings: Variation in inhibition and recovery of net photosynthesis. Environ. Pollut. 3, 273-289 (1972).

Brandt, C.S., Heck, W.W.: Effects of air pollutants on vegetation. In: Air Pollution, 2nd ed., Vol. I (ed. A.C. Stern), p. 401-443. New York-London: Academic Press 1968.

Bredemann, G., Radeloff, H.: Fluorine absorption through the bark of shoots and its effects. Angew. Botany 19, 172-181 (1937).

Bremanis, E., Deering, J.R., Meade, C.F., Keyworth, D.A.: Elimination of nitrogen and chloride interference in the iodometric determination of sulfur as sulfur dioxide. Mater. Res. Stand. 7, 459-460 (1967).

Brennan, E., Leone, I., Daines, R.H.: Ammonia injury to apples and peaches in storage. Plant Dis. Rep. 46, 792-795 (1962).

Brennan, E., Leone, I., Daines, R.H.: Investigation of SO_2 effects on rubber trees as a means of forestalling injury to Malayan plantations from refinery emissions. J. Air. Pollut. Control Assoc. 14, 229-233 (1964).

Brewer, R.F.: Effects of hydrogen fluoride gas on seven citrus varieties. Am. Soc. Hortic. Sci. Proc. 75, 236-243 (1960).

Carlson, R.M., Keeney, R.M.: Specific ion electrodes: techniques and uses in soil, plant and water analysis. In: Instrumental Methods for Analysis of Soils and Plant Tissue (ed. L.M. Walsh), p. 39-66. Madison, Wisc.: Soil Sci. Soc. Am. 1971.

Closuit, M.: Foret et Sante publique. Schweiz. Z. Forstw. 117, 809-817 (1966).

De Cormis, L., Bonte, C.: Sur la variation de la teneur en soufre des plantes ayant séjourné en atmosphère polluée par l'anhydride sulfureux. Phytiatr. Phytopharm. 2, 151-155 (1968).

Costonis, A.C., Linzon, S.N.: Methods for measuring photochemical air pollution in forest trees. Mitt. Forstl. Bundesversuchsanstalt 92, 57-70 (1971).

Czaja, A.Th.: Plant damage resulting from dust-grained silicate of soda. Z. Pflanzenkr. Pflanzenschutz 58, 54-61 (1951).

Daines, R.H.: Sulfur dioxide and plant response. J. Occup. Med. 10, 516-524 (1968).

Dochinger, L.S.: The impact of air pollution on eastern white pine: the chlorotic dwarf disease. J. Air Pollut. Control Assoc. 18, 814-816 (1968).

Dochinger, L.S.: Can trees cleanse the air of particulate pollutants? Intern. Shade Tree Conf. Proc. 48, 45-48 (1972).

Dochinger, L.S., Bender, F.W., Fox, F.L., Heck, W.W.: Chlorotic dwarf of eastern white pine caused by an ozone and sulphur dioxide interaction. Nature 225, 476 (1970).

Dochinger, L.S., Jensen, K.F.: Effects of chronic and acute exposure of sulfur dioxide on the growth of hybrid poplar cuttings. Environ. Pollut. 9, 219-229 (1975).

Dochinger, L.S., Seliskar, C.E.: Air pollution and the chlorotic dwarf disease of eastern white pine. Forest. Sci. 16, 46-55 (1970).

Dochinger, L.S., Townsend, A.M., Seegrist, D.W., Bender, F.W.: Responses of hybrid poplar trees to sulfur dioxide fumigation. J. Air Pollut. Control Assoc. 22, 369-371 (1972).

Donaubauer, E.: Methoden zur Erkennung und Beurteilung forstschädlicher Luftverunreinigungen. Mitt. Forstl. Bundesversuchsanstalt 92 (1971)

Evans, L.S., Miller, P.R.: Ozone damage to ponderosa pine: a histological and histochemical appraisal. Am. J. Botany 59, 297-304 (1972).

Evans, L.S., Ting, I.R.: Ozone-induced membrane permeability changes. Am. J. Botany 60, 155-162 (1973).

Flemming, G.: In welchem Fall können Waldstreifen die Rauchgaskonzentration vermindern? Luft u. Kältetech. 6, 255-258 (1967).

Gabelman, W.H.: Alleviating the effects of pollution by modifying the plant. HortScience 5, 250-254 (1970).

Gerhold, H.D., Palpant, E.H.: Prospects for breeding ornamental Scotch pines resistant to air pollutants. Proc. 6th Cent. States Forest Tree Improv. Conf. 6, 34-36 (1968).

Godzik, S.: Absorption of $^{35}SO_2$ and its localization in the leaves of certain trees. Polish Acad. Sci., Zabrze, Res. Inst. Upper Silesian Ind. Reg., Mater. Miedzynorodowej Konf., Wplyw. Zanieczyszczen Powietrza na Lasy, Katowice, Poland. 6, 63-77 (1968).

Godzik, S.: Vergleichende Untersuchungen über die Aufnahme von Schwefeldioxyd aus der Atmosphäre durch einige Pinus-Arten. Mitt. Forstl. Bundesversuchsanstalt 97, 319-334 (1972).

Greszta, J.: Die Beschädigung des Assimilationsapparates der Kiefer sowie die dadurch entstandenen Verluste der Holzmasse. Mitt. Forstl. Bundesversuchsanstalt 97, 431-442 (1972).

Greszta, J., Olszowski, J., Godzik, S.: The effect of air pollution on wood volume increment in the common pine (*Pinus silvestris* L.). Ekologia Polska 42 (A), 827-845 (1969).

Grill, D., Haertel, O.: Zellphysiologische und biochemische Untersuchungen an SO_2-begasten Fichtennadeln, Resistenz und Putterkapazität. Mitt. Forstl. Bundesversuchsanstalt 97, 367-386 (1972).

Guderian, R.: Untersuchungen über quantitative Beziehungen zwischen dem Schwefelgehalt von Pflanzen und dem Schwefeldioxidgehalt der Luft. I. Teil. Z. Pflanzenkr. Pflanzenschutz 77, 200-220 (1970).

Hanson, G.P., Thorne, L.: Vegetation to reduce air pollution. Lasca Leaves 22, 60-65 (1972).

Heagle, A.S., Body, D.E., Heck, W.W.: An open-top field chamber to assess the impact of air pollution on plants. J. Environ. Qual. 2, 365-368 (1973).

Heagle, A.S., Body, D.E., Pounds, E.K.: Effect of ozone on yield of sweet corn. Phytopathology 62, 683-687 (1972).

Heck, W.W.: Factors influencing expression of oxidant damage to plants. Ann. Rev. Phytopathology 6, 165-188 (1968).

Heck, W.W., Dunning, J.A., Johnson, H.: Design of a simple plant exposure chamber. U.S. Dep. H.E.W. Natl. Cent. Air Pollut. Control Publ. APTD-68-6 (1968).

Hedgecock, G.G.: Winter killing and smelter injury in forests of Montana. Torreya 12, 25-30 (1912).

Heggestad, H.E.: How plants fight "man-made" pollution. Sci. Teach. 39, 21-24 (1972).

Heggestad, H.E., Heck, W.W.: Nature, extent, and variation of plant response to air pollutants. Adv. Agron. 23, 111-145 (1971).

Hepting, G.H.: Diseases of forest and tree crops caused by air pollutants. Phytopathology 58, 1098-1101 (1968).

Hill, A.C.: A special purpose plant environmental chamber for air pollution studies. J. Air Pollut. Control Assoc. 17, 743-748 (1967).

Hill, A.C.: Vegetation: A sink for atmospheric pollutants. J. Air Pollut. Control Assoc. 21, 341-346 (1971).

Hill, A.C., Heggestad, H.E., Linzon, S.N.: Ozone. In: Recognition of Air Pollution Injury to Vegetation: A Pictorial Atlas (eds. J.S. Jacobson, A.C. Hill), Inf. Rep. No. 1, Sect. B, p. 1-22. Pittsburgh: Herbick and Held 1970.

Houston, D.B.: Response of selected *Pinus strobus* L. clones to fumigation with sulfur dioxide and ozone. Can. J. Forestry Res. 4, 65-68 (1974).

Houston, D.B., Stairs, G.R.: Genetic control of sulfur dioxide and ozone treatment in eastern white pine. Forest Sci. 19, 267-271 (1973).

Huey, N.A.: The lead dioxide estimation of sulfur dioxide pollution. J. Air Pollut. Control Assoc. 18, 610-611 (1968).

Jensen, K.F.: Absorption, translocation, and soil deposition of aerial pollutants by vegetation. Soil Conserv. Soc. Am Proc. 28, 26-29 (1973a).

Jensen, K.F.: Response of nine forest tree species to chronic ozone fumigation. Plant Dis. Rep. 57, 914-917 (1973b).

Jensen, K.F.: Sulfur content of hybrid poplar cuttings fumigated with sulfur dioxide. USDA For. Serv. Res. Note NE-209. Northeast. For. Exp. Stn., Upper Darby, Pa. (1975).

Jensen, K.F., Dochinger, L.S.: Responses of hybrid poplar cuttings to chronic and acute levels of ozone. Environ. Pollut. 6, 289-295 (1974).

Jensen, K.F., Kozlowski, T.T.: Effect of SO_2 on photosynthesis of quaking aspen and white ash seedlings. Proc. 3rd North Am. For. Biol. Workshop, p. 359 (1974).

Jensen, K.F., Kozlowski, T.T.: Absorption and translocation of sulfur dioxide by seedlings of four forest tree species. J. Environ. Qual. 4, 379-382 (1975).

Jones, H.C., MacLaughlin, S.B.: A simple method for measuring fluorides in foliage using the fluoride specific in electrode. Mitt. Forstl. Bundesversuchsanstalt 92, 2-7 (1971).

Kadota, M.: Taiki osen to jumoku toshirin no zosei. J. Jap. Soc. Air Pollut. 6, 31-38 (1971).

Kadota, M., Yamada, H.: Ozon no akamatsu oyobi kuromatsu ba ni oyobosu eikyo. J. Jap. Soc. Air Pollut. 5, 158 (1970).

Karpen, D.N.: Ozone and sulfur dioxide synergism: Foliar injury to a ponderosa pine geographic race plantation in the Puget Sound region. Plant Dis. Rep. 54, 945-948 (1970).

Katz, M.: Effect of sulfur dioxide on vegetation. Can. Natl. Res. Counc. Publ. 815 (1939).

Katz, M.: Sulphur dioxide in the atmosphere and its relation to plant life. Ind. Eng. Chem. 41, 2450-2465 (1949).

Katz, M., McCallum, A.W.: The effect of sulfur dioxide on conifers. In: Air Pollution (ed. L.C. McCabe). New York: McGraw-Hill 1952.

Keller, T.: Auswirkungen der Luftverunreinigungen auf die Vegetation. Städtehygiene 22, 130-136 (1971).

Keller, T.: Translocation of fluoride in woody plants. Fluoride 7, 31-35 (1974).

Keller, T., Schwager, H.: Der Nachweis unsichtbarer ("physiologischer") Fluor-Immissionsschädigungen an Waldbäumen durch eine einfache koloimetrische Bestimmung der Peroxidase-Aktivität. Eur. J. For. Pathol. 1, 6-18 (1971).

Knabe, W.: Natürliche Abnahme des aus Immissionen aufgenommenen Fluors in Fichtennadeln. Staub, Reinhalt. Luft 30, 384-385 (1970).

Knabe, W.: Wirkungen von Luftverunreinigungen auf Waldbäume. Mitt. Forstl. Bundesversuchsanstalt 97 (1972).

Knabe, W.: Zur Ausweisung von Immissionsschutzwaldungen. Forstarchiv 44, 21-27 (1973).

Krutikow, A.: Changes in the biocenosis of forests in the Cracow industrial area. Las Polski 13/14, 23-24 (1969).

Lacasse, N.L., Moroz, W.J.: Handbook on effects assessment: Vegetation damage. University Park, Penn.: Cent. Environ. Stud. 1969.

Lampadius, F.: Die lufthygienische Bedeutung des Waldes in ihrer Abhängigkeit von schädlichen Raucheinwirkungen auf den Wald. Angew. Meteorol. 4, 248-249 (1963).

Lampadius, F.: Die Bedeutung der SO_2-Filterung des Waldes im Blickfeld der forstlichen Rauchschadentherapie. Wiss. Z. Tech. Univ. Dresden 17, 503-511 (1968).

Langer, G.: Particle deposition and reentrainment from coniferous trees. Part II. Experiments with individual leaves. Kolloid Z. Z. Polym. 204, 119-124 (1965).

Lihnell, D.: Sulphate contents of tree leaves as an indicator of SO_2 air pollution in industrial areas. In: Air Pollution. Proc. 1st Eur. Congr. on the Influence of Air Pollut. on Plants and Anim., pp. 341-352, 1969.

Linzon, S.N.: Economic effects of sulfur dioxide on forest growth. J. Air Pollut. Control Assoc. 21, 81-86 (1971).

Lux, H.: Forstliche Rauchschadprobleme in der Slowakei. Soz. Forstwirtsch. 21, 60-62 (1971).

Majernik, O., Mansfield, T.A.: Direct effect of SO_2 pollution on the degree of opening of stomata. Nature 227, 377-378 (1970).

Majernik, O., Mansfield, T.A.: Stomatal responses to raised atmospheric CO_2 concentrations during exposure of plants to SO_2 pollution. Environ. Pollut. 3, 1-7 (1972).

Mamajev, S.A., Shkarlet, O.D.: Effects of air and soil pollution by industrial waste on the fructification of Scotch pine in the Urals. Mitt. Forstl. Bundesversuchsanstalt 97, 443-450 (1972).

Mansfield, T.A., Majernik, O.: Can stomata play a part in protecting plants against air pollutants? Environ. Pollut. 1, 149-154 (1970).

Martin, A., Barber, F.R.: Some measurements of loss of atmospheric sulphur dioxide near foliage. Atmos. Environ. 5, 345-352 (1971).

Materna, J., Kohout, R.: The absorption of sulphur dioxide by spruce and pine trees. Naturwissenschaften 50, 407 (1963).

Materna, J., Kohout, R.: Uptake of sulphur dioxide into the leaves of some tree species. Comm. Inst. For. 6, 39-47 (1969).

Matsuoka, Y., Takasaki, T., Udagawa, O.: Shokubatsu yochu ni shuseki suru io seibun to taikichu no aryusan gasu ni tsuite. J. Jap. Soc. Air Pollut. 8, 538 (1973).

Meyer, M.W.: Absorption and release of ammonia from and to the atmosphere by plants. No. 73-28881. Ann Arbor, Mich.: Univ. Microfilms, Inc. 1973.

Miller, P.R., McBride, J.R.: Study on the impact of oxidant air pollution on the mixed conifer forest ecosystem. In: Oxidant Air Pollutant Effects on a Western Coniferous Forest Ecosystem. Task B Rep. p. A1-A36 (1973).

Miller, P.R., Parmeter, J.R., Jr., Flick, B.H., Martinez, C.W.: Ozone dosage response of ponderosa pine seedlings. J. Air Pollut. Control Assoc. 19, 435-438 (1969).

Miller, P.R., Parmeter, J.R., Jr., Taylor, O.C., Cardiff, E.A.: Ozone injury to the foliage of *Pinus ponderosa*. Phytopathology 53, 1072-1076 (1963).

Nikolaevskiy, V.S.: Effect of sulfur dioxide on the enzymatic activity of tree leaves. In: AICE Survey of USSR Air Pollution Literature (ed. M.Y. Nuttonson), Vol. III, p. 109-114. Silver Spring, Md.: Am. Inst. Crop Ecol. 1970.

Nowak, G., Czapla, J.: Badania nad pobieraniem i przemieszczaniem siarki znakowanej izotopem ^{35}S w siewkach sosny pospolitej (*Pinus silvestris*). Sylwan 10, 43-49 (1971).

Pelz, E.: Gaseous air pollution and choice of species in districts with industrial smoke hazards. Forst. u. Jagd. 6, 347-349 (1956).

Podgorow, N.W.: Die staubfilternde Rolle von Anpflanzungen. Lesn. Khoz. (Moscow) 20, 39-40 (1967).

Popov, B.V.: Making a city verdant. In: AICE Survey of USSR Air Pollution Literature (ed. M.Y. Nuttonson), Vol. II, p. 60-75. Silver Spring, Md.: Am. Inst. Crop Ecol. 1969.

Pryakhin, V.D.: Recommended procedures for planting sanitary - protective green belts in areas of industrial enterprises. In: AICE Survey of USSR Air Pollution Literature (ed. M.Y. Nuttonson), Vol. II, p. 28-31. Silver Spring, Md.: Am. Inst. Crop Ecol. 1969.

Pyatt, F.B.: Plant sulfur content as an air pollution gauge in the vicinity of a steelworks. Environ. Pollut. 5, 103-115 (1973).

Rich, S.: Effects of trees and forests in reducing air pollution. Symp. on Trees and Forests in an Urbanizing Environ. Plan. Resour. Dev. Ser. 17, 29-33 (1971).

Rich, S., Turner, N.C.: Importance of moisture on stomatal behavior of plants subjected to ozone. J. Air Pollut. Control Assoc. 22, 718-721 (1972).

Ricks, G.R., Williams, R.J.H.: Effects of atmospheric pollution on deciduous woodland. Part 2: Effects of particulate matter upon stomatal diffusion resistance in leaves of *Quercus petraea* (Mattuschka) Leibl. Environ. Pollut. 6, 87-109 (1974).
Roberts, B.R.: Tree as air purifiers. Arborist's News 36, 23-25 (1971a).
Roberts, B.R.: Foliar absorption of gaseous air pollutants. Am. Nurseryman 133 (9), 44-45 (1971b).
Roberts. B.R.: Foliar sorption of atmospheric sulphur dioxide by woody plants. Environ. Pollut. 7, 133-140 (1974).
Roberts, B.R.: The influence of different sulfur dioxide concentrations on growth of white birch and pin oak seedlings in the field. J. Am. Soc. Hortic. Sci. 100, 640-642 (1975).
Roberts, B.R., Townsend, A.M., Dochinger, L.S.: Photosynthetic response to SO_2 fumigation in red maple. Plant Physiol. 47, 30 (Supplement) (1971).
Rohmeder, E.: Die Wirkung von Staub und Ruß auf den Zuwachs der Fichte. Forst. u. Holzwirtsch. 15, 245-248 (1960).
Rohmeder, E., Merz, W., Schonborn, von, A.: Suchtung von gegen Industriealgase relativ resistenten Fichten- und Kiefernsorten. Forstwiss. Centralblatt 81, 321 (1962).
Santamour, F.S., Jr.: Air pollution studies on Platanus and American elm seedlings. Plant Dis. Rep. 53, 482-484 (1969).
Sato, T.: Taiki osendo to jumoku no yochu ioganyuryo no kankei. J. Jap. Soc. Air Pollut. 5, 153 (1970).
Scheffer, T.C., Hedgcock, G.G.: Injury to northwestern forest trees by sulfur dioxide from smelters. Washington, D.C.: USDA Tech. Bull. 1117 (1955).
Scurfield, G.: Air pollution and tree growth. Forestry Abstr. 3, 339-347, 517-528 (1960).
Sinclair, W.A.: Polluted air: Potent new selective force in forests. J. Forestry 67, 305-309 (1969).
Skelly, J.M., Moore, L.D., Stone, L.L.: Symptom expression of eastern white pine located near a source of oxides of nitrogen and sulfur dioxide. Plant Dis. Rep. 56, 3-6 (1972).
Smith, W.H.: Technical review: Trees in the city. J. Am. Inst. Plan. 36, 429-436 (1970).
Smith, W.H.: Lead contamination of roadside white pine. Forest Sci. 17, 195-198 (1971).
Smith, W.H.: Lead and mercury burden of urban woody plants. Science 176, 1237-1239 (1972).
Smith, W.H.: Metal contamination of urban woody plants. Environ. Sci. Technol. 7, 631-636 (1973).
Smith, W.H.: Air pollution - Effects on the structure and function of the temperate forest ecosystem. Environ. Pollut. 6, 111-129 (1974).
Spierings, F.: Method for determining the susceptibility of trees to air pollution by artificial fumigation. Atmos. Environ. 1, 205-210 (1967).
Stone, L.L., Skelly, J.M.: The growth of two forest tree species adjacent to a periodic source of air pollution. Phytophathology 64, 773-778 (1974).
Thomas, H.L.: Control what you breathe. Ind. Res. 16 (13), 53-58 (1974).
Thomas, M.D.: Gas damage to plants. Ann. Rev. Plant Physiol. 2, 293-322 (1951).
Thomas, M.D.: Effects of air pollution on plants. In: Air Pollution. World Health Organ. Monogr. Ser. 46, 233-275 (1961).
Thomas, M.D., Hendricks, R.H., Hill, G.R.: Some impurities in the air and their effects on plants. In: Air Pollution (ed. L.C. McCabe), p. 41-47. New York: McGraw-Hill 1952.

Thompson, R.J., McMullen, T.B., Morgan, C.B.: Fluoride concentration in the ambient air. J. Air Pollut. Control Assoc. 21, 484-487 (1971).
Thorne, L., Hanson, G.P.: Species differences in rates of vegetal ozone absorption. Environ. Pollut. 3, 303-312 (1972).
Townsend, A.M.: Breeding shade trees for disease resistance. Arborist's News 37, 126-129 (1972).
Townsend, A.M.: Sorption of ozone by nine shade tree species. J. Am. Soc. Hortic. Sci. 99, 206-208 (1974).
Townsend, A.M., Dochinger, L.S.: Relationship of seed source and developmental stage to the ozone tolerance of *Acer rubrum* seedlings. Atmos. Environ. 8, 957-964 (1974).
Treshow, M.: Symptomatology of fluoride injury on vegetation. In: Handbook of Effects Assessment: Vegetation Damage (ed. N.L. Lacasse, W.J. Moroz). Pa. State Univ. Sect. 7, 1-41 (1969).
Treshow, M.: Environment and plant response. New York: McGraw-Hill 1970.
Treshow, M., Anderson, F.K., Harner, F.: Responses of Douglas fir to elevated atmospheric fluorides. Forest Sci. 13, 114-120 (1967).
Treshow, M., Pack, M.R.: Fluoride. In: Recognition of Air Pollution Injury to Vegetation. A Pictorial Atlas (ed. J.S. Jacobson, A.C. Hill), pp. D1-D17. Pittsburgh: Herbick and Held 1970.
Unsworth, M.H., Biscoe, P.V., Pinckney, H.R.: Stomatal responses to sulphur dioxide. Nature 239, 458-459 (1972).
Vandegrift, A.E., Shannon, L.J., Sallee, E.E., Gorman, P.G., Park, W.R.: Particulate air pollution in the United States. J. Air Pollut. Control Assoc. 21, 321-329 (1971).
Vigorov, L.I.: Combining protective and ornamental characteristics of woody plants. In: AICE Survey of USSR Air Pollution Literature (ed. M.Y. Nuttonson), Vol. II, p. 90-95. Silver Spring, Md.: Am. Inst. Crop. Ecol. 1969.
Weidensaul, T.C.: Are trees efficient air purifiers? Arborist's News 38, 85-89 (1973).
Weidensaul, T.C.: Importance of plants in oxygen production, carbon dioxide consumption, and air purification. Hort. Science 9, 8-12 (1974).
Weinstein, L.H., McCune, D.C.: Field surveys, vegetation sampling, and air and vegetation injury to vegetation. In: Recognition of Air Pollution Injury to Vegetation. A Pictorial Atlas (ed. J.S. Jacobson, A.C. Hill), pp. G1-G4. Pittsburgh: Herbick and Held 1970.
Wentzel, K.F.: Susceptibility of plants to air pollution, and differences in resistance. Forstarchiv 39, 189-194 (1968).
White, E.J., Turner, F.: A method for estimating income of nutrients in a catch of airborne particles by a woodland canopy. J. Appl. Ecol. 7, 441-461 (1970).
Williams, R.J.H., Lloyd, M.M., Ricks, G.R.: Effects of atmospheric pollution on deciduous woodland I: Some effects on leaves of *Quercus petraea* (Mattuschka) Leibl. Environ. Pollut. 2, 57-68 (1971).
Wood, F.A., Drummon, D.B., Wilhour, R.G., Davis, D.D.: An exposure chamber for studying the effects of air pollutants on plants. Pa. State Univ. Agric. Exp. Stn. Prog. Rep. 335 (1973).
Zahn, R.: The influence of various environmental factors on plant sensitivity towards sulphur dioxide. Z. Pflanzenkr. Pflanzenschutz 70, 81-95 (1963).
Zinke, P.J.: Forest interception studies in the United States. Intern. Symp. Forestry Hydrol., pp. 137-161. Oxford, England: Pergamon Press 1967.

CHAPTER 10

Indirect Selection for Improvement of Desired Traits

K. VON WEISSENBERG

Introduction

All genetic traits are related to physiological processes that interact with the environment and result in phenotypic expression. Traits such as yields of turpentine, tall oil, and other valuable extractives are more directly related to biochemical syntheses than are bole straightness or fibre length, which are indirectly dependent. Diseases caused by biotic agents are examples of very intricate physiological interactions between host and agent cells (Strobel, 1975).

Physiological processes underlying the expression of a multitude of traits are often studied by biochemical methods. The genetic correlation between a phenotypic trait and some biochemical properties may be of such a magnitude that indirect selection becomes an attractive alternative to direct selection and it should not be overlooked in breeding strategies. This chapter presents the application of some aspects of indirect selection to breeders and biochemists cooperating in the field of tree breeding.

The Theory of Indirect Selection

Nelson and Birkeland (1929) suggested that rust resistance, hardiness, and yield in wheat (*Triticum aestivum* L.) could be increased by selection for globulin fractions in the grain. The theoretical foundations of indirect selection have been treated in detail by Falconer (1967) and the following presentation is based, in part, on his work.

Indirect selection is defined as a selection procedure by which improvement of a desired trait X is achieved by selecting the breeding material on the basis of another trait Y which is genetically correlated with the desired trait. Environmental causes of correlation affect phenotypic correlations which cannot be utilized for indirect selection and are sources of error in breeding programs. Such correlations are valuable for phenotypic selection only. The term indirect selection has often been used rather broadly in recent forestry literature and phenotypic correlations have been used as arguments for this method of selection, without distinguishing whether phenotypic or genotypic selection is implied. The above definition will be used in this presentation. When cited authors have not made this distinction it has been difficult, however, to adhere to the definition.

The genetic gain that can be expected from one cycle of indirect selection, assuming that the desired trait is quantitative rather than qualitative, depends on four factors:

$$G_x = i_y h_y \sigma_{ax} r_a \quad \text{where} \tag{1}$$

G_x = gain in the desired trait X by indirect selection for a marker trait Y

i_y = selection intensity achieved when selecting for the marker trait

h_y = square root of the heritability of the marker trait

σ_{ax} = square root of the additive genetic variance of the desired trait

r_a = additive genetic correlation between the desired trait and the marker trait.

Genetic correlation between two traits occurs when: 1. two different genes controlling the traits are located close together on the same chromosome (linkage), 2. one gene affects both traits (pleiotropy), or 3. linkage and pleiotropy occur simultaneously. If the relationship between two traits is dependent on linkage, and assuming no espistasis between the genes, the genetic correlation between the two traits will be decreased by repeated cycles of selection due to breaking up of the linkage block during the breeding process. If the relationship is due to pleiotropy, however, the genetic correlation will remain unchanged over several cycles of selection. The most useful markers are traits that have a high genetic correlation based predominantly on pleiotropy.

In order to justify the efforts involved in the development of indirect selection for a trait, its gain should be compared to the predicted gains achieved by direct selection. The comparison is made by using the following relationship:

$$\frac{G_x}{D_x} = r_a \cdot \frac{i_y}{i_x} \cdot \frac{h_y}{h_x} \qquad \text{where} \tag{2}$$

G_x = gain in trait X by one cycle of indirect selection using the marker Y

D_x = gain in trait X by one cycle of direct selection

r_a = additive genetic correlation between trait X and a marker Y

h_y, h_x = square root of the heritability for the marker Y and trait X, respectively.

Assuming that a potential marker Y has the same heritability as the desired trait the formula is reduced to:

$$\frac{G_x}{D_x} = r_a \cdot \frac{i_y}{i_x} \tag{3}$$

Under this assumption the relative efficiency of indirect vs. direct selection is dependent on: 1. the genetic correlation between the marker trait and the desired trait and 2. the selection intensities that can be achieved by the direct and indirect selection procedures, respectively.

If the marker has a heritability higher than that of the desired trait, the genetic correlation or selection intensity can be smaller without decreasing the predicted gain.

Since selection intensity is a function of the ratio of number of plants selected to total number of plants tested, any trait that can be measured more quickly or at less cost than by direct tests for the desired trait, will permit a higher selection intensity to be applied with the same investment of time and effort. Thus, the relatively high selection intensity required for indirect selection can be achieved by measuring a suitable marker by means of any of a large number of routine biochemical or other analytical procedures, some of which are discussed in other chapters of this book.

Considering the modest selection intensities presently achieved by large-scale direct mass selection in many breeding programs throughout the world, doubling and in some cases even tripling such intensities appears possible by means of efficient indirect selection (Table 1).

Table 1. Proportions to be selected for normally distributed traits from a large base population at various multiples of selection intensities, i_x (based on Falconer, 1967; Namkoong, 1969)

i_x	$1 \times i_x$	$2 \times i_x$	$3 \times i_x$
	(Proportion of population to be selected)		
0.50	0.700	0.380	0.166
0.75	0.540	0.166	0.031
1.00	0.380	0.058	0.004
1.25	0.260	0.016	0.0002
1.50	0.166	0.004	-
1.75	0.100	0.0006	-

Due to the increased capacity of the selection procedure, the breeder may also be able to provide a sufficiently large number of selected plants to achieve further gains from subsequent cycles of selection for both the desired trait and other important traits as well.

Dickerson and Hazel (1944) have pointed out that genetic gain per unit of time is the critical factor to be considered when comparing breeding procedures. Thus, if one cycle of selection is completed twice as fast by indirect selection compared to direct selection, the relative efficiency of the two methods may be considered doubled (Lewis, 1973).

The need for a moderate to high genetic correlation required for successful indirect selection can often be met, e.g. if the marker is a component of the desired trait. A component is defined as a trait directly or indirectly contributing to the expression of the desired trait. The genetic correlation between two such traits (Stonecypher et al., 1973) is often very high. High genetic correlations also, however, are found between traits that have no apparent or well-understood causal relationship to each other. Examples of such correlations are those found between monoterpene percentages and resistance to fusiform rust (*Cronartium fusiforme* Hedgc. and Hunt ex Cumm.), in loblolly (*Pinus taeda* L.) and slash (*Pinus elliottii* var. *elliottii* Engelm.) pines (Rockwood, 1973, 1974). On the other hand, if the genetic correlation is 1.00 or nearly that, the two traits may be considered to be the same. If the correlation between, e.g. juvenile and mature growth of a tree is 1.00, the two traits are the same; the early test for growth may not be considered a case of indirect selection, which it otherwise would be.

Indirect Selection in Forest Tree Breeding

Disease Resistance. Indirect selection is likely to be superior to direct selection when large technical difficulties exist in selecting for the desired trait directly. Such difficulties include large errors in measurement of the desired trait and very expensive, as well as time-consuming methods of measurement. Since these difficulties occur frequently in breeding programs, the possibilities of using indirect selection have been explored both in plant and animal breeding. Related literature has been briefly reviewed by Weissenberg (1973).

Eleven reports have been found suggesting or encouraging indirect selection in forest trees. Seven of these are concerned with breeding for disease resistance (Table 2). Nearly all have relied on biochemical methods for measuring the marker trait.

In 1961 a research program was initiated aiming at discovering markers for resistance to blister rust (*Cronartium ribicola* J.C. Fisch ex Rabenh.) in white pines (Hanover, 1963). The results (Hanover and Hoff, 1966; Hoff, 1968, 1970) of the program did not, however, lead to discovery of markers for indirect selection and therefore Bingham (1970) reported that further physiological studies to identify markers should be deferred by the group until more complete understanding of resistance had been obtained. Subsequently several different mechanisms of resistance in white pines to blister rust have been discovered with specific modes of inheritance (McDonald and Hoff, 1970, 1971; Kinloch et al., 1970). Such various mechanisms may partly, among other reasons, cause difficulties in finding markers correlated with several types of resistance.

Efforts to discover markers for disease resistance in other pines have been more successful. Although Schütt's (1964) report on a mycological method to determine resistance of Scotch pine (*Pinus sylvestris* L.) to needle cast (*Lophodermium pinastri* Schrad ex Hook.) did not reach pratical applications, it stimulated further research (Scholz, 1973). Scholz and Stephan (1974) were able to show that the phenotypic correlation was significant between the buffer capacity of press-juice of needles and resistance of the clone. They suggested that the buffer capacity might be used as a marker for indirect methods of selection. The development of this biochemical method is a prime example of logical reasoning based on sound knowledge of the pathogen's physiological requirements.

In developing methods for indirect selection for resistance to fusiform rust in southern yellow pines, the results have not been reached on such a sound basis. The efforts by Hare (1970), Weissenberg (1973), and Lewis (1973) have not reached results leading to applied selection procedures. Weissenberg calculated, however, that gains from indirect mass selection reached in a comparatively short time may be equal or superior to gains presently predicted from two direct selection methods (Blair, 1970; Stonecypher et al., 1973) provided that: 1. the heritability of the marker is equal to that of resistance to fusiform rust (0.22 according to Blair, 1970), 2. the genetic correlation between the marker and resistance is moderate (0.6-0.7), and 3. the selection intensity is two or three times greater, respectively, than that achieved when the direct methods are applied.

Rockwood (1972) discovered strong genetic and phenotypic correlations between certain monoterpenes and rust resistance, in connection with studies of general patterns of variation of monoterpene composition in loblolly pine. Rockwood (1973) then calculated gains in resistance to fusiform rust in loblolly pine by selecting trees on the basis of their monoterpene content. Compared to a simple direct mass selection scheme, indirect selection for branch cortex β-phellandrene resulted in gain in resistance surpassing that from direct mass selection. This supported Weissenberg's (1973) calculations. The gain in time, however, would have been considerable since data required for direct selection were obtained in 6 growing seasons while indirect selection for monoterpenes could have been completed in the short time required for raising seedlings able to produce the oleoresin.

Rockwood (1974) concluded that indirect selection for resistance to fusiform rust for slash pine is justified provided that well-substantiated genetic correlations for the marker are found. Resistant slash pine would be selected and subsequently tested with direct methods, and if these trees remain resistant, they would be used in breeding programs. A more conservative approach would utilize monoterpene contents of families rather than individual trees. Monoterpene analysis could serve as a supplement to actual performance of the family in rust tests to further verify resistance. These suggestions by Rockwood indicate one of the advantages of indirect selection: the number of plants to be tested in costly, time- and space-consuming field tests could be reduced, thus increasing the capacity of the selection program to screen large numbers of plants for resistance.

The resistance of Scotch pine to the root and butt rot (*Fomes annosus* (Fr.) Karst.) has been found to be genetically correlated with the amount of Δ-3-carene and α-pinene to such an extent that indirect selection has been proposed in the Soviet Union (Chudnyi, 1973, 1974). The correlation was suspected to be due to pleiotropy. Very similar relationships have been indicated by preliminary results obtained in Finland from studies on Norway spruce (*Picea abies* (L.) H. Karst.) resistant to spread of the same fungus (Weissenberg, 1974 and unpublished data).

Some of the problems related to the use of indirect selection for disease resistance should be considered: 1. if different mechanisms of resistance occur within large portions of the host population, markers have to be found which are highly correlated with these mechanisms, 2. variation in pathogenicity within pathogen populations may occur; such variation may reduce the probability of finding markers sufficiently correlated with resistance to races of the pathogen ocurring in the region where the improved cultivar will be grown (Weissenberg, 1973), and 3. if the marker trait is highly and genetically correlated with a specific gene for, or is a causal agent of, resistance, indirect selection may lead to vertical resistance (Rockwood, 1974); this may be avoided by using field tests sufficiently replicated in time and environment for measurement of the genetic correlation between marker and resistance. Thus, the measured resistance remains horizontal.

Insect Resistance. The volatile fraction of wood resin has traditionally received attention as a source of olfactory attractants to insects (cf. rev. Borden and Stokkink, 1971; Hanover, 1975). A group in Finland has, since 1957, published a series of papers (cf. Perttunen, 1957; Kangas et al., 1965, 1970, 1971) on their investigations on the effect of α-pinene and other monoterpenes in the bark of Scotch pine on the orientation of the bark beetles *Blastophagus minor* Hart. and *Blastophagus piniperda* L. The attractant effect on *B. piniperda* is due to α-terpineol synergistically enchanced by cis- and trans-carveol. *B. minor* is more sensitive to these attractants than *B. piniperda*.

Table 2. Suggested applications of and encouraged research for indirect selection in forest trees

Organism	Desired trait	Marker trait	Remarks[a]	Author
Picea abies	Resistance to *Adelges abietis*	Total phenol content in foliage	Suggested	Tjia and Houston, 1975
P. elliottii	Yield of tall oil	Yield of ethanol-benzene extractives	Suggested; $r_p = 0.70$	Franklin et al., 1970
P. elliottii	Yield of oleoresin	Yield of ethanol-benzene extractives	Suggested; $r_a = 0.58$	Franklin et al., 1970
P. elliottii	Resistance to *Cronartium fusiforme*	Branch cortex; percentage of β-phellandrene	Suggested; $r_p = 0.43$	Rockwood, 1974
P. radiata	Resistance to *Sirex noctilio*	Ethylene production	Suggested	Shain and Hillis, 1972
P. radiata	Resistance to *Sirex noctilio*	Reaction to mucus of *Sirex noctilio*	Suggested	Kile et al., 1974
P. radiata	Production of phenols for increased resistance to diseases	Ethylene production	Suggested	Shain and Hillis, 1972
P. sylvestris	Resistance to *Lophodermium pinastri*	Inhibited growth of pathogen on needle-juice-agar	Suggested; consistent ranking of over 20 clones	Schütt, 1964
P. sylvestris	Resistance to *Lophodermium pinastri*	Buffer capacity of needle-juice	Suggested; $r_p = 0.84$	Scholz and Stephan, 1974
P. sylvestris	Resistance to *Fomes annosus*	Percentage of Δ-3-carene and α-pinene	Suggested; thought to be due to pleiotropy	Chudnyi, 1973; ref. 1974
P. taeda	Resistance to *Cronartium fusiforme*	Primary cortex; amount of the fatty acid 18:2	Further research encouraged; $r_p = 0.95$	Weissenberg, 1973

Table 2 (continued)

Organism	Desired trait	Marker trait	Remarks[a]	Author
P. taeda	Resistance to *Cronartium fusiforme*	Growth inhibitors	Further research encouraged; $r_p = -0.83$	Weissenberg, 1973
P. taeda	Resistance to *Cronartium fusiforme*	Branch cortex; percentage of β-phellandrene and α-pinene	Suggested; $r_a = 0.54$ and 0.78 $r_a = -0.81$ and -0.94	Rockwood, 1973
P. taeda	Resistance to *Cronartium fusiforme*	Shoot samples; peroxidase and poly-phenoloxidase activity	Further research encouraged	Lewis, 1973
Pseudotsuga menziesii	Resistance to deer browsing	Total monoterpene content in foliage	Suggested	Radwan and Ellis, 1975
Ulmus americana	Resistance to *Ceratosystis ulmi*	Vessel diameter index	Suggested	Sinclair et al., 1975

[a] r_p = coefficient of phenotypic correlation, r_a = coefficient of genotypic correlation

Smith (1970) ranked the monoterpenes of several pine species responsible for resistance to three species of *Dendroctonus* as follows: limonene > 3-carene > α-pinene = standard. A standard tree was one without a high percentage of any of these monoterpenes.

Fractions isolated from Scotch pine phloem containing monoterpene alcohols and sesquiterpenes were attractant to the weevil *Hyolbius abietis* L., a destructive insect on pine plantations in the Nordic countries. These terpenoids possibly acted synergistically (Selander et al., 1974).

The possibility of using these attractants produced by the host for the bark beetles and the weevil for indirect selection have not yet, to my knowledge, been explored.

A rapid and simple ethylene analysis indicated the capacity of radiata pine (*Pinus radiata* D. Don) in Australia to synthesize polyphenols, which possibly are responsible for inhibition of growth of *Amylostereum areolatum* (Fr.) Boldin, and thus cause lack of nutrients for larvae of the wood wasp (*Sirex noctilio* F.). The ability to produce sufficient quantities of ethylene may be a valuable marker for indirect selection (Shain and Hillis, 1972). Kile et al. (1974) have suggested that resistance to artificially applied mucus of the wood wasp also may provide a measurement of resistance to attack by the insect itself. Consequently, breeding for resistance to the insect in radiata pine by indirect selection may be an approach worth further investigation.

Although the above-mentioned examples of some tentative possibilities of indirect selection for resistance to insects may be encouraging, some complications have to be considered: 1. insects, unlike fungi, actively orientate themselves preferentially to available food sources and breeding materials. Depending on the availability of material, insects usually are able to use inferior food sources, suggesting that a resistance mechanism has to have a profound effect on orientation of the insect in order to divert attacks during times of limited supply of food and breeding material. 2. Influential attractants are produced by pioneering insects and associated microorganisms subsequent to attack on the host. Consequently, efforts to identify compounds suitable as markers should include studies on the effect of the compound(s) on the pheromone-producing capacity of the pioneering insects and associated microorganisms (Vité and Pitman, 1968; Bedard et al., 1969; Pitman and Vité, 1969; Berryman, 1972; Smith, 1972; Brand et al., 1975). 3. The attractants produced by the host may act synergistically with those produced by the insect population (Borden and Stokkink, 1971).

Other Traits. Throughout the history of forest tree breeding it has been recognized that several traits are very difficult to measure directly. The concept of "Frühtests", early tests for traits occurring at later phases of the long life span of the tree, grew out of this recognition. Many breeders will have to wait for final and conclusive evidence of high genetic correlation between juvenile and mature growth or juvenile growth and total yield during one rotation. Only if sufficiently high genetic correlations are found can the early test be used as a basis for genotypic selection.

Breeding for yield of tall oil is hampered by the fact that tall oil is a product of the pulping process. Measurement of this trait results in destruction of the genotype. Consequently, Franklin et al. (1970) studied the phenotypic and genotypic correlation between tall oil yield and yield of oleoresin and ethanol-benzene extractives. They concluded that indirect selection for increased yield of tall oil would be possible by using the yield of ethanol-benzene extractives as a marker.

The yield of levopimaric acid from slash pine is of commercial interest but time consuming to measure compared with measurement of oleoresin yield from commercial chipping. Squillace et al. (1971) found the genetic correlation between the two traits high enough (0.80) to justify breeding for oleoresin yield and expectations of concomitant response in yield of levopimaric acid.

Velling (1974) speculated that the earlywood percentage may be a suitable marker for wood density, but conclusive evidence was not presented. Other examples of the possibilities of using indirect selection to improve traits other than disease and insect resistance have not been found in the literature. Since most other traits of interest are finally assessed at the end of the rotation, breeders have to wait for maturation of the genetic experiments established around the world. These experiments will in time, and if properly analyzed, give valuable information on genetic correlations suggesting choices of markers which may facilitate indirect selection for traits difficult to select for directly.

Concluding Remarks

Although indirect selection may be an attractive alternative to direct selection for many traits, some general problems have to be considered in addition to those treated in the disease and insect resistant sections.

1. Finding suitable marker traits is extremely difficult and perhaps more fortuitous than expected. Since markers may be causally related to the desired trait, a thorough knowledge of the biochemical and physiological mechanisms controlling the desired trait is of invaluable help (Scholz, 1973; Scholz and Stephan, 1974). On the other hand, since no apparent causal relationship may exist between the two traits (Rockwood, 1973, 1974), other approaches may also lead to results. It is wise, however, not to embark upon costly and frustrating searches for markers with that aim as the sole objective of the study. The advice is, in cases of noncausal relationships, look for markers as byproducts from general studies of variation and genetic parameters in breeding populations (Rockwood, 1972).

2. The determination of the potential value of a marker trait is predicted on accurate estimates of its heritability and genetic correlation with the ultimately desired trait. Genetic studies satisfactory for this purpose are costly to establish and require considerable time to complete. Particularly the estimates of genetic correlation are often inaccurate (Falconer, 1967; Rockwood, 1973, 1974), and efforts to increase accuracy and precision may be costly. For detailed treatment of problems associated with confidence limits of genetic correlations see Mode and Robinson (1959); Falconer (1967); and Grossman (1970). As long as the estimate of the correlations are inaccurate or have low precision, indirect selection should be used cautiously and/or as a supplement to direct selection. For the same reason, the relative efficiency of direct and indirect selection can seldom be measured accurately and should rather serve as an indicator to possible advantages of alternative procedures.

3. Only genetic correlations due to pleiotropy will remain unchanged for several cycles of selection. Thus, the genetic control of the correlation has to be understood before long-range programs are initiated. On the other hand, genetic correlations, regardless of their control, may be useful for one or a few cycles of selection.

4. Since pleiotropic genes affecting both the marker and the desired trait will be rapidly stabilized due to efficient selection, a selection limit may be reached. The gain simultaneously obtained, however, may justify the program. How rapidly stabilization is reached when applying selection to heterozygous populations of forest trees is not known.

5. Indirect selection is most advantageous when only one trait is to be rapidly improved. Disease and insect resistance for cultivars to be used in specific problem areas are, therefore, among traits suitable for indirect selection. Since, however, most breeding programs aim at simultaneous improvement of several traits, the most effective use of correlated characters is in combination with desired characters as additional sources of information about the breeding value of the individuals. This leads to the use of selection indices and to topics outside the realm of this chapter and book.

References

Bedard, W.D., Tilden, P.E., Wood, D.L.: Western pine beetle: field response to its sex pheromone and a synergistic host terpene, myrcene. Science 164, 1284-1285 (1969).

Berryman, A.A.: Resistance of conifers to invasion by bark beetle-fungus associations. Biol. Sci. 22, 598-602 (1972).

Bingham, R.T.: Rust Resistance in Conifers - Present Status, Future Needs. Second World Consultation on Forest Tree Breeding, FAO, Rome 1, 483-497 (1970).

Blair, R.L.: Quantitative inheritance of resistance to fusiform rust in loblolly pine. Ph.D. Diss., North Carolina State Univ., Raleigh, North Carolina, 87 pp. (1970).

Borden, J.H., Stokkink, E.: Secondary attraction in the *Scolytidae*: An annotated bibliography. Dept. Fish. For., C.F.S., For. Res. Lab., Victoria, B.C. Inf. Rep. BC-X-57, 77 pp. (1971).

Brand, J.M., Bracke, J.W., Markovetz, A.J., Wood, D.L., Brown, L.E.: Production of verbenol pheromone by a bacterium isolated from bark beetles. Nature 254, 136-137 (1975).

Chudnyi, A.V.: O nekotorykh printsipakh izucheniya struktury populyatsii drevesnykh porod. (On some principles of tree species population structure). VNIILM, Pushkino, 13 p. (1973).

Chudnyi, A.V.: Metsäpuiden populaatiorakenteen tutkimusmenetelmistä. (Investigation methods in forest tree population studies). Silva Fenn. 8, 89-95 (1974).

Dickerson, G.E., Hazel, L.N.: Effectiveness of selection on progeny performance as a supplement to earlier culling in livestock. J. Agr. Res. 69, 459-476 (1944).

Falconer, D.S.: Introduction to Quantitative Genetics, 365 pp. Glasgow: Robert MacLehose and Co. 1967.

Franklin, E.C., Taras, M.A., Volkman, D.A.: Genetic gains in yield of oleoresin, wood extractives, and tall oil. TAPPI 53, 2302-2304 (1970).

Grossman, M.: Sampling variance of the correlation coefficients estimated from analysis of variance and covariance. Theor. Appl. Genet. 40, 357-359 (1970).

Hanover, J.W.: New developments in breeding western white pine. II. Biochemistry of rust resistance. Proc. Forest Genetics Workshop: SAF Committee on Forest Tree Improvement, Southern Forest Tree Improvement Committee Meeting, Oct. 25-27, 1962, pp. 77-78. Macon, Ga. (1963).

Hanover, J.W.: Physiology of tree resistance to insects. Ann. Rev. Ent. 20, 75-95 (1975).

Hanover, J.W., Hoff, R.J.: A comparison of phenolic constituents of *Pinus monticola* resistant and susceptible to *Cronartium ribicola*. Phys. Plant. 19, 554-562 (1966).

Hare, R.C.: Physiology and biochemistry of pine resistance to the fusiform rust fungus, *Cronartium fusiforme*. Ph.D. Diss., 155 pp. Univ. of Florida, Gainesville, Florida (1970).

Hoff, R.J.: Comparative physiology of *Pinus monticola* Dougl. resistant and susceptible to *Cronartium ribicola* J.C. Fich. ex Rabenh. Ph.D. Diss., 76 pp. Washington State University, Pullman, Washington (1968).

Hoff, R.J.: Inhibitory compounds of *Pinus monticola* resistant and susceptible to *Cronartium ribicola*. Can. J. Bot. 48, 371-376 (1970).

Kangas, E., Perttunen, V., Oksanen, H.: Responses of *Blastophagus minor* Hart. (*Col., Scolytidae*) to the pine phloem fraction known to be attractant to *Blastophagus piniperda* L. Ann. Ent. Fenn. 36, 120-122 (1970).

Kangas, E., Perttunen, V., Oksanen, H.: Physical and chemical stimuli affecting the behaviour of *Blastophagus piniperda* L. and *B. minor* Hart. (*Col., Scolytidae*). Acta Ent. Fenn. 28, 120-126 (1971).

Kangas, E., Perttunen, V., Oksanen, H., Rinne, M.: Orientation of *Blastophagus piniperda* L. (*Col., Scolytidae*) to its breeding material. Attractant effect of terpineol isolated from pine rind. Ann. Ent. Fenn. 31, 61-73 (1965).

Kile, G.A., Bowling, P.J., Dolezal, J.E., Bird, T.: The reaction of *Pinus radiata* twigs to the mucus of *Sirex noctilio* in relation to resistance to *Sirex* attack. Aust. For. Res. 6, 25-34 (1974).

Kinloch, B.B., Jr., Parks, G.K., Fowler, G.W.: White pine blister rust: simply inherited resistance in sugar pine. Science 167, 193-195 (1970).

Lewis, R.A.: Quantitative Assessment and Possible Biochemical Indicators of Variation in Resistance to Fusiform Rust in Loblolly Pine, pp. 90. School of Forest Resources, North Carolina State Univ. Cooperative programs (1973).

McDonald, G.I., Hoff, R.J.: Resistance to *Cronartium ribicola* in *Pinus monticola*: Early shedding of infected needles. Res. Note INT-124, 8 pp. Ogden, Utah: USDA For. Serv. 1970.

McDonald, G.I., Hoff, R.J.: Resistance to *Cronartium ribicola* in *Pinus monticola*: Genetic control of needle-sports-only resistance factors. Can. J. For. Res. 1, 197-202 (1971).

Namkoong, G.: Accurate values for selection intensities. Silvae Genet. 18, 172 (1969).

Nelson, C.I., Birkeland, J.M.: A serological ranking of some wheat hybrids as an aid in selecting for certain genetic characters. J. Agric. Res. 38, 169-181 (1929).

Perttunen, V.: Reactions of two bark beetle species, *Hylurgops palliatus* Gyll. and *Hylastes ater* Payk. (*Col., Scolytidae*) to the terpene α-pinene. Ann. Ent. Fenn. 23, 101-110 (1957).

Pitman, G.B., Vité, J.P.: Aggregation behaviour of *Dendroctonus ponderosae* (*Coleoptera:Scolytidae*) in response to chemical messengers. Can. Entomol. 101, 143-149 (1969).

Radwan, M.A., Ellis, W.D.: Clonal variation in monoterpene hydrocarbons of vapors of Douglas-fir foliage. For. Sci. 21, 63-67 (1975).

Rockwood, D.L.: Aspects of monoterpene composition in loblolly pine. Ph.D. Diss., 129 pp. North Carolina State Univ., Raleigh, North Carolina (1972).

Rockwood, D.L.: Monoterpene-fusiform rust relationship in loblolly pine. Phytopath. 63, 551-553 (1973).

Rockwood, D.L.: Cortical monoterpene and fusiform rust resistance relationships in slash pine. Phytopath. 64, 976-979 (1974).

Scholz, F.: Biochemical methods in forest genetics. Forstarchiv 44, 171-174 (1973).

Scholz, F., Stephan, B.R.: Physiologische Untersuchungen über die unterschiedliche Resistenz von *Pinus sylvestris* gegen *Lophodermium pinastri*. Eur. J. For. Path. 4, 118-126 (1974).

Schütt, P.: Eine mykologische Methode der Resistenzprüfung, entwickelt für den Lophodermium-Befall der Kiefer. Phytopath. Z. 51, 62-72 (1964).

Selander, J., Kalo, P., Kangas, E., Perttunen, V.: Olfactory behaviour of *Hylobius abietis* L. (*Col., Curculonidae*). I. Response to several terpenoid fractions isolated from Scotch pine phloem. Ann. Ent. Fenn. 40, 108-115 (1974).

Shain, L., Hillis, W.E.: Ethylene production in *Pinus radiata* in response to *Sirex-Amylostereum* attack. Phytopath. 62, 1407-1409 (1972).

Sinclair, W.A., Zahand, J.P., Melching, J.B.: Anatomical marker for resistance of *Ulmus americana* to *Ceratocystis ulmi*. Phytopath. 65, 349-352 (1975).

Smith, R.H.: Xylem resin as a factor in the resistance of pines to forced attacks by bark beetles. Proc. 2nd World Consultation For. Tree Breeding. FAO, Rome 1, 537-553 (1970).

Smith, R.H.: Xylem resin in the resistance of the *Pinacea* to bark beetles. USDA, For. Serv. Gen. Tech. Rep. PSW-1, 7 pp. (1972).

Squillace, A.E., Hendrick, G.W., Green, A.J.: Variation and inheritance of levopimaric acid content and its relationship to oleoresin yield in slash pine. Silvae Genet. 20, 90-91 (1971).

Stonecypher, R.W., Zobel, B.J., Blair, R.: Inheritance patterns of loblolly pines from a nonselected natural population. North Carolina Agirc. Exp. Station, Tech. Bull. No. 220, 60 pp. (1973).

Strobel, G.A.: A mechanism of disease resistance in plants. Sci. Amer. 232, 81-88 (1975).

Tjia, B., Houston, D.B.: Phenolic constituents of Norway spruce resistant or susceptible to eastern spruce gall aphid. For. Sci. 21, 180-184 (1975).

Velling, P.: Summary: Phenotypic and genetic variation in the wood basic density of Scotch pine (*Pinus sylvestris* L.). Acta For. Fenn. 188, 1-29 (1974).

Vité, J.P., Pitman, G.B.: Bark beetle aggregation: effects of feeding on the release of pheromones in *Dendroctonus* and *Ips*. Nature 218, 169-170 (1968).

Weissenberg, K. von: Indirect selection for resistance to fusiform rust in loblolly pine. Acta For. Fenn. 134, 1-46 (1973).

Weissenberg, K. von: Kuusen kestävyydestä juurikääpäsienen leviämistä vastaan. Metsänviljelyn koeaseman tiedonantoja 12, 38-45 (1974).

CHAPTER 11

Pollen Handling Techniques in Forest Genetics, with Special Reference to Incompatibility

E. G. KIRBY and R. G. STANLEY

Introduction

Several new pollen treatment and application techniques have been introduced in recent years. Field and laboratory techniques that insure and assess pollen viability and, in some cases, overcome incompatibility barriers are now fairly standard.

This chapter reviews some of the old and presents some of the new techniques in pollen biology which are applicable to forest genetics. Collection, storage, viability, treatment and application are the five pollen handling methods covered.

Methods

Pollen Management

Collection. The production of genetically improved trees depends very much on knowledge of pollination of the species under study and proper handling of pollen. Most tree species are wind-pollinated and the quantities of pollen produced in a flowering season varies with the time of pollen shedding, the species, location and climate. Year-to-year variations of as much as one month in the time of pollen shedding for a particular species is fairly common. The length of pollen shedding within a stand of trees may cover six weeks in a mild climate, such as Australia (Fielding, 1957). On the other hand, dehiscence may take place within a few days in northern climates. This occurs when low temperatures directly precede the period of pollen maturation. Temperature is the most critical factor controlling shedding of pollen. In addition, dehiscence in pine species generally displays a diurnal pattern with early afternoon shedding and little shedding at night. The approximate shedding times in northern temperate zones of a few genera are: Corylus, February; Salix, April; Pinus, May; Acer, late May (Stanley and Kirby, 1973).

Pollen from angiosperm and gymnosperm tree species is usually collected from similar locations. Catkins on the trees are enclosed with a bag in which the dehisced pollen is accumulated. Catkins or strobili while still on the branches, may be removed and brought into a greenhouse or laboratory where pollen shedding can be hastened. A controlled environment eliminates contamination by foreign pollen species as demonstrated by Barner and Christiansen (1958), who designed such a pollen house.

A simple technique for pollen collection is used in our laboratory. Catkins, or male cones, are collected in the field just prior to shedding, brought into the laboratory, and placed in polyethylene bags that are tied at the top but not completely sealed. The bags are then placed in a small room with a dehumidifier where the temperature is between 35° and 40°C. The bags are periodically checked, and after dehiscence the bags are removed and pollen is strained through a series of sieves to insure a pure pollen sample. The viability of the pollen is then determined by *in vitro* germination tests, or by other viability assays.

Typical pollen yields are given in Table 1. The amount of pollen produced by a tree depends on the age and nutritional state of the tree. Strobili containing the greatest amounts of pollen grow near the top and outside of pine trees (Chira, 1966).

Table 1. Pollen yields. (After Snyder and Clausen, 1974)

Genus	Source	Quantity (ml)
Angiosperms		
Alnus	100 catkins	4
Betula	100 catkins	5 - 20
Fagus	100 flowers	1 - 1.6
Populus	100 catkins	50 - 100
Ulmus	100 flowers	0.1 - 0.4
Gymnosperms		
Larix	100 strobili	1.0 - 1.4
Pinus	100 strobili	7 - 33
Pseudotsuga	100 strobili	0.7 - 3.7

As mentioned previously, young strobili and catkins can be removed from trees and dehiscence accelerated. However, while male flowers of ash (Fraxinus) can be removed 15 weeks prior to maturity and development and dehiscence induced, pine and others are limited to considerably shorter periods (Table 2). If branches with developing male flowers are removed too early, chromosome aberrations may occur, dehiscence may be inhibited, or immature, nonviable pollen grains may be formed. Reviews of pollen collections methods are found in Snyder and Clausen (1974) and Stanley and Linskens (1974).

<u>Storage</u>. Pine pollen is recognized as the easiest and most stable pollen to store as compared with other forest species. Stability during storage is related to the nature of the outer layer of the pollen wall (exine) and pollen protoplast metabolism during storage. Pollen which metabolizes fewer of its soluble carbohydrates and organic acids during storage can retain its viability for longer periods (Stanley and Poostchi, 1962).

Temperature and water content of pollen are two major factors influencing viability after storage. Reduced moisture content is essential before storing pollen below 0°C. It is possible to freeze-dry (lyophilize) pollen of Douglas fir and maintain good seed set after storage under vacuum at room temperature for up to two years (Livingston and Ching, 1967).

Pollen of most species can be dried to about 5% moisture content at room temperature or under controlled conditions with desiccant (such as silica gel) and then placed with the desiccant at -20°C in a deep freezer or in the freezing compartment of a refrigerator. Generally,

Table 2. Acceleration of pollen dehiscence under controlled conditions

Genus	Temperature	Light Power (W)	Light Distance (m)	Light Time (h)	Weeks detached before natural shedding
Angiosperms					
Acer	22°C	500	1	15	12
Alnus	27°C	600	1	22	-
Corylus	10°C	-	1	-	3-4
Fagus	27°C	600	1	22	-
Fraxinus	22°C	500	1	15	12-15
Quercus	20-22°C	500	1	15	1-3
Gymnosperms					
Abies	27°C	600	1	20	3
Chamaecyparis	27°C	600	1	20	3
Larix	20°C	200	1.5	3	3
Picea	27°C	600	1	20	8
Pinus	27°C	600	1	20	4
Pseudotsuga	27°C	600	1	20	4

Data from Worsley (1960), Larsen (1958), and Johnson (1945).

this is the least expensive and easiest method for storing pollen. Acer pollen is, however, very sensitive to moisture during storage and becomes nonviable below 30% moisture content.

Two precautions are essential with low moisture content. First, minimize repeated exposure to room temperature followed by refreezing. Second, in using pollen that has been freeze-dried or desiccated, the pollen must be rehydrated with water vapor prior to germination, pollination or viability testing.

Pollen storage in organic solvents, for example acetone, benzene or petroleum ether, maintains viability, while more polar solvents, such as alcohol or chloroform, may inactivate pollen (Iwanami, 1972; Iwanami and Nakomura, 1972). Storage in organic solvents may improve viability of Acer pollen and pollen from other species after storage. In Gainesville, storage of pine and pear pollen in organic solvents has proved unsuccessful to date. As must be expected, each species of pollen has its own response characteristics. Additional access to the literature on pollen storage may be found in Stanley and Linskens (1974).

Viability. Fresh pollen usually germinates satisfactorily when placed in proper artificial media, or when placed on receptive stigmas. However, as noted earlier, many factors present in fresh pollen may be altered during storage. Therefore, *in vitro* germination tests, or other ways of assaying pollen viability are required prior to field application of stored pollen.

In vitro germination tests may be quite simple. Jovancevic (1962) observed under the low power microscope that the number of pollen grains of *Picea omorica*, *Salix alba* and *Quercus robur* displaying a shriveled appearance, were abnormally small, or had a modified exine color, equally the number of nonviable pollen grains. The conditions for *in vitro* germination tests for some woody species are given in Table 3.

Table 3. Optimum germinating conditions for *in vitro* viability tests

Species	Temperature °C	Nutrients	Media	Reference
Angiosperms				
Eucalyptus sp.	30	20% sucrose	1.5% agar	Boden (1958)
Corylus avellana	30	18% sucrose	no buffer water, pH 5.5	Hoekstra (1970)
Juglans nigra	27+	20% sucrose	water	Hall and Farmer (1971)
Fraxinus excelsa	21	15% sucrose	2.0% agar	Nikolaeva (1962)
Fraxinus ornus	21	5% sucrose	2.0% agar	Nikolaeva (1962)
Gymnosperms				
Pinus canariensis	27	None	water	Duffield (1954)
Pinus ponderosa	27	None	water	Duffield (1954)
Pinus elliottii	25	EDTA-FE	1.0% agar	Echols and Mergen (1956)
Pinus edulus (2n, 4n)	30	2% sucrose 0.01% B	1.0% agar	Chira (1967)
Larix sibirica	27	B,Ca,Mg,KNO_3	water	Ho and Rouse (1970)

Table sugar is usually sufficiently pure to affect good germination if sucrose is required in the germination medium. However, in pine and other species with slow rates of pollen tube growth, it is preferable to avoid using table sugar because of bacterial contamination. Pine pollen is often so heavily contaminated with fungal hyphae after 72 h of germination that it cannot be effectively counted. With slow-growing pollen, where fungal infection can be a problem, additional replicate samples are desirable; it is likely that some of the samples will not become infected. A fungistat, for example "Nystatin" (mycostatin), can be used at very low concentrations (0.01%), but such fungistats are generally expensive. Removing or killing contaminating microorganisms is possible, but time consuming and seldom done by forest geneticists.

The observation that boron is required for the germination of pollen of some species opened a new area of study. Ash pollen and pollen of most other angiosperm species and a few gymnosperm species, will respond to the addition of boron to the germination medium. Boron or calcium requirements should be determined for *in vitro* germination studies when confronted with "new" pollen.

Different techniques for pollen germination *in vitro* include: 1. the hanging drop, 2. spot tests, 3. well tests, 4. agar or gelatin media and 5. membrane substrates (Fig. 1). The concentration of pollen is critical in inducing specific interactions, the so-called "mutual stimulation effect". The pollen concentration must be established for each species. Below or above the optimal concentration, germination is usually restricted providing a false index of viability.

Care must be taken when counting a sample of pollen under the microscope. The "mutual stimulation effect" will cause localized areas of high germination frequency. Such areas are valid places to count for maximum pollen viability, but in these areas it may be difficult to determine pollen tube length. Addition of water or sucrose buffer can dilute the sample more evenly to facilitate counting. Another source of error in pollen counting may arise if the cover slip is dropped suddenly, moving ungerminated pollen grains to the outside edge of the cover slip, thus giving an incorrect count. In general for woody species, normal seed formation occurs when using pollen with at least 40% viability (Callaham and Duffield, 1961).

Although *in vivo* seed set assays of pollen viability are the most desirable viability scale, such tests are rarely performed. However, *in vivo* estimations of pollen germination ability are often determined. Pollen is directly applied to the stigma and pollen tube growth observed after a determined time interval by the use of the analine blue stain (stains callose in pollen tube walls) and a fluorescence microscope (Jensen, 1962; Berlyn and Miksche, 1976).

Simple *in vitro* staining reactions are often sufficient to provide an adequate assessment of pollen viability. Specific stains are not as accurate as germination tests; however, pollen demonstrating at least 50% viability by staining tests usually yields an acceptable seed set level. Tetrazolium reagents that react with enzymes are the best known staining assays for pollen viability.

Stains for specific cellular constituents may not yield as good as index of viability as redox dyes that react with enzymes. For example, simple staining reactions may lead to the conclusion that starch is absent from oak pollen; however, it may be still viable. Acridine orange gives a characteristic fluorescence with DNA (greenish-yellow)

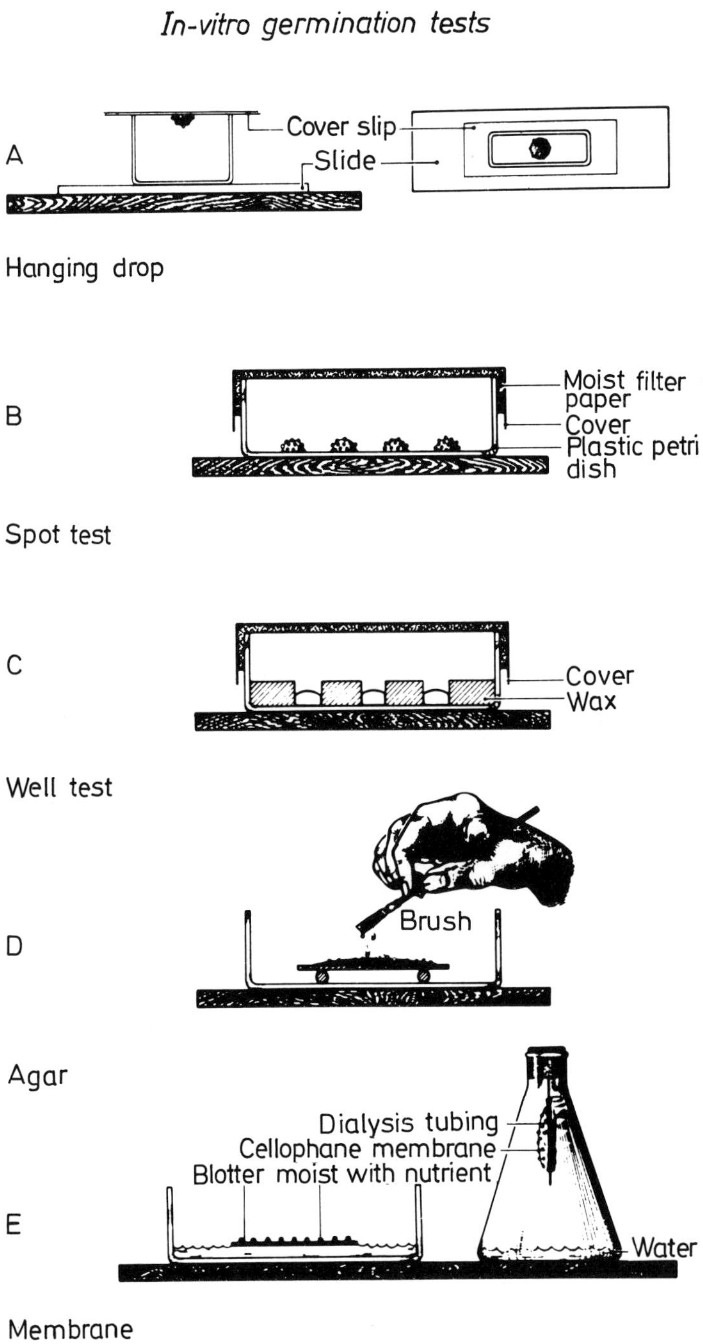

Fig. 1. Methods for germinating pollen *in vitro*. (From Stanley and Linskens, 1974)

and RNA (bright orange to red) of viable pollen grains of pine and birch when observed under the fluorescence microscope. Dead pollen fluoresces a dull yellow-green (Kozubov, 1967). Another fluorescent dye, fluorescein diacetate, when used at a concentration of 10^{-6} M enters the pollen and is hydrolyzed by esterase to yield a compound that fluoresces at 490 nm. Other compounds reacting with enzymes commonly used as viability indicators include those that react in an oxidation-reduction manner with specific substrates. A few examples of these topographical redox reactions are given in Table 4. The preferred compound is 2,3,5-triphenyltetrazolium chloride (TTC) (Cook and Stanley, 1960).

All pollen is not equally sensitive to the TTC assay. Some chemical reactions may reduce TTC and yield a false color reaction. However, Picea and Pinus pollen test fairly accurately by TTC, or at least with sufficient agreement with *in vitro* germination and seed set to suggest wider wise use of this color test. Other redox dyes, such as nitroblue TTC or resazurin, may also be effective in assaying pollen viability.

Pollen Treatment and Application

Promoters of Pollen Tube Growth. Pollen contains many organic substrates and constituents which are altered or undergo slow respiration during storage. If soluble sugars decrease during storage, tube growth can be enhanced if sugars are supplied by the female tissue or in the *in vitro* medium. Pollen from many species grow slowly, or not at all, without added sugars or cofactors, such as boron, calcium, manganese, magnesium or a growth regulator such as indole acetic acid or gibberellic acid.

Auxins or growth regulators released by pollen can influence the formation of the abscission layer and prevent cone drop. Michalski (1967) and Sweet and Lewis (1969) showed that pine pollen contains and releases considerable amounts of these substances. Addition of growth substances to the female tissue may permit a slow-growing incompatible pollen to overcome incompatibility and set seed.

Matsubara (1973) reported that 1.0% or 0.1% kinetin, applied to styles after pollination as a paste mixed with lanolin, is successful in overcoming self-incompatibility in a cultivar of *Lilium longiflorum*. Partial success in obtaining seed from incompatible crosses in Lilium has resulted from the addition of naphthalene acetamide (an auxin) (Emsweller and Stuart, 1948) and potassium gibberellate (Emsweller et al., 1960). Similar treatments may be helpful in aiding pollen growth and overcoming incompatibility in forest species.

Barriers to pollen tube growth *in vitro* can be overcome by adding stimulatory chemicals to the growth medium and *in vivo* by adding the factor directly to the pollen grains. However, when pollen is placed in aqueous solutions, substances located in the pollen walls are easily eluted. Such compounds include enzymes, carbohydrates, amino acids, proteins and others (Mäkinen and Brewbaker, 1967; Stanley and Search, 1971; Kirby and Smith, 1974). These compounds may effect germination capacity and incompatibility (Knox et al., 1972). Re-drying of pollen after such treatments does not restore germination capacity. Pine pollen is one of the few species releasing only slight amounts of elutable proteins (Stanley and Search, 1971).

Probably the easiest and most effective method of chemical pretreatment is to mix chemical additives as powders with dry pollen and then apply the dry mix to female flowers or test it for germination *in vitro*. Such added chemicals stimulate pollen tube growth and, thus, facilitate seed set. They can also function as fillers to stretch pol-

Table 4. A few redox dyes that can be used in determining pollen viability

Reagent	Formula	Color Change	Enzyme Involved
2,3,5 triphenyltetrazolium chloride (TTC)	$C_6H_5CN_2C_6H_5N_2C_6H_5Cl$	Colorless to red	Reductase
Triphenyl tetrazolium red		Colorless to red	Reductase
Sodium biselenite	$NaHSeO_3$	Colorless to red	Not known
Diaminobenzidine	$NH_2C_6H_4C_6H_4NH_2$	Ligh yellow to blue	Peroxidase
Guaiacol	$OHC_6H_4OCH_3$	Colorless to brown-red	

len supplies. The growth response of pollen of angiosperm species to chemical pretreatment is generally greater than that of slow-growing gymnosperm pollen species.

Mentor Pollen. Mentor pollen is dead or non-functioning pollen which contributes certain factors influencing the growth of functioning pollen. Mentor pollen can contribute to the formation of a seed even though the male cell of the mentor does not fuse with the egg cell. Polyakov (1964) reviewed the effects of mentor pollen on the fertilization and developing zygotes of hybrids. Li (1960) applied multiple pollinations to *Populus nigra* at 12-h intervals with *Populus berolian-sis* pollen mixed with the *P. nigra* mentor, resulting in increased seed set and seedling vigor.

Use of pollen mixes and repeated pollinations suggests that the induction of changes facilitated by mentor pollen can occur during 1. progamic phase, when pollen first interacts with the female tissue; 2. gamic phase, when fertilization occurs; 3. postgamic phase, when pollen may influence the developing embryo. Dead pollen can also serve as mentor pollen, helping to give rise to hybrids.

Sufficient amounts of gamma rays and X-rays can kill pollen. However, such treatments do not effect diffusable pollen wall compounds. Irradiation at certain levels in some species can stimulate pollen tube growth. Livingston and Stettler (1973) found that Co^{60} or X-rays up to 63 Krad stimulated pollen tube elongation in Douglas fir. Mentor pollen which has been killed by radiation can be mixed with untreated pollen to influence pollen tube growth and seed set. Studies by Stettler (1968) demonstrated that the use of radiation killed mentor pollen is a good method for obtaining hybrids in black cottonwood (*Populus trichocarpa*). Branches from female individuals of *P. trichocarpa* were transferred to the greenhouse in the dormant state. When dormancy was broken, inflorescences were treated with a pollen mix consisting of a ratio of 2:1 (by weight) of foreign pollen with pollen of *P. trichocarpa* irradiated at 100 Krad. Pollen from five different secondary species was used and hybrids were obtained from all crosses. Fertilization is effected by the pollen of the foreign incompatible species, since genetically inactive irradiated pollen cannot fertilize the ovule. Stettler (1968) theorized that mentor pollen functions in providing a physiological stimulus which results in catkin retention.

The key to the mechanism by which mentor pollen acts in combination with foreign, usually incompatible, pollen may lie in compounds contained within the walls of pollen grains. Because of the presence of proteins and enzymes within pollen walls, Tsinger and Petrovskaya-Baranova (1961) proposed that pollen walls are physiologically active, living structures. Mäkinen and Brewbaker (1967) and Stanley and Search (1971) have shown that compounds contained within pollen walls are very mobile. Furthermore, cytological studies have shown that enzymes are associated with the intine layer of the pollen wall (Knox and Heslop-Harrison, 1969, 1970) and proteins are found in the cavities of the exine as well (Heslop-Harrison et al., 1973, 1974). There appears to be some genetic control of the quantities of proteins and carbohydrates contained within pollen walls of binucleate and trinucleate pollen species (Kirby and Smith, 1974). The function of pollen wall compounds as recognition substances in interspecific incompatibility in poplars has been demonstrated by Knox et al. (1972). This group was successful in obtaining fertilization and hybrid seed from previously incompatible crosses by adding elutable protein components from compatible pollen grains. This work is strong support for the participation of pollen wall compounds in control of incompatibility as well as control of pollen germination and tube growth.

(A) DIRECT TECHNIQUE

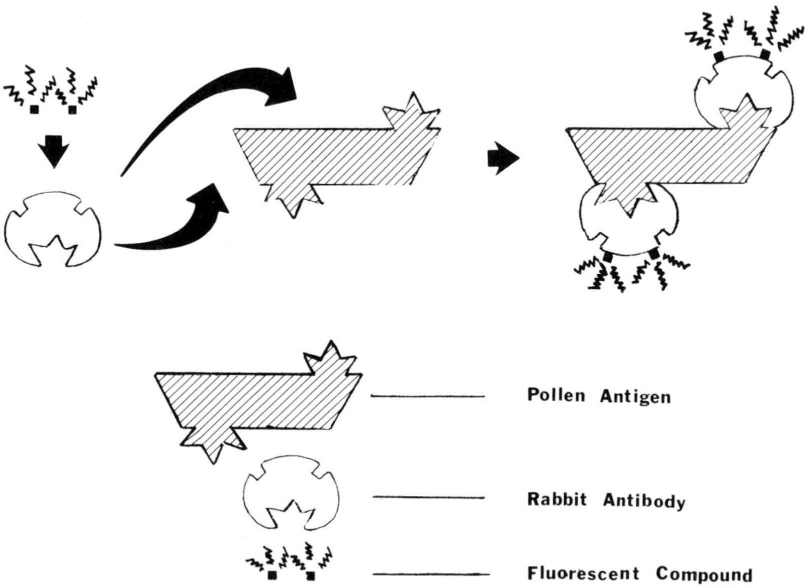

(B) INDIRECT TECHNIQUE

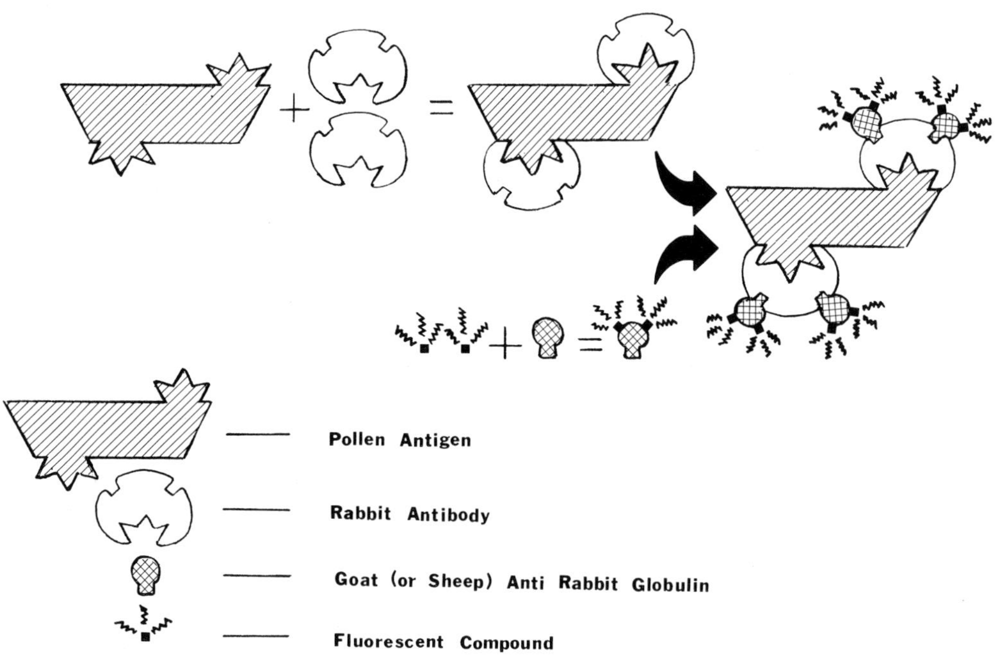

Legends see opposite page.

Precise localization of pollen wall compounds has been a topic of considerable interest. Knox et al. (1970) adapted the immunofluorescence technique in localizing pollen wall proteins. Briefly described, the technique involves extracting large samples of pollen with buffered saline and immunizing rabbits with freshly prepared extracts. Serum from sensitized rabbits is collected. Two techniques for coupling serum antibodies with fluorescent labeled molecules have been employed. In the direct technique rabbit antibodies are conjugated directly with a fluorescent compound, such as fluorescein isothiocyanate (FITC) or rhodamine isothiocyanate (RITC) (Fig. 2A). When applied to sections or whole pollen grains, the labeled antibody binds to the pollen antigen and precipitates wherever antigens appear. The indirect technique involves preparation of a fluorescent labeled globulin against rabbit serum (Fig. 2B). Such globulins are usually produced in sheep or goats and then conjugated with the fluorescent label, FITC or RITC. Rabbit antibodies are allowed to bind to pollen antigens. Fluorescent labeled anti-rabbit globulins are placed over the preparation resulting in the fluorescent labeled fraction combining with the rabbit antibody-pollen antigen complex. Fluorescent label appears wherever the rabbit antibody-pollen antigen complex is found. Under most circumstances the indirect technique for localization is preferred; the conjugation of fluorescent label need only be done once, to sheep or goat globulins, since the specificity of the technique lies with the rabbit antibodies. Fortunately, fluorescent labeled goat and sheep anti-rabbit globulins are commercially available.

Preparations made by using the immunofluorescence technique are observed under the ultra-violet microscope, with the light source adjusted to the exciting wavelength of the labeled fluorescent molecule, fluorescein or rhodamine. For an excellent description of fluorescent antibody techniques, the reader is referred to Goldman (1968).

Application of Pollen. In forest tree breeding programs pollen can be applied to trees in micro and macro quantities. It can also be applied as pure pollen or mixed with inert diluents, such as talc, dead pollen or pollen from another species (mixed pollination). By diluting pollen with genetically inactive substances, one can stretch the supply to cover many more trees and still set seed at a reasonable level. As little as 30% viable pine pollen sets normal quantities of seed (Callaham and Duffield, 1961; Callaham, 1966).

Common methods for applying pollen in tree breeding programs include hand brush, syringe, pole dusting, power spray dusting, water spray, shotgun shells and even airplane dusting.

Fig. 2. (A) The direct technique for immunofluorescent localization of pollen antigen. The fluorescent compound is conjugated directly to the specific rabbit antibody produced in response to the pollen antigen. The labeled antibody will bind to sites containing pollen antigen. (B) The indirect technique for immunofluorescent localization of pollen antigens. A general globulin against rabbit serum is prepared in a second animal (sheep or goat). This globulin is conjugated with fluorescent compound. Specific rabbit antibodies are allowed to bind to pollen antigens. Fluorescent labeled anti-rabbit globulins are placed over the preparation, resulting in the fluorescent labeled fraction combining with the rabbit antibody-pollen antigen complex. (After Goldman, 1968)

A simple pollination device can be made from a 10 or 25 ml hypodermic syringe by removing the plunger and placing a rubber bulb over the open end (Fig. 3A). Pollen is placed in the syringe and sprayed on female flowers by squeezing the bulb (Fig. 3B). Pole dusters extend the spray action of the syringe type over the length of the pole and, thus, into the crown of trees. The United States Forest Service has adapted a portable air compressor to supply power to pole dusters. Power sprays and shotgun shells loaded with pollen disperse pollen into the tops of trees where most conelets or flowers are located. Tree pollens which are not sticky or clumped together are most easily handled.

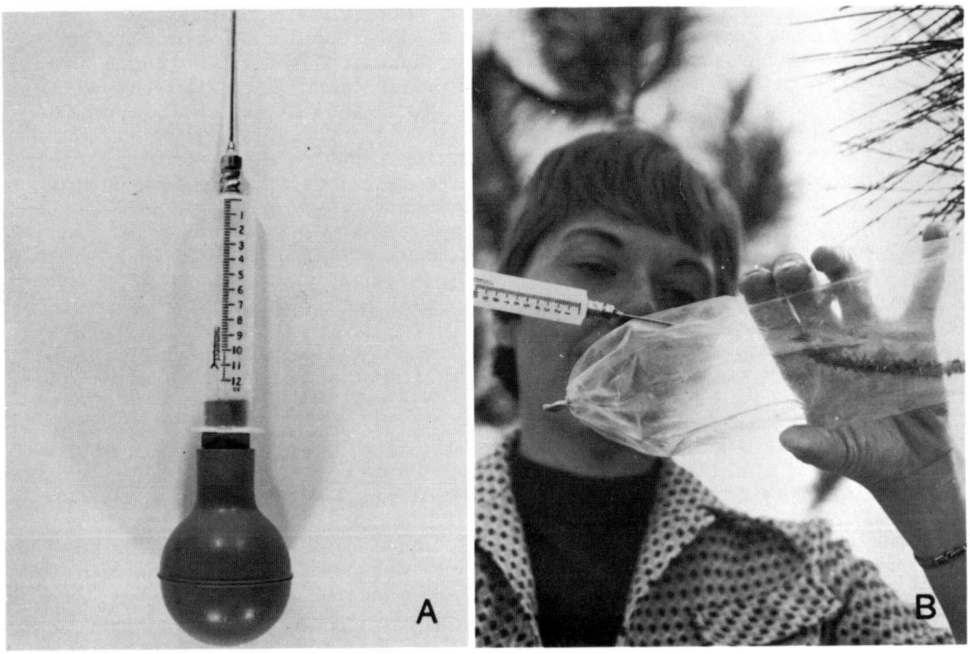

Fig. 3 A and B. A simple device for application of pollen in the field. (A) 10 ml syringe fitted with rubber ball. (B) Field application. Female cones have been isolated by the plastic bag

Pollen dispersed in a water mist applied from a back pack or pump truck can facilitate fertilization in some species, particularly gymnosperms. However, in general, viability of pollen in angiosperm forest tree species is decreased when pollen is placed in an aqueous solution, in which wall-held compounds can diffuse from pollen surfaces, as mentioned earlier.

Application of pollen by shotgun shells, airplanes or helicopters is expensive and such methods are uneconomical in normal seed orchard management.

References

Barner, H., Christiansen, H.: On the extraction of forest tree pollen from inflorescences forced in a specially designed house. Silvae Genetica 7, 19-21 (1958).

Berlyn, G.P., Miksche, J.P.: Botanical microtechnique and cytochemistry. Iowa State University Press (1976).

Boden, R.W.: Handling and storage of pollen in Eucalyptus breeding. Aust. Forestry 22, 73-81 (1958).

Callaham, R.Z.: Hybridizing pines with diluted pollen. In: Proc. 8th Southern Conf. of Forest Tree Improvement, Macon, Georgia, pp. 110-111 (1966).

Callaham, R.Z., Duffield, J.W.: Stretching the pollen supply. J. Forestry 59, 204-207 (1961).

Chira, E.: The effect of age and position in the crown on pollen quality of *Pinus sylvestris*. Sb. Vys. Sk. Zemed. Brne. 35, 339-343 (1966).

Chira, E.: Acceleration of the pollen development of *Picea excelsa* under laboratory conditions. Biologia (Bratislv.) 22, 260-265 (1967).

Cook, S.A., Stanley, R.G.: Tetrazolium chloride as an indicator of pine pollen germinability. Silvae Genetica 9, 134-136 (1960).

Duffield, J.W.: Studies on extraction, storage, and testing of pine pollen. Z. Forstgenet. 3, 39-45 (1954).

Echols, R.M., Mergen, F.: Germination of slash pine pollen *in vitro*. Forestry Sci. 2, 321-327 (1956).

Emsweller, S.L., Stuart, N.W.: Use of growth regulating substances to overcome incompatibilities in Lilium. Proc. Amer. Soc. Hort. Sci. 51, 581-589 (1948).

Emsweller, S.L., Urhing, J., Stuart, N.W.: The roles of naphthalene acetamide and potassium gibberellate in overcoming self-incompatibility in *Lilium longiflorum*. Amer. Soc. Hort. Sci. 75, 720-725 (1960).

Fielding, J.W.: Notes on the dispersal of pollen by Monterey pine. Aust. Forestry 21, 17-21 (1957).

Goldman, J.: Fluorescent antibody methods. New York, London: Academic Press 1968.

Hall, G.C., Farmer, R.E.: *In vitro* germination of black walnut pollen. Can. J. Botany 49, 799-802 (1971).

Heslop-Harrison, J., Heslop-Harrison, Y., Knox, R.B., Howlett, B.: Pollen wall proteins: "gametophytic" and "sporophytic" fractions in the pollen walls of the Malvaceae. Ann. Botany 37, 403-412 (1973).

Heslop-Harrison, J., Knox, R.B., Heslop-Harrison, Y.: Pollen wall proteins: exine and intine held fractions associated with the incompatibility response in the Cruciferae. Theoret. Appl. Genetics 44, 133-137 (1974).

Ho, R.H., Rouse, G.E.: Pollen germination of Siberian larch (*Larix sibirica*). Can. J. Botany 48, 213-215 (1970).

Hoekstra, F.A.: Respiration and vitality of bi- and trinucleate pollen. Inc. Newsletter 3, 52-54 (1970).

Iwanami, Y.: Retaining viability of *Camelia japonica* pollen in various organic solvents. Plant Cell Physiol. 13, 1139-1141 (1972).

Iwanami, Y.: Acceleration of the growth of *Camelia sasangua* pollen by soaking in organic solvent. Plant Physiol. 52, 508-509 (1973).

Iwanami, Y.: Storage in organic solvent as a means for preserving viability in pollen grains. Stain Tech. 47, 137-139 (1972).

Jensen, W.A.: Botanical Histochemistry. San Francisco, London: W.H. Freeman 1962.

Johnson, L.P.V.: Development of sexual and vegetative organs on detached forest tree branches cultures in the greenhouse. Forestry Chron. 21, 130-136 (1945).

Jovancevic, M.: Determining the germinative potential of forest tree pollen by the size, form and color of pollen grains. Nar. Sumar. 16, 493-502 (1962).

Kirby, E.G., Smith, J.E.: Elutable substances of pollen grain walls. In: Fertilization in Higher Plants (ed. H.F. Linskens), pp. 127-130. Amsterdam: North-Holland 1974.

Knox, R.B., Heslop-Harrison, J.: Cytochemical localization of enzymes in the wall of the pollen grain. Nature 223, 92-94 (1969).

Knox, R.B., Heslop-Harrison, J.: Pollen wall proteins: location and enzymic activity. J. Cell Sci. 6, 1-27 (1970).

Knox, R.B., Heslop-Harrison, J., Reed, C.E.: Localization of antigens associated with the pollen grain wall by immunofluorescence. Nature 225, 1066-1068 (1970).

Knox, R.B., Willing, R.R., Ashford, A.E.: Role of pollen wall proteins as recognition substances in interspecific hybridization in poplars. Nature 237, 381-383 (1972).

Kozubov, G.M.: Luminescent method for studying the pollen of plants. Botan. Zhur. 52, 1156-1157 (1967).

Larson, P.R.: Effect of gibberellic acid on forcing hardwood cuttings for pollen collection. USDA, For. Serv. Lake States For. Expt. Sta. Tech. Note 538 (1958).

Li, S.W.: Repeated pollinations in poplar breeding. Vestn. Sel'skokhoz. Nauk., Moskova 5, 1940-1943 (1960).

Livingston, G.K., Ching, K.K.: The longevity of freeze-dried Douglas-fir pollen. Silvae Genetica 16, 89-129 (1967).

Livingston, G.K., Stettler, R.F.: Radiation-induced stimulation of pollen tube elongation in Douglas fir. Radia. Botany 13, 65-72 (1973).

Mäkinen, Y., Brewbaker, J.L.: Isoenzyme polymorphism in flowering plants. I. Diffusion of enzymes out of intact pollen grains. Physiol. Plant. 20, 477-482 (1967).

Matsubara, S.: Overcoming self-incompatibility by cytokinins treatment on *Lilium longiflorum*. Botan. Mag. 86, 43-46 (1973).

Michalski, L.: Growth regulators in the pollen of pine. Acta Soc. Botan. Pol. 36, 475-481 (1967).

Nikoleava, Z.V.: The pollen of ash (Faxinus). Tashkert. Gos. Univ. Lenin. Nauk Trudy 204, 115-127. (Transl. U.S. Dept. Comm. II No. 68-50430)(1962).

Polyakov, I.M.: New data on the use of radioactive isotopes in studying fertilization of plants. In: Pollen Physiology and Fertilization (ed. H.F. Linskens), pp. 194-199. Amsterdam: North-Holland 1964.

Snyder, E.B., Clausen, K.E.: Seeds of woody plants of the United States. Chap. 3, U.S. Dept. Agri., Agric. Handbook 450 (1974).

Stanley, R.G., Kirby, E.G.: Shedding of pollen and seeds. In: Shedding of Plant Parts (ed. T. Kozlowski), pp. 295-340. London, New York: Academic Press 1973.

Stanley, R.G., Linskens, H.F.: Pollen: Biology, Biochemistry, Management. Berlin-Heidelberg-New York: Springer 1974.

Stanley, R.G., Poostchi, I.: Endogenous carbohydrates organic acids and pine pollen viability. Silvae Genetica 11, 1-3 (1962).

Stanley, R.G., Search, R.W.: Pollen protein diffusates. In: Pollen Development and Physiology (ed. J. Heslop-Harrison), pp. 174-176. London: Butterworths 1971.

Stettler, R.F.: Irradiated mentor pollen: its use in remote hybridization in black cottonwood. Nature 219, 746-747 (1968).

Chapter 12
Tissue Culture of Trees

L. Winton and O. Huhtinen

Introduction

Callus production, cell cultures, and protoplast fusion offer exciting possibilities for augmenting current tree-improvement programs. One of the main limitations of tree breeding is the relatively slow life cycle of trees and the minimum of 15-20 years between generations. Recent improvement in selection, hybridization and fertilizer application will undoubtedly increase the production of wood for timber and pulp; however, the advantage of callus and cell culture is the rapid vegetative propagation of trees that are difficult to root or graft by conventional methods. Selection of 3-4 year old vigorous seedlings with high photosynthetic efficiency, rapidly propagated by culturing methods, may possibly reduce by as much as 75% the time to obtain improved planting stock for reforestation.

The economic potential of the aseptic culture of forest trees is still at the dawn of exploitation. However, significant advances have been made during the past decade in regenerating a few hardwood tree species, and currently, workers are attempting to regenerate conifer trees. A glossary of tissue culture terms is given in Table 1, to help the new student in his understanding of this chapter.

Table 1. Glossary of tissue culture terms

Adventitious:	Roots, shoots, embryos, or other organs or tissues developing in an abnormal position.
Allotetraploid:	Plant with four sets of chromosomes, usually with two sets from each parent.
Anther:	Microsporangium bearing microspores which develop into pollen grains (microgametophytes).
Anther culture:	The aseptic culture of anthers for the production of haploid plants from microspores.
Aseptic culture:	Surface sterilization of parental explants, free from pathogens, but not necessarily free of internal symbionts.
Asexual propagation:	Vegetative, somatic nonsexual propagation of plant parts without fertilization.
Autotetraploid:	Plant with four identical sets of chromosomes, originating from duplications of one set from one parent.

Auxins:	Plant growth hormones of several types which cause cell enlargement, apical dominance and root initiation. One natural auxin is IAA (indoleacetic acid).
Axenic culture:	Free of symbionts: not possible with surface sterilization, and incorrectly used for aseptic culture.
Callus culture:	Proliferation from a parental explant of many cells in protoplasmic continuity, but having no equivalence with any normal tissue. Same as tissue culture.
Cell differentiation:	Internal chemical or ultrastructural changes preceding or accompanying specialization of function.
Cell suspension:	Culture of single cells in moving liquid medium, often used incorrectly to describe suspension cultures of cells and cell aggregates.
Clonal propagation:	Asexual propagation from one individual (ortet) of many new plants (ramets) all with the same genotype.
Cytokinin:	A class of plant growth hormones associated with cell division and cell differentiation.
Diploid:	Having twice the number of chromosome sets found in normal sex cells, and reconstituted during fertilization or the somatic hybridization of haploid cells as the normal somatic (diploid) number of chromosome sets per cell.
Embryo:	The young plant developing in the megagametophyte from the fertilization of an egg cell, or without fertilization. In aseptic cultures, adventitious embryos show polarization followed by the growth of a shoot from one end and a root from the other end.
Embryoid:	A cell group approximating an embryo, but with a more random arrangement of cells.
Excise:	Cut or isolate callus tissue from its parental explant, or to remove adventitious shoots from callus tissue for rooting.
Explant:	A plant part excised and prepared for aseptic culture by surface sterilization followed by the exposure of live cells to a nutrient medium.
Fertilization:	The normal union of two gametes during sexual reproduction.
Gamete:	A sex cell, or nucleus, derived from a gametophyte and containing a haploid set of chromosomes, or one-half the number in somatic cells.
Genotype:	The genetic makeup of an individual carried in the chromosomes.

Haploid: One-half the number of chromosome sets found in somatic cells, forming the typical chromosome number in sex cells, or gametes.

Hybridization: The production of offspring of genetically different parents, normally from sexual reproduction but also asexually by the fusion of protoplasts or by transformation.

Inoculum: A small piece of tissue cut from callus, or a small amount of cell material from a suspension culture placed in contact with fresh medium for continued growth of the culture. Inocula (plural).

Karyotype: The characteristic number, size and shape of the chromosomes in a haploid set.

Megagametophyte: Produces the female egg, or sex cell. In gymnosperms, it also produces a nutritive tissue composed of haploid cells.

Meristem: A localized group of cells, actively dividing and undifferentiated but ultimately giving rise to permanent tissue such as shoots, roots, wood or bark.

Meristimoid: A localized group of cells in callus tissue, characterized by an accumulation of starch, RNA and protein, and giving rise to advantitious shoots or roots.

Microspore: A haploid cell ultimately developing into a pollen grain (microgametophyte) that produces male gametes.

Nutrient medium: A solid or liquid combination of major and minor salts, an energy source (sucrose), vitamins, hormones, and occasionally other defined or undefined supplements. Usually made up from previously prepared stock solution, then sterilized by autoclaving or filtering through a micropore filter. Media (plural).

Organ culture: The aseptic culture of intact plant parts or organs that do not proliferate but maintain their original form.

Organized tissue: Composed of regularly differentiated cells.

Organogenesis: The initiation of adventitious shoots or roots from callus tissue, or embryos from single cells or cell aggregates.

Parasexual hybridization: Asexual fusion of protoplasts from genetically different parents. Also called somatic, or vegetative, hybridization.

Parenchymatous: Generally spherical and undifferentiated cells with primary cell walls, capable of division and differentiation.

Passage: The duration of growth of callus or cell material from one subculture to another.

Plantlet:	A small rooted shoot. Also, a piece of callus having both roots and shoots that are not connected together inside the callus.
Ploidy:	The level of polyploidy, or the number of sets of chromosomes per cell over two sets.
Polyploid:	Containing three or more sets of chromosomes per cell, and also referred to specifically as triploid, tetraploid, etc.
Protoplast:	A spherical cell without a wall, surrounded by a membrane.
Somatic:	Referring to vegetative, or nonreproductive, tissue.
Subculture:	Cutting solid callus into small cubes (inocula) for transfer to fresh medium. Sometimes used to denote the adding of fresh liquid medium to a suspension culture.
Suspension culture:	Cells or cell aggregates dispersed and growing in moving liquid medium.
Tissue culture:	Now used interchangeably with callus culture, but originally used to describe what is now called organ culture.
Transformation:	Specifically, the transfer of genetic information to a plant cell by DNA isolated from another cell, but now a general term for genetic transfer by different means, also called transgenosis.
Vacuole:	A storage area in a cell surrounded by a membrane.
Vegetative propagation:	Same as asexual, or nonsexual, propagation.

We will present techniques that will enable the new student to establish and maintain various types of cultures in the laboratory. Tree species which have been cultured by aseptic methods are listed in recent reviews by Durzan and Campbell (1974), Murashige (1974), and Winton (1972a, b, 1973, 1974a). Tree species that have been regenerated as rooted shoots or trees are listed in Table 2.

The aseptic culturing of somatic, or vegetative, tissues will be emphasized for the primary purpose of vegetatively propagating forest tree species. Some attention will be given to the culture of anthers for the production of haploid plants, but the culturing of pollen, megagametophytic tissue, stem tips, apical meristems, buds, or seed embryos will not be presented. These have been dealt with to some degree in the reviews cited above, and in greater detail in specific chapters in recent books edited by Street (1973a, 1974) and Kruse and Patterson (1973). We will also preview the future uses of protoplasts for somatic, or asexual, hybridization and for gene introduction through transformation.

Table 2. Tree species regenerated from somatic callus cultures[a]

Betula pendula	European birch	Huhtinen and Yahyaoglu, 1974
Broussonetia kazinoki	Paper mulberry	Oka and Ohyama, 1972
Citrus maxima (grandis)	Shaddock	Chaturvedi and Mitra, 1974
Citrus sinensis	Sweet orange	Chaturvedi and Mitra, 1974
Elaeis guineensis	Oil palm	Jones, 1974
Eucalyptus citriodora	Lemon-scented gum	Aneja and Atal, 1969
Populus canescens	Gray poplar	Chalupa, 1974
Populus deltoides	Eastern cottonwood	Berbee and Hildebrandt, 1972
Populus euroamericana	Carolina poplar	Berbee and Hildebrandt, 1972
Populus nigra	Black poplar	Berbee and Hildebrandt, 1972; Chalupa, 1974
Populus nigra var. *typica*	Black poplar	Chalupa, 1974
Populus nigra var. *italica*	Lombardy poplar	Venverloo, 1973
Populus tremula	European aspen	Chalupa, 1974
Populus tremula, tetraploid	European aspen	Winton, 1971
Populus tremuloides	Quaking aspen	Wolter, 1968
Populus tremuloides, triploid	Quaking aspen	Winton, 1968, 1970; Mathes, 1968
Populus tremuloides x *alba*	(hybrid)	Winton, 1972b
Prunus amygdalus	Almond	Mehra and Mehra, 1974
Ulmus americana	American elm	Durzan and Lopushanski, 1975

[a]Regeneration is defined here as the production of shoots which rooted directly while still attached to callus or after they were excised. Not included are callus shoots that were not directly connected to a root, sometimes referred to as plantlets.

The purpose, therefore, of this presentation is to briefly discuss the latest methods and techniques used to grow vegetative tree parts in aseptic culture, and suggest uses of such cultures in both present and future tree-breeding programs.

Callus Cultures from Somatic Tissues

The terms tissue culture and callus culture have been used interchangeably to describe the proliferation from a parental explant of many cells in protoplasmic continuity (Street, 1974). The derived tissue, or callus, has no equivalence with any normal plant tissue. Such cultures have been referred to by many workers as being composed of uniform and undifferentiated cells; when, in fact, some or many cells have been highly differentiated. Murashige (1974) cautions against characterizing callus tissue as undifferentiated when one really means unorganized. Callus may proliferate from parental cells in the cambium, cortex or pith regions of stem sections, but in culture the callus cells are mainly parenchymatous, with primary walls and large storage vacuoles, and are generally spherical in newly initiated cultures.

Callus can be subcultured (cut into small cubes and transferred to fresh medium) at intervals of 3-5 weeks, and can thus be maintained indefinitely in the laboratory on a nutrient medium balanced for each particular species. The tissue or cells can be used in biochemical studies of metabolism, enzyme systems, carbohydrate and lignin production and deposition, as well as for pathogen-host interactions, the production of secondary products for the drug industry, and for the vegetative propagation of selected trees. The callus remains unorganized in all but the last of the above uses. For the production of adventitious shoots or roots, Torrey (1966) proposed that groups of cells within the callus organize into meristemoids that eventually give rise to organs. Thorpe and Murashige (1968, 1970) showed that heavy starch accumulations occur in cells in the region of shoot initiation, before any organized structure (meristemoid) is visible microscopically. Protein and RNA (but not DNA) also accumulate in smaller amounts, and by the time meristemoids are formed their cells change from having large vacuoles to having none. Adventitious shoots usually root easily when excised from the callus and placed in a rooting medium, and occasionally shoots will grow their own roots while still attached to callus (Winton, 1970).

The main factors in establishing a callus culture are the collection and storage of material until use, surface sterilization, preparation of parental explants, nutrient medium, incubation, growth and isolation of callus from explants, subcultures, and the selection and establishment of the final stock culture. Fig. 1 outlines the main steps in tree regeneration from callus cultures.

Seedlings. Huhtinen and Yahyaoglu (1974) developed a method of callus initiation for both Norway spruce and European birch. They collect the top few centimeters of green and succulent growth and remove the leaves from the firmer 2-3 cm of stem. The stem segment is surface sterilized in 70% ethanol for 3-5 min, then rinsed 3-5 times in sterile water. All of the outer cortical tissue is removed, leaving only the woody cylinder 2-3 mm in diameter. This cylinder is cut into discs 2-3 mm thick, which are placed flat on the surface of nutrient medium at 5-7 discs per Petri dish.

Each dish is sealed with Parafilm (which is cheaper and as functional as Teflon tape), and the discs are incubated in the dark at 25 ± 2°C, on Murashige and Skoog (1962) medium containing 25.0 mg/l (milligrams per liter) IAA (indoleacetic acid) and 0.5 mg/l kinetin. Table 3 gives the composition of Murashige and Skoog (1962) medium without hormones (auxins or cytokinins), and will be referred to in this chapter as MS medium.

After callus is initiated in the dark, it is isolated and transferred to fresh medium in continuous fluorescent light of 1000-2000 lux. Shoots grew from birch callus, and were excised and rooted in the same medium without kinetin, but containing 0.1 mg/l 2,4-D (2,4-dichlorophenoxyacetic acid). From the callus of Norway spruce there were only found three normal-appearing needle-shoots that grew from

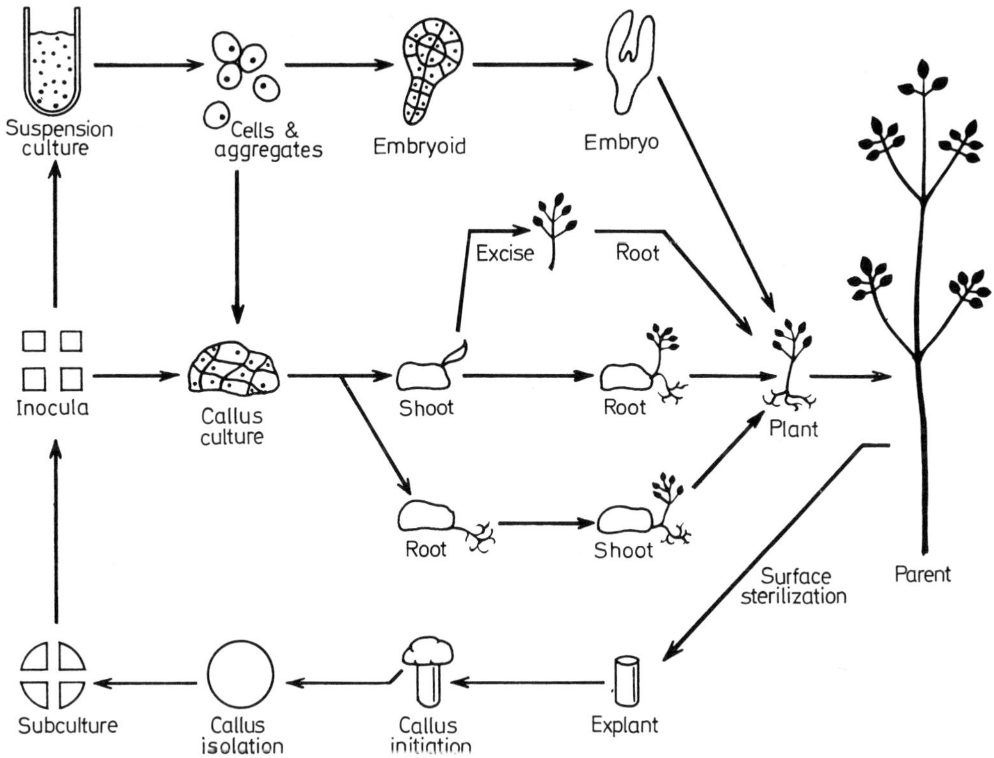

Fig. 1. Regeneration of trees from somatic callus or suspension cultures. (Drawing courtesy of Mr. Steve Newcombe)

three separate callus masses (each in a separate flask) in 1973. The shoots elongated to 2-3 cm in length, but all eventually died. However, they are the first shoots to grow from subcultured, somatic callus tissue of a conifer tree species since Ball (1950) grew the first conifer buds from redwood callus.

Root Sprouts. Subsurface roots 1 inch or less in diameter are collected from aspen trees, scrubbed in soap and water, cut to sections 10-15 cm long, and planted in a 1:1 sand-vermiculite mixture and watered sparingly (Winton and Mathes, 1973). Root sprouts are excised from the root sections when 3-5 cm tall, and planted in the same mixture in flat dishes covered with plastic bags. Rooted sprouts are then transplanted to uncovered 3-inch pots and grown to 20-40 cm in height.

When stems are collected for callus initiation, the basal 5-10 cm are discarded to reduce contamination from endogenous bacteria and the top 3-5 cm are discarded if the growth is too succulent. Leaves are removed and each stem is cut into sections 2-3 cm long. These are surface sterilized in bleach containing 5-6% sodium hypochlorite, plus a few drops of a surfactant such as Tween-20, and shaken occasionally in a flask for 12-15 min.

After the bleach is decanted and the stem sections rinsed in sterile water 3-4 times, internode segments are cut 5-7 mm long and plunged

Table 3. Components of nutrient media in milligrams per liter[a]

Component	WS Wolter and Skoog, 1966	MS Murashige and Skoog, 1962	LS Linsmaier and Skoog, 1965	BL Brown and Lawrence, 1968
NH_4NO_3	50	1,650	1,650	1,650
KNO_3	170	1,900	1,900	1,900
$MgSO_4 \cdot 7H_2O$	764	370	370	370
KH_2PO_4	-	170	170	170
$CaCl_2 \cdot 2H_2O$	-	440	440	440
Na_2SO_4	425	-	-	-
$Ca(NO_3)_2 \cdot 4H_2O$	425	-	-	-
KCl	140	-	-	-
$NaH_2PO_4 \cdot H_2O$	35	-	-	-
$MnSO_4 \cdot H_2O$	9	22.3	22.3	16.9*
$ZnSO_4 \cdot 7H_2O$	3.2	8.6	8.6	10.6*
H_3BO_3	3.2	6.2	6.2	6.2
KI	1.6	0.83	0.83	0.83
$Na_2MoO_4 \cdot 2H_2O$	-	0.25	0.25	0.25
$CuSO_4 \cdot 5H_2O$	-	0.025	0.025	0.025
$CoCl_2 \cdot 6H_2O$	-	0.025	0.025	0.025
Fe(EDTA)[b]	5.6	5.6	5.6	5.6
Nicotinic acid	0.5	0.5	- *	0.5
Thiamine · HCl	0.1	0.1	0.4*	0.1
Pyridoxine · HCl	0.1	0.1	- *	0.1
Inositol	100	100	100	100
Asparagine	-	-	-	100 *
Glycine	-	2	- *	- *
Sucrose	20,000	30,000	30,000	30,000
Agar	10,000	10,000	10,000	8,000 *

[a]Medium is made with double-distilled water, adjusted to pH 5.8, then autoclaved for 20 min at 130°C and 15 psi.

[b]5.6 mg/l Fe are contained in 5 ml/l of a stock solution made with 5.57 g $FeSO_4 \cdot 7H_2O$ and 7.45 g Na_2-EDTA (disodium ethylenedinitrilo-tetraacetate) per liter of water.

*Changes from MS medium.

basal end up into Wolter and Skoog (1966) medium made with ammonium nitrate omitted (see Table 3 for WS medium), but containing 0.04 mg/l 2,4-D and 1.0 mg/l kinetin (Winton, 1970, 1971). Five explants per dish are incubated in the dark at 27 ± 1°C, in unsealed Petri dishes, kept moist by the presence of open pans of water in the incubator.

After 6-8 weeks, whole buttons of firm white callus can be isolated from the parental explants and subcultured monthly to fresh medium in the dark. This callus will turn red and die in light unless ammonium nitrate is added or the callus is grown on MS medium. However, shoots have not been initiated by us from aspen callus grown in the light.

For shoot initiation, inocula 3 mm^3 are placed on the same WS medium without 2,4-D, but having 0.05 mg/l BAP (benzylaminopurine) instead of kinetin. Shoots are visible on the lower sides of the inocula in 2-3 weeks. After an additional 1-2 weeks in the dark, the callus pieces with the shoots are transferred to original WS medium containing hormones and placed in 1000-2000 lux of fluorescent-tungsten light. Roots grow from the callus and occasionally from shoots in 2-3 weeks. Shoots can also be excised and rooted in WS medium containing 3% IBA (indolebutyric acid) and no BAP for rooting of 50-60% of the shoots (Winton, 1970).

One-Year-Old Branches from Hardwoods. Chalupa and Durzan (1973) and Chalupa (1974) reported better callus growth from poplar explants grown on Linsmaier and Skoog (1965) medium than on Wolter and Skoog (1966) medium. LS medium (Table 3) is MS medium without glycine, nicotinic acid and pyridoxine, and having 0.4 instead of 0.1 mg/l of thiamine. Added to the LS medium were 2 mg/l NAA (naphthaleneacetic acid), 1 mg/l BAP, and 100 mg/l arginine. Callus was firm white in the dark and firm green in 16 h/day of 10,000 lux of fluorescent-tungsten light at 25 ± 0.05°C.

The callus was subcultured every 3-4 weeks by Chalupa (1974), and inocula were then transferred to the same medium without NAA and having 0.15-0.7 mg/l BAP. Shoots appeared 10-14 days after subculture, and were excised when 4-8 weeks old and rooted in WS medium containing 0.4 mg/l NAA and no BAP.

Venverloo (1973) grew callus from several poplar species on MS medium containing supplements (in mg/l) of 0.5 2,4-D, 100 inositol, and 1 each pyridoxine, nicotinic acid and thiamine. She also found that for shoots to form, the medium had to be auxin free and contain 0.2 mg/l BAP. Similar conditions for shoot initiation were also reported by Bawa and Stettler (1972) for black cottonwood callus cultures. Venverloo (1973) excised the shoots and could root them in medium containing low levels (0.01-0.05 mg/l) of NAA or 2,4-D, but no rooting occurred when a high level of 0.5 mg/l of 2,4-D was used.

One-Year-Old Branches from Conifers. Chalupa and Durzan (1973) surface sterilized Norway spruce twigs by immersing three times, for five min each, in a calcium hypochlorite solution with 4% available chlorine, followed by immersion in a 5% solution of hydrogen peroxide. Details for the formulation of the solutions were not given.

The method of Harvey and Grasham (1968) and Grasham and Harvey (1970) was used successfully by Winton (1972c) to surface sterilize terminal branch sections from 5-6 year old trees of Douglas-fir, loblolly pine and other conifer species. Store the collected terminal branches (10-20 cm long) in a wet towel in a plastic bag for several days in the refrigerator before using. This allows the drying sap to seal the conductive elements in the cortical tissue, which prevents penetration of the sterilant that could result in severe damage or possible death of the cambial tissue.

For the preparation of explants, needles are carefully cut off at their base, and whole branch sections are washed in cool tap water with Alconox detergent, rinsed in tap water, dipped for 10 sec in 95% ethanol, 60 sec in bleach (5-6% NaOCl), and rinsed three times in sterile water. Using aseptic techniques, branch sections can be cut to segments 1-2 cm long with a sterile razor blade and all tissue outside the cambium is removed. The woody cylinder can be cut to segments 1 cm long or into discs 1-2 mm thick and the explants placed on the surface of the agar medium. Dishes are sealed with Parafilm tape and incubated in the dark at 27 ± 1°C, on Brown and Lawrence (1968) medium modified by Winton (1972c) and containing 5 mg/l NOAA (naphthoxyacetic acid) and 0.1 mg/l BAP. (See Table 3 for BL medium.)

After 1-2 weeks in the dark, the explants and their developing callus from Douglas-fir are moved to 16 h/day in 3000-5000 lux of cool-white fluorescent light at a constant temperature of 23°C. The soft callus turns green, and after another 1-2 months can be isolated and subcultured to fresh medium containing 1-5 mg/l NOAA, where it becomes softer in texture and darker green in color.

In 3000 lux of fluorescent light for 16 h/day, the callus will form occasional embryoids. However, a uniformly unorganized callus will develop on the same medium in 500 lux for 16 h/day at 24°C, alternating with 8 h of darkness at 15°C. The same tissue will also grow in both environments as a firmer and lighter-green callus, if BAP and all vitamins are omitted and 0.5 mg/l 2,4-D is substituted for NOAA.

Suspension Cultures from Somatic Callus Cultures

A suspension culture consists of cells and cell aggregates dispersed and growing in moving liquid medium (Street, 1973b). The culture may be grown for only one passage of from one to several weeks, or may be continuous and provisions made for periodic withdrawal of cell material and the addition of fresh liquid medium. In most suspensions, cells often remain in small aggregates, and true cell suspensions composed of single cells are almost impossible to establish and maintain in culture. Fig. 1 outlines the establishment and use of suspension cultures for regeneration.

Numerous methods have been devised to both aerate and disperse cell material in suspension cultures. These include shaking and rotating the culture vessels, as well as bubbling sterile air through stationary vessels. Recent reviews by Hildebrandt (1973) and Street (1973b) contain detailed instructions for setting up the laboratory, making stock solutions, and mixing and sterilizing the nutrient media.

Chalupa and Durzan (1973) cultured cells and aggregates of Norway spruce callus in suspension cultures. However, instead of using tumble tubes introduced by Steward et al. (1952) they used nipple flasks (Steward and Shantz, 1956). Callus 20-25 days old from initiation was placed in LS medium (Table 3) with 1 mg/l each of NAA and kinetin, and rotated slowly at 1 rpm in 3000-6000 lux of fluorescent-tungsten light at 25°C. The suspension was composed of both single cells and cell aggregates of various sizes. The dry weight of the cell material increased 10-14 times during the first two-week passage, and after repeated transfers to fresh medium some cells formed small embryoids of 5-10 cells, presumably originating from single cells.

Suspension cultures were established by Winton (1972c) from soft-green callus of Douglas-fir, as well as from soft-yellow callus of pine species. Cells and aggregates were dispersed in liquid BL medium (Table 3) containing 0.008 mg/l 2,4-D and 0.05 mg/l BAP, as well as in medium containing 25.0 mg/l IAA and 0.5 mg/l kinetin. Douglas-fir and pine cell aggregates produced small embryoid structures in simplified liquid medium containing only inorganic salts, iron and sucrose (Winton, 1974a).

Suspension cultures were also established from white spruce callus by Durzan et al. (1973) and Chafe and Durzan (1973), and from callus of jack pine and white spruce by Durzan and Steward (1968) by methods similar to that of Chalupa and Durzan (1973) for Norway spruce. Small embryoids were found in all suspensions, along with cells and cell aggregates.

To date, no tree species has been reproduced directly from cells or aggregates growing in liquid suspension cultures. However, Durzan and Lopushanski (1975) transferred American elm callus from agar medium to suspension culture and back to agar medium. Some of the cells or aggregates in the suspension grew into unorganized callus tissue, followed by the initiation of shoots and their growth into plants. The original callus was initiated from seed hypocotyl segments, and it was subcultured to fresh medium three times during 105 days before it was placed in liquid medium. The suspension of cells and aggregates was given fresh liquid medium every 20 days for 157 days, then the cell material was placed back to agar medium where it grew into callus and subsequently produced shoots.

Regeneration from Somatic Callus or Suspension Cultures

So far we have emphasized the vegetative propagation of selected forest trees from callus or suspension cultures. In future recurrent selection programs, designed to improve planting stock for reforestation, these culture techniques will prove most valuable if they can be used to propagate juvenile trees 3-4 years old, long before gametophytic (reproductive) tissue is normally produced.

Space has not permitted the exhaustive description of methods used to culture the 200 or more tree species so far reported. The methods cited are representative of the most recent techniques that have proven successful in establishing somatic callus or suspension cultures from several hardwood and conifer trees. We hope that among these, the new student will find a method that works or can be adapted to your particular species of interest.

Neither has time permitted the inclusion of the myriad of details that are essential to the successful establishment of tree callus or suspension cultures. For cook-book instructions on laboratory organization, the preparation of nutrient stock solutions and final media, methods of sterilization by autoclaving or filtering, as well as other essential points, the new student should acquire a recent book on basic tissue culture techniques such as Kruse and Patterson (1973) or Street (1973a).

In the sections that follow, relatively new methods are examined that have been under study for only the past 5-10 years using herbaceous plant species, and for a much shorter time with tree species. They are frontier methods, and much less is known of these than of callus and suspension cultures. However, what information is available on tree species is presented, with the hope of stimulating new studies in this field.

Protoplasts and Somatic Hybridization

A protoplast is a plant cell that has had its cellulosic wall removed, leaving a spherical cell surrounded by a cytoplasmic membrane. Protoplasts can be prepared from the cells of many different plant tissues including leaf mesophyll, meristems, microspores and megagametophytes. But, because the cells must first be separated from each other, wall removal probably is easier from cells of callus or suspension cultures. Some workers have removed walls by mechanical means, but most workers use a combination of the enzymes pectinase and cellulase, often in the same solution.

Protoplasts may prove useful for many different kinds of study such as the production and deposition of cellulose, enzyme systems, cytological and ultrastructural investigations, cell differentiation, organogenesis, and transformation (see next section), as well as for its originally intended purpose of somatic, or nonsexual hybridization between plant species. Without walls, protoplasts may fuse together to form new cells in a manner similar to the fusion of two sex cells during fertilization. However, there are still many unsolved problems in this field, where most workers are mainly using herbaceous plant species. A few workers have isolated tree protoplasts, but for the most part the big advances in tree research will be expected to come after major breakthroughs occur in protoplast research with herbaceous plants.

For a general background and for details of techniques, recent reviews are suggested which are found in the same books that detail callus and suspension cultures. These include chapters by Ruesink (1971), Cocking and Evans (1973), Miller et al. (1973), and Eriksson et al. (1974). Specific problems in protoplast culture are treated in the 49 papers presented at the International Colloquium in Versailles in 1972, and published in book form by l'Institut National de la Recherche Agronomique, 149 rue de Grenelle, 75007 Paris in 1973, under the title Protoplastes et Fusion de Cellules Somatiques Végétales. All papers are in English or French with French and English summaries.

Fig. 2 compares the somatic hybridization of tree species in the same genus by protoplast fusion, with hybridization between genera by transformation.

Tree Protoplasts. Relatively little work has been done and even less reported on the isolation of protoplasts from tree species. Neither has there been any report of the successful aseptic culture, wall regrowth, cell division, callus formation, or regeneration from tree protoplasts. Among forest trees, Huhtinen (Huhtinen and Winton, 1973) and Chalupa (1974) isolated protoplasts from Norway spruce callus, and Winton (1974a, b) isolated protoplasts from Douglas-fir, loblolly pine, shortleaf pine, and western hemlock.

Winton et al. (1975) placed about 1 gram of soft-green Douglas-fir callus in a test tube containing 5 ml of an enzyme-salt solution. The solution was composed of 2% each of commercial preparations of Macerase-TM (pectinase and hemicellulases) and Cellulysin-TM (cellulase and polysaccharidases) from the CalBiochem Company, plus 0.2 *M* mannitol and the inorganic salts from MS medium. In this preliminary study, the total osmolality was roughly estimated at 0.3 *M*, by comparing the very slight degree of plasmolysis of the callus cells in the enzyme-salt solution to the same degree of plasmolysis exhibited by similar cells in a standard 0.3 *M* solution of sucrose. By the same method,

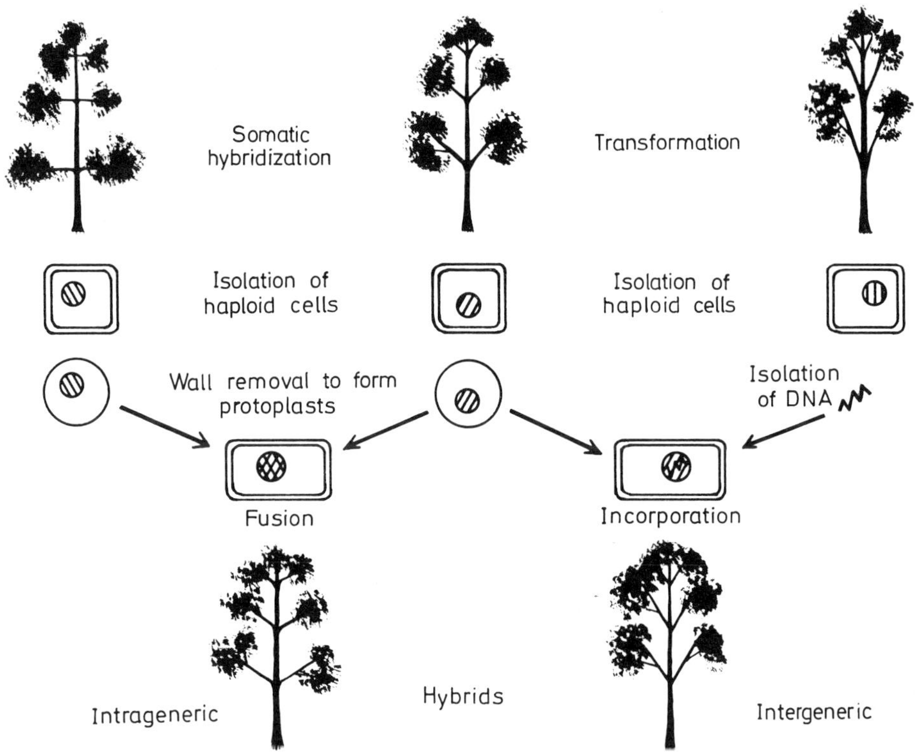

Fig. 2. Intrageneric hybridization by protoplast fusion and intergeneric hybridization by transformation. (Drawing by courtesy of Mr. Steve Newcombe)

the contribution of the two enzymes together was estimated at 0.05 *M*, so the wash solution used for removing the enzymes was made to 0.3 *M* by adding 0.25 *M* mannitol to MS salts estimated at 0.05 *M*.

Spherical protoplasts began forming after 1-2 h on a drum rotating at 20 rpm at 24°C in 1000 lux of fluorescent light. After 3 h, protoplasts were formed from 80-90% of the cells in the original suspension of cells and aggregates. The enzyme-salt solution was decanted and the protoplasts were washed 3 times in the wash solution without the enzymes.

The protoplasts were not cultured, but some were fixed by standard methods (along with cells in a control suspension without enzymes) and were examined with a transmission electron microscope. Photomicrographs showed that the enzyme treatment removed all traces of the cellulosic wall, leaving only the spherical protoplasts surrounded by a cytoplasmic membrane.

One of the problems we encountered in isolating conifer protoplasts were the small, but apparently significant differences in osmolality of the vacuolar sap between cells from different species. Only the solutions used for Douglas-fir were closely adjusted, but our observations indicate that cells from loblolly pine might require an osmoticum of 0.35-0.4 *M*. Another problem is the cold-filtration necessary for the enzyme-salt solution if the protoplasts are to be cultured aseptically. Normal filtering causes rapid clogging of the 0.3 μm pores, but Ruesink (pers. comm.) suggested that the unfiltered enzyme solution be allowed to stand after vigorous shaking, followed by decanting the clear liquid and centrifuging it for a few minutes at 500 xg. The supernatant should still contain all of the enzyme activity and easily pass through a micropore filter.

After protoplasts are formed, they must be washed to remove the enzymes and debris. Some workers use fine sieves with 8 μm pores, but Huhtinen (Huhtinen and Winton, 1973) centrifuges the enzyme-protoplast suspension at 600 xg for 10 min to form a soft pellet, then decants and replaces the clear enzyme solution with wash solution and resuspends the solid material. After washing the material three times, the suspension is filtered and washed through sterile glass wool on a Buchner funnel without suction. The glass wool catches the debris but allows the protoplasts to pass through. The filtrate is then centrifuged at 1000 xg for 10 min and the washed and cleaned protoplasts form a soft pellet for resuspension and culture.

The main problem of culturing protoplasts seems to be aeration. Small drops of wash solution containing protoplasts can be placed on the glass covers of sterile Petri dishes and grown as a hanging-drop culture. Whatever method is used, the osmoticum can be changed at intervals of 1-2 days, with decreasing osmolality as the wall forms, until normal nutrient medium is used entirely after walls are completed around all cells.

<u>Protoplast Fusion</u>. No information is available on the fusion of tree protoplasts, so workers should start first with methods that work for herbaceous species, then make modifications as they are needed. There are three main systems available for fusing protoplasts: 1. adding a high concentration of sodium nitrate to the osmoticum (Power et al., 1970), 2. incubating isolated protoplasts for a short time of 30-60 min at a high pH of 9.5-10.5 in the presence of calcium ions (Keller and Melchers, 1973), and 3. the addition of polyethylene glycol and calcium ions to the osmoticum (Kao and Michayluk, 1974).

Natural agglutins can be used to induce protoplasts to adhere to each other, but true fusion is seldom observed (Eriksson et al., 1974). Also, the use of sodium nitrate alone has inhibited cell division after fusion, thus eliminating this method for somatic hybridization from protoplast fusion. Past results with polyethylene glycol in the osmoticum have shown that aggregates form, in many cases, composed of many protoplasts that fuse together. Unfortunately, plants are neither desired or expected from these multiple aggregates because of their high ploidy levels. However, by carefully choosing the optimum molecular weight and adjusting the concentration of ethylene glycol, as well as varying the plating densities of the protoplasts, a combination might be found that will give a high proportion of fusion products that are each composed of only two protoplasts.

The use of high pH and calcium ions is a relatively new method, and so far there is only one report of successful protoplast fusion by this technique (Keller and Melchers, 1973). From these same fused protoplasts, Melchers (1974) selected callus of somatic hybrid cells and one plant has been regenerated that is identical to the sexual hybrid between two varieties of *Nicotiana tabacum*. In both parental varieties the protoplasts were chlorophyll deficient, light sensitive, and would not grow under high light intensities. However, the fused hybrid cells were complementary and light resistant. Melchers (1974) simply incubated the fused protoplasts under intense light and selected the green calluses and regenerated a plant from these. Plastid color of parental cells has also been used as a marker for somatic hybridization by Kao et al. (1974). In the first reported case of somatic hybridization, Carlson et al. (1972) selectively screened hybrid callus from parental callus by its faster growth after the fusion of tobacco protoplasts.

The selection of hybrid fusion products will certainly also be a problem among tree species. One solution may be the use of haploid microspores for protoplasts, and select parental species having recessive marker genes that are expressed in the fusion products differently than in the parental cells. The use of a cytophotometer might also be used to determine differences in DNA levels among cells, which would directly reflect the ploidy levels. Diploid cells could be selected as fusion products and haploid cells discarded as parental cells (Miksche, pers. comm.; Berlyn and Cecich, this volume).

Genetic Transformation

New genetic information can be transmitted to bacterial cells by isolated DNA (transformation), by intact bacteria (transcession), and by viral particles (transduction). All of these were grouped together in the term "transgenosis" by Doy et al. (1973), to describe the genetic transfer of information to plant cells when the precise mechanism is unknown. However, most plant studies to date have involved isolated DNA, so the term transformation has been reinstated.

The most recent reviews are by Holl et al. (1974) and Reinhard (1974), also in the book edited by Street (1974). They explore the transformation by isolated DNA of several herbaceous plant species, citing both their own work and that of others. Pioneering work on transformation in plant cells was carried out by Hess (1972, 1973), using whole cells surrounded by walls. However, the cell wall effectively limits the intrusion into the cell of viral particles (Cocking, 1970) and possibly also of DNA. Thus, the use of protoplasts was suggested by Ohyama et al. (1973), who reported the uptake of isolated, bacterial DNA by protoplasts of soybean and carrot.

However, after cells have been transformed they must be regenerated into plants, and so far only a few plants have been regenerated from unfused protoplasts. These include tobacco (Takebe et al., 1971), carrot and brome grass (Gamborg et al., 1973), as well as haploid plants from mesophyll cells of a haploid tobacco plant (Ohyama and Nitsch, 1972). But even with these problems of regeneration of plants, as well as the isolation, incorporation and expression of the transferred DNA, many workers remain optimistic that transformation may eventually permit the transfer of genetic material across existing natural barriers between plant genera. This could become an important method of hybridizing species in different genera, especially if somatic hybridization by means of protoplast fusion is found to be limited to hybridization between species within the same genera. We thus see in these two systems the potential means of increasing genetic variability among herbaceous plants and tree species, by a successive expansion of available parental material first within, then between genera (Fig. 2).

Cultures of Gametophytic Cells and Tissues

The preceding sections have dealt with the aseptic culture of somatic, or vegetative, tree parts and the use of the callus or cells for regeneration, somatic hybridization and transformation. In addition, many other problems of tree breeding could potentially be solved by using cultures from gametophytic, or reproductive, tree parts.

The main advantage of culturing either micro- or megagametophytic cells is that each is haploid, having only one-half the number of chromosomes found in somatic cells. Haploid cells can be used to study chromosome morphology and karyotypes, recover recessive markers useful in further breeding studies, and investigate mutation rates. Their main value in tree breeding, however, lies in two distinct areas. First, haploid cells can be used for protoplast production, so that the fused hybrid cells will reconstitute the normal diploid condition found in somatic cells. And secondly, plants grown from haploid cells might be induced to double their one set of chromosomes, either naturally or spontaneously, producing diploid plants that have homozygous alleles at each gene locus. This would fix valuable traits in self-fertile trees, for their expression in succeeding generations (Fig. 3).

Polyploidy may also be an important area of tree-breeding that would benefit from the use of haploid plants. Conifer seedlings having more than the normal two sets of chromosomes per cell are polyploids, and generally are stunted and do not survive. Tetraploids have four sets of chromosomes, two from each parent, and are thus allotetraploids. If further doubling of chromosome sets occurred in homozygous diploid plants, the resulting autotetraploids would be homozygous at each gene locus and might have surprisingly vigorous characteristics. The effects of autotetraploidy might be even greater among hardwood tree species.

Reviews of haploidy in forest tree-breeding have been published by Huhtinen (1972) and Winton and Stettler (1974). The two main sources of haploid cells are from reproductive tissue (microspores) in anthers (microsporangia) and from the nutritive cells in the megagametophyte of conifer trees (the nutritive endosperm of hardwood trees is triploid). Megagametophytic tissue is generally isolated from immature seeds and grown independently in aseptic culture (Bonga, 1974). However, microspores (which eventually develop into pollen grains, or microgametophytes) are generally left intact in the anther, and so this method is referred to as anther culture. Nitsch (1972) and Sunderland (1973) present general techniques in anther culture of herbaceous plant species, and the recent book edited by Kasha (1974) includes a wide range of papers on special problems in haploidy.

Reviews of culturing callus and regenerating plants from pollen and megagametophytic tissue have been published by Winton and Stettler (1974) and Bonga (1974) and will not be discussed in this presentation.

Tree Anther Cultures. Harn and Kim (1972) induced callus from apricot anthers, cultured on MS medium supplemented with 2 mg/l each of 2,4-D and kinetin. Six calluses emerged from inside the anthers and were considered as originating from microspores, although chromosomes counts were not then attempted because of the slow growth of the callus. Santamour (pers. comm.) was unable to grow either plants or callus from elm and magnolia anthers cultured on Nitsch (1972) medium. Sommer and Sharp (pers. comm.) obtained callus from anthers of American elm, but they were unable to determine whether it originated from microspores or from somatic cells of the anther wall.

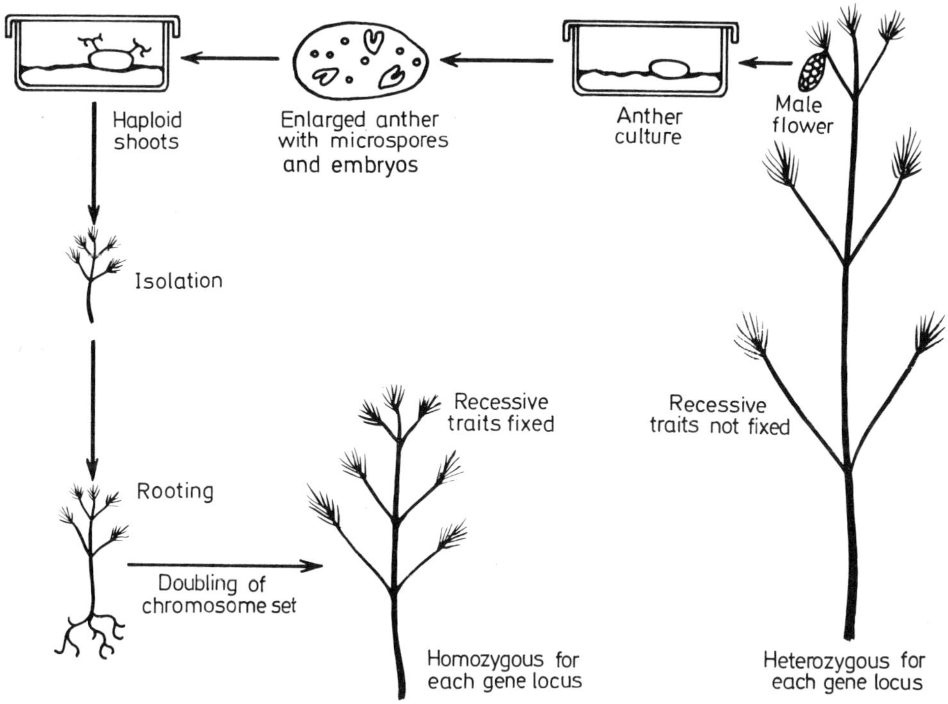

Fig. 3. Production of homozygous diploid trees from haploid plants regenerated from microspores cultured inside of anthers. (Drawing courtesy of Mr. Steve Newcombe)

Likewise, Sato (1971, 1972, 1974) cultured anthers from several tree species on WS medium and obtained callus, but was unable to determine whether the callus originated from microspores. Roots were initiated from anther callus of two species of alder, three of Prunus, Cryptomeria, and three hybrids and two species of aspen. Shoots grew from many aspen calluses and were excised, rooted and grown into normal plants. Unfortunately, chromosome counts were not made of the callus to determine if it was haploid, and only diploid counts were observed in root tips of the regenerated plants.

Jordan (1974) observed multicellular microspores in anther cultures of sweet cherry, cultured on Nitsch (1972) medium supplemented with 1 mg/l each of NAA and BAP. Some callus grew from anther walls, which is normal somatic tissue, and no organogenesis has been observed. Michellon et al. (1974) cultured anthers from almond and peach on Miller (1963) medium containing 1 mg/l 2,4-D, 0.1 mg/l IAA and 0.2 mg/l kinetin. Callus grew from the anthers, and microscopic examinations showed that the callus cells in culture were haploid, diploid or triploid. No roots or shoots were formed.

For conifer species, Bonga (1974) has presented an excellent cytological study on callus formation and the development of embryo-like structures from microspores of several pine species. However, he points out that microspores in the same microsporangium have variable genomes, and the mixing of callus from several microspores would result in a nonisogenic callus. Haploid cells could be separated and cultured individually, but Bonga (1974) prefers the simpler route of using the uniformly haploid cells from megagametophyte tissue for the growth of haploid plants and subsequent diploidization.

Winton (1974a) cultured anthers from Douglas-fir and quaking aspen, but only diploid counts were observed in the few callus masses recovered from both species. In an effort to selectively grow possible haploid cells in the callus, 5 mg/l of PFP (parafluorophenylalanine) were added to BL medium. However, only diploid counts were found after several months in the slow growing callus.

Gupta and Carlson (1972) first suggested the preferential growth with PFP of haploid tobacco cells over diploid cells, in callus growing on LS medium containing 4 mg/l IAA and 1 mg/l kinetin. Dix and Street (1974) also cultured haploid and diploid tobacco callus on LS medium, made with 0.4 mg/l 2,4-D, 0.03 mg/l kinetin and 1000 mg/l casein hydrolysate. They reported that both types of cells were insensitive to 10 mg/l PFP in the medium, and also that the growth of both types of cells was inhibited at levels of 25-75 mg/l PFP. Niizeki (1973) grew haploid and diploid tobacco callus on Miller (1963) medium with 2 mg/l each of IAA and kinetin, plus PFP at levels of 0, 50, 100 or 200 mg/l. After one month, diploid callus growth was inhibited 50-60% on medium with 50 or 100 mg/l PFP, but haploid callus growth was inhibited only 12-40% on the same media. At 200 mg/l PFP, both diploid and haploid cells evantually died. To date these results are inconclusive, and PFP should not be used routinely without further testing.

Discussion

The techniques presented here represent only the most recent work in the fields of callus, cell, protoplast and anther cultures of tree species. Several significant papers have been published within the past five years, and we should not be surprised to witness an even greater upsurge of understanding during the next five years.

In our estimation, the most important and immediate problem is to find an efficient method of regenerating both hardwood and conifer trees from aseptic, somatic callus and suspension cultures. Such a technique would have immediate application in propagating selected juvenile trees in current tree-breeding programs. Regeneration is also necessary to grow into trees the results of somatic hybridization (fused protoplasts) and transformed cells.

Many workers throughout the world are now trying to solve this problem of regeneration. Empirical methods have produced limited success in reproducing a relatively few species of hardwood trees, but so far there have been no conifer trees grown from subcultured callus or suspension cultures. As a result, a team effort was started in 1974 at The Institute of Paper Chemistry, Appleton, Wisconsin, U.S.A., combining the efforts of the biochemistry, electron microscopy, tissue culture and genetics laboratories (Winton et al., 1974). Other team approaches will undoubtedly start at other laboratories in the near future, illustrating the importance placed on this problem by so many researchers and administrators alike. Not only do we want to know how to propagate trees from aseptic cultures, but we must also understand the mechanisms involved in order to apply the method to as many tree species as possible.

Acknowledgements. The authors wish to thank Drs. Toshio Murashige, Donald Durzan, Harry Sommer, Jerome Miksche, and Mrs. Shirley Verhagen for reading the manuscript and for their suggestions.

References

Aneja, S., Atal, C.: Plantlet formation in tissue cultures from lignotubers of *Eucalyptus citriodora* Hook. Current Sci. (India) 38, 69 (1969).

Ball, E.: Differentiation in callus cultures of *Sequoia sempervirens*. Growth 14, 295-325 (1950).

Bawa, K., Stettler, R.: Organ culture with black cottonwood: morphogenetic response of female catkin primordia. Can. J. Botany 50, 1627-1631 (1972).

Berbee, J., Hildebrandt, A.: Induction of callus and trees from stem tip cultures of a hybrid poplar. In Vitro 7, 269 (1972).

Bonga, J.: *In vitro* culture of microsporophylls and megagametophyte tissue of Pinus. In Vitro 9, 270-277 (1974).

Brown, C., Lawrence, R.: Culture of pine callus on defined medium. Forest Sci. 14, 62-64 (1968).

Carlson, P., Smith, H., Dearing, R.: Parasexual interspecific plant hybridization. Proc. Natl. Acad. Sci. (Wash.) 69, 2292-2294 (1972).

Chafe, S., Durzan, D.: Tannin inclusions in cell suspension cultures of white spruce. Planta (Berlin) 113, 251-262 (1973).

Chalupa, V.: Control of root and shoot formation and production of trees from poplar callus. Biol. Plantarum (Praha) 16, 316-320 (1974).

Chalupa, V., Durzan, D.: Growth of Norway spruce (*Picea abies* (L.) Karst.) tissue and cell cultures. Commun. Inst. Forestalis Chechoslov. 8, 111-125 (1973).

Chaturvedi, H., Mitra, G.: Clonal propagation of Citrus from somatic callus cultures. Hort. Science 9, 118-120 (1974).

Cocking, E.: Virus uptake, cell wall regeneration, and virus multiplication in isolated plant protoplasts. Intern. Rev. Cytol. 28, 89-124 (1970).

Cocking, E., Evans, P.: The isolation of protoplasts. In: Plant Tissue and Cell Culture (ed. H. Street), pp. 100-120. Berkeley-Los Angeles: Univ. California Press 1973.

Dix, P., Street, H.: Effects of p-fluorophenylalanine (PFP) on the growth of cell lines differing in ploidy and derived from *Nicotiana sylvestris*. Plant Sci. Letters 3, 283-288 (1974).

Doy, C., Gresskoff, P., Rolfe, B.: Biological and molecular evidence for the transgenosis of genes from bacteria to plant cells. Proc. Natl. Acad. Sci. (Wash.) 70, 723-726 (1973).

Durzan, D., Campbell, R.: Prospects for the mass production of improved stock of forest trees by cell and tissue culture. Can. J. Forest Res. 4, 151-174 (1974).

Durzan, D., Chafe, S., Lopushanski, S.: Effects of environmental changes on sugars, tannins, and organized growth in suspension cultures of white spruce. Planta (Berlin) 113, 241-249 (1973).

Durzan, D., Lopushanski, S.: American elm from cell suspension culture. Can. J. Forest Res., in press (1975).

Durzan, D., Steward, F.: Cell and tissue culture of white spruce and jack pine. Bimon. Res. Notes, Can. Forest Serv. 24, 30-31 (1968).

Eriksson, T., Bonnett, H., Glimelius, K., Wallin, A.: Technical advances in protoplast isolation, culture and fusion. In: Tissue Culture and Plant Science. Proc. 3rd Intern. Cong. Plant Tissue and Cell Culture, Leicester, England (ed. H. Street), pp. 213-231. London, New York: Academic Press 1974.

Gamborg, O., Kao, K., Miller, R., Fowke, L., Constabel, F.: Cell regeneration, division and plant development from protoplasts. In: Protoplastes et Fusion de Cellules Somatiques Végétales, pp. 155-173. Paris: Centre Natl. Rech. Sci. 1973.
Grasham, J., Harvey, A.: Preparative techniques and tissue selection criteria for *in vitro* culture of healthy and rust-infected conifer tissues. USDA, Forestry Serv. Res. Pap. INT-82 (1970).
Gupta, N., Carlson, P.: Preferential growth of haploid plant cells *in vitro*. Nature New Biol. (London) 239, 86 (1972).
Harn, C., Kim, M.: Induction of callus from anthers of *Prunus armeniaca*. Korean J. Breeding 4, 49-53 (Korean with Engl. summary)(1972).
Harvey, A., Grasham, J.: Procedures and media for obtaining tissue cultures of 12 conifer species. Can. J. Botany 47, 547-549 (1968).
Hess, D.: Transformation an höheren Organismen. Naturwissenschaften 59, 348-355 (German with Engl. summary)(1972).
Hess, D.: Transformationsversuche an höheren Pflanzen: Untersuchungen zur Realisation des Exosomen-Modells der Transformation bei *Petunia hybrida*. Z. Pflanzenphysiol. 68, 432-440 (1973).
Hildebrandt, A.: Plant cell suspension cultures. In: Tissue Culture (eds. P. Kruse, M. Patterson), pp. 215-219. New York, London: Academic Press 1973.
Holl, F., Gamborg, O., Ohyama, K., Pelcher, L.: Genetic transformation in plants. In: Plant Tissue and Cell Culture (ed. H. Street), pp. 301-327. Berkeley, Los Angeles: Univ. California Press 1973.
Huhtinen, O.: Production and use of haploids in breeding conifers. In: IUFRO Genetics - SABRAO Joint Symposia, Tokyo, D-3 (I), 1-8 (1972).
Huhtinen, O., Winton, L.: Cell and tissue culture for production of haploid, polyploid and clonal propagation. In: IUFRO Workshop for Methods in Biochemical Genetics of Forest Trees, Göttingen, Germany, Section 2, pp. 1-12 (1973).
Huhtinen, O., Yahyaoglu, Z.: Das Frühen von aus Kalluskulturen herangezogenen Pflänzchen bei der Birke (*Betula pendula* Roth.). Silvae Genetica 23, 32-34 (1974).
Jones, L.: Plant cell culture and biochemistry: studies for improved vegetable oil production. In: Industrial Aspects of Biochemistry (ed. B. Spender), pp. 813-833. Fed. Eur. Biochem. Soc. 1974.
Jordan, M.: Multizelluläre Pollen bei *Prunus avium* nach *in vitro* Kulture. Z. Pflanzenzüchtung 71, 358-363 (1974).
Kao, K., Constabel, M., Michayluk, M., Gamborg, O.: Plant protoplast fusion and growth of intergeneric hybrid cells. Planta (Berlin) 120, 215-227 (1974).
Kao, K., Michayluk, M.: A method for high-frequency intergeneric fusion of plant protoplasts. Planta (Berlin) 115, 355-367 (1974).
Kasha, K. (ed.): Haploids in higher plants. Proc. 1st Intern. Symp. Univ. Guelph, Canada, 1974.
Keller, W., Melchers, G.: The effect of high pH and calcium on tobacco leaf protoplast fusion. Z. Naturforsch. 28, 737-741 (1973).
Kruse, P., Patterson, M. (eds.): Tissue culture methods and application. New York, London: Academic Press 1973.
Linsmaier, E., Skoog, F.: Organic growth factor requirement of tobacco tissue cultures. Physiol. Plantarum (Cph.) 18, 100-127 (1965).
Mehra, A., Mehra, P.: Organogenesis and plantlet formation *in vitro* in almond. Botan. Gaz. 135, 61-73 (1974).
Melchers, G.: Summation: Haploid research in higher plants. In: Haploids in Higher Plants Advances and Potential. Proc. 1st Intern. Symp., Univ. Guelph, Canada, June 10-14, 1974 (ed. K. Kasha), pp. 393-401. Univ. Guelph Press 1974.
Michellon, R., Hugard, J., Jonard, R.: The isolation of tissue cultures of peach (*Prunus persica*) and almond (*Prunus amygdalus*) by anthers cultured *in vitro*. C.R. Acad. Sci. (Paris) 278-D, 1719-1722 (Fr.)(1974).

Miller, C.: Kinetin and kinetin-like compounds. In: Moderne Methoden der Pflanzenanalyse (eds. H. Linskens, M. Tracey), Vol. VI, pp. 192-202. Berlin-Heidelberg-New York: Springer 1963.
Miller, R., Kao, K., Gamborg, O.: Plant protoplast culture. In: Tissue Culture (eds. P. Kruse, M. Patterson), pp. 500-505. New York, London: Academic Press 1973.
Murashige, T.: Plant propagation through tissue cultures. Ann. Rev. Plant Physiol. 25, 135-166 (1974).
Murashige, T., Skoog, F.: A revised medium for rapid growth and bioassays with tobacco tissue cultures. Physiol. Plantarum 15, 473-497 (1962).
Niizeki, M.: Studies on plant cell and tissue culture. IV. Effect of parafluorophenylalanine on haploid and diploid cells of tobacco plant *in vitro*. J. Fac. Agr. Hokkaido Univ. (Sapporo) 57, 349-356 (1974).
Nitsch, J.: Haploid plants from pollen. Z. Pflanzenzüchtung 67, 3-18 (1972).
Ohyama, K., Gamborg, O., Shyluk, J., Miller, R.: Studies on transformation: uptake of exogenous DNA by plant protoplasts. In: Protoplastes et Fusion de Cellules Somatiques Végétales, pp. 423-428. Paris: Centre Natl. Rech. Sci. 1973.
Ohyama, K., Nitsche, J.: Flowering haploid plants obtained from protoplasts of tobacco leaves. Plant and Cell Physiol. 13, 229-236 (1972).
Oka, S., Ohyama, K.: Adventitious bud initiation on the cultured callus of *Broussonetia kazinoki*. 3rd Symp. Plant Tissue Culture Japan 31 (Japanese)(1972).
Power, J., Cummins, S., Cocking, E.: Fusion of isolated plant protoplasts. Nature 225, 1016-1018 (1970).
Reinhard, E.: Biotransformations by plant tissue cultures. In: Tissue Culture and Plant Science 1974, Proc. 3rd Intern. Congr. Plant Tissue Cell Culture, Leicester, England (ed. H. Street), pp. 433-459. London, New York: Academic Press 1974.
Ruesink, A.: Protoplasts of plant cells. In: Methods in Enzymology. Vol. 23, pp. 197-209. New York: Academic Press 1971.
Sato, T.: Haplont generation in anther cultures and present status on anther cultures of forest trees. Forest Tree Breed. 70, 2-5 (Japanese)(1971).
Sato, T.: Studies on the anther cultures in the research and experimental stations of the Ministry of Agriculture and Forestry of Japan. Res. News (Kenkyu Zyoho) 17, 35 (Japanese)(1972).
Sato, T.: Callus induction and organ differentiation in anther culture of poplars. J. Jap. Forestry Soc. 56, 55-62 (Japanese with Engl. summary)(1974).
Steward, F., Caplin, S., Millar, F.: Investigations of growth and metabolism of plant cells. I. New techniques for the investigation of metabolism, nutrition and growth in undifferentiated cells. Ann. Botany 16, 58-77 (1952).
Steward, F., Shantz, E.: The chemical induction of growth in plant tissue cultures (eds. R. Wain, F. Wightman), pp. 165-186. London: Butterworths 1956.
Street, H. (ed.): Plant tissue and cell culture. Botany Monographs II. Berkeley, Los Angeles: Univ. California Press 1973a.
Street, H.: Cell (suspension) cultures - techniques. In: Plant Tissue and Cell Structure (ed. H. Street), pp. 59-99. Berkeley, Los Angeles: Univ. California Press 1973b.
Street, H. (ed.): Tissue culture and plant science. Proc. 3rd Intern. Congr. Plant Tissue and Cell Culture, Leicester, England. London, New York: Academic Press 1974.
Sunderland, N.: Pollen and anther culture. In: Plant Tissue and Cell Culture (ed. H. Street), pp. 205-239. Berkeley, Los Angeles: Univ. California Press 1973.

Takebe, I., Labib, G., Melchers, G.: Regeneration of whole plants from isolated mesophyll protoplasts of tobacco. Naturwissenschaften 58, 318-320 (1971).
Thorpe, T., Murashige, T.: Starch accumulation in shoot-forming tobacco callus cultures. Science 160, 421-422 (1968).
Thorpe, T., Murashige, T.: Some histochemical changes underlying shoot initiation in tobacco callus cultures. Can. J. Botany 48, 277-285 (1970).
Torrey, J.: The initiation of organized development in plants. Adv. Morphogenesis 5, 39-91 (1966).
Venverloo, C.: The formation of adventitious organs. I. Cytokinin-induced formation of leaves and shoots in callus cultures of *Populus nigra* L. 'Italica'. Acta Botan. Neerl. 22, 390-398 (1973).
Winton, L.: Plantlets from aspen tissue cultures. Science 160, 1234-1235 (1968).
Winton, L.: Shoot and tree production from aspen tissue cultures. Am. J. Botany 57, 904-909 (1970).
Winton, L.: Tissue culture propagation of European aspen. Forest Sci. 17, 348-350 (1971).
Winton, L.: Annotated bibliography of somatic conifer callus cultures. Genetics and Physiology Notes No. 16. Appleton, Wisconsin: Inst. Paper Chem. 1972a.
Winton, L.: Bibliography of somatic callus cultures from deciduous trees. Gen. and Physiology Notes No. 17. Appleton, Wisconsin: Inst. Paper Chem. 1972b.
Winton, L.: Callus and cell cultures of Douglas-fir. Forest Sci. 18, 151-154 (1972c).
Winton, L.: Bibliographic addendum of tree callus cultures. Gen. and Physiology Notes No. 18. Appleton, Wisconsin: Inst. Paper Chem. 1973.
Winton, L.: Second addendum to the bibliography of tree callus culture. Gen. and Physiology Notes No. 19. Appleton, Wisconsin: Inst. Paper Chem. 1974a.
Winton, L.: The use of callus, cell and protoplast cultures for tree improvement. In: U.S.R.O.C. Cooperative Science Program Seminar on Plant Tissue and Cell Culture, Taipei 1974b.
Winton, L., Mathes, M.: Aspen callus. In: Tissue Culture (eds. P. Kruse, M. Patterson), pp. 161-165. New York, London: Academic Press 1973.
Winton, L., Parham, R., Johnson, M., Einspahr, D.: Tree improvement by callus, cell, and protoplast culture. TAPPI 57, 151-152 (1974).
Winton, L., Parham, R., Kaustinen, H.: Isolation of conifer protoplasts. Gen. and Physiology Notes No. 20. Appleton, Wisconsin: Inst. Paper Chem. 1975.
Winton, L., Stettler, R.: Utilization of haploidy in tree breeding. In: Haploids in Higher Plants Advances and Potential. Proc. 1st Intern. Symp., Univ. Guelph, Canada, June 10-14 (ed. K. Kasha), pp. 259-273. Univ. Guelph Press 1974.
Wolter, K.: Root and shoot initiation in aspen callus cultures. Nature 219, 509-510 (1968).
Wolter, K., Skoog, F.: Nutritional requirements of Fraxinus callus cultures. Am. J. Botany 53, 263-269 (1966).

Chapter 13

Manipulation of Flowering in Conifers Through the Use of Plant Hormones

R. P. Pharis

Introduction

The rationale for attempting to induce flowering in juvenile-phase seedlings (i.e., "that phase in ontogeny during which seedlings cannot be induced to form flowers by any means" (Doorenbos, 1965)) needs no explanation to those engaged in forest tree breeding. However, not only is termination "at will" of juvenility a goal of importance, but regularization of cone crops on sexually mature seed orchard trees is imperative in any program of reforestation where genetically superior stock are used. In the past these two goals have been treated as separate problems and Doorenbos (1965), Robinson and Wareing (1969) have urged that a clear distinction be drawn between treatments which induce the seedling to reach the "ripeness-to-flower" state (Klebs cited in Lang, 1965) and treatments which promote flowering in an already adult plant. However, I am not certain that such a clear distinction should be made since many of the cultural treatments which promote flowering in adult, but otherwise non-flowering trees, will also promote, on occasion, precocious flowering in juvenile-phase seedlings.

It is the purpose of the present paper then, to discuss in some detail, the methodology by which strobili may be repeatably induced in both young and sexually mature trees, and further to delve into some of the factors affecting strobilus differentiation and development, including the means by which the differentiation and development processes have been, and can be critically analyzed.

Discussion

Strobilus Induction

Here, the term induction is used in the broadest sense to include any treatment which promotes strobilus initiation or the differentiation of previously initiated (but undetermined) primordia into strobili. Cultural treatments known to promote strobilus induction have been split into "hormonal" and "non-hormonal", and manipulation of the sex of the strobilus will be dealt with briefly.

Non-Hormonal Cultural Treatments

Over the years many physical techniques have been developed to either hasten the ripeness-to-flower stage, or to enhance flowering in sexually mature trees. These include girdling, strangulation, root pruning, branch pruning, allowing the plant to become "pot bound", water stress, low nitrogen nutrition, nitrogen fertilization, especially with nitrate, geotropic stimulation, growth acceleration and changes in the amount of radiant energy received (Wareing, 1959; Longman, 1960; Doorenbos, 1965; Robinson and Wareing, 1969; Jackson and Sweet, 1972; Puritch, 1972; Zimmerman, 1972). As well, high temperature treatments (30°C day; 23°C night during the last period of the growing season) first utilized for beech by Holmgård and Olsen (1966) have shown good success with Norway spruce (*Picea abies* (L.) Karst.)[1,2] and Sitka spruce (*Picea sitchensis* (Bong.) Carr.)[3]. Unfortunately many of these techniques are expensive, extremely time consuming, highly destructive, variable with regard to response, or possess a combination of many or all of these faults. Even so, they have in the past offered the only hope to geneticists wishing to cross-breed superior phenotypes and/or their genetically superior progeny.

The techniques noted above will not be discussed further except as relates to their possible use as adjunct treatments to hormonal applications. However, in retrospect, on the basis of what is known (Owens, 1969; Ebell, 1971, 1972; Owens and Molder, 1975, and [4]) about the timing of sexual differentiation (in the Pinaceae at least), more consistent results in strobilus induction should have been possible with many of the above treatments through judicious timing and duration of treatment. It now appears that these treatments probably affect differentiation of previously initiated primordia, the period of time during which the primordia are receptive to treatment being finite (Ebell, 1972) and somewhat variable, depending upon the species and geographic location.

Hormonal Treatments

The Cupressaceae and Taxodiaceae. The very striking response of strobilus initiation to exogenous application of a plant hormone (gibberellin A_3) was first noted by Kato et al. (1958) with sugi (*Cryptomeria japonica*). This work has been extended to at least 28 other species in these families (see review by Pharis and Kuo, 1976) and GA_3 applications are now being used effectively in breeding programs with sugi (Ohba et al., 1971). In at least one species, Arizona cypress (*Cupressus arizonica* Greene) staminate strobili can be induced as early as age 88 days (GA_3 application beginning at age 55 days from germination) and viable pollen results (Pharis et al., 1965).

The usual technique for application is to spray an aqueous GA_3 solution (100 to 500 ml/l) to "drip-off" once or twice weekly, a surfactant such as Tween 20 (0.05% solution) being used. Alternatively, aqueous solutions of GA_3 (100 to 500 mg/l) can be used as a soil drench for potted seedlings, volumes of 50 to 250 ml being applied once every two weeks, or even less frequently (Glenn, 1973). This latter technique gives very uniform results with no toxicity (toxicity is sometimes experienced with aqueous sprays of GA_3 at high concentrations, especially under high light intensity). A third possibility is the injection of from 10 to 600 μg of GA_3 into the woody portion of the branch (using up to 10 μl of 80% ethanol as a carrier). With this technique, dose : response ratios of 1 strobilus : 0.002 μg GA_3 have been obtained on 1 to 2 year-old Arizona cypress seedlings (Pharis et al., 1969b).

Footnotes see on page 278.

The effect of GA_3 on strobilus initiation is greater on older seedlings and increasing concentrations of GA_3 at any one age will produce additional strobili in a given period of time (Kato, 1959; Hashizume, 1959, 1963; Saito and Hashizume, 1959; Shidei et al., 1960; Pharis et al., 1965). It has thus been postulated that in younger seedlings of Cupressaceae and Taxodiaceae species the endogenous level of GAs is adequate for shoot elongation and other growth processes, but below the level required to induce differentiation of strobili, and that in nature the juvenile phase exists because the concentration of endogenous GA is used preferentially for shoot elongation and/or other processes. Only when other factors begin to limit vegetative growth does the level of endogenous GA become sufficient for initiation of cone development (Pharis and Owens, 1966; Pharis and Morf, 1967).

Firm evidence for this hypothesis does not exist. Arizona cypress sprayed with low dosages (10 mg/l) of GA_3 will produce significant internode elongation without the induction of strobili, increased levels (100 mg/l) of GA_3 accomplishing both (Glenn, 1973). Also, the amount of GA-like substances extractable from seedlings of Arizona cypress treated with increasing dosages of GA_3 is proportional to the number of strobili produced (Kuo, 1973).

Gibberellins other than GA_3 are also effective in strobilus initiation within the Cupressaceae and Taxodiaceae (Pharis and Morf, 1968 and [6]), and for those GAs tested on Arizona cypress, a relative order of efficacy is shown in Table 1.

Table 1. Relative efficacy[a] of 7 gibberellins in the Cupressaceae[b]

GA_3	> or =	GA_9	> or =	$GA_{4/7}$[c]	= GA_4	> GA_1	> or =	GA_7	> GA_5[d]	> GA_{13}
100		100		80-100	80-100	65-80		15-80	50	10

[a]Based on a combination of "time to first flowering" and number of strobili. Data adapted from Pharis and Morf (1968) and additional unpublished research results[6].

[b]For 3 species of Cupressus and 1 of Thuja.

[c]ca. 70% GA_7.

[d]Injected in 75% ethanol. GA_5 almost ineffective in aqueous spray.

<u>The Pinaceae</u>. Despite the good results obtained using GA_3 and other GAs on the Cupressaceae and Taxodiaceae, repeatable "promotion" of flowering in species of Pinaceae has, until very recently, eluded most research workers. However, as early as 1960, GA_3 was noted to have a slight promotive effect on strobilus formation in seedlings of *Pinus densiflora* (Kato and Fukuhara, 1960; Hashizume, 1966, 1973) and *Pinus mugo* (Shidei and Yoshikawa, 1963), and Hashizume (1967) was able to promote flowering on girdled saplings of *Larix leptolepis* with GA_3 or naphthaleneacetic acid (NAA). But, this work with Pinus was not readily repeatable at the University of Calgary[6] and other successful published reports have not come to my attention.

Many of the non-hormonal cultural treatments discussed in the above are effective in promoting flowering in species of all three families (Pharis and Kuo, 1975), yet GA_3, a potent flowering hormone for species in the Cupressaceae and Taxodiaceae, appeared to be essentially ineffective for the Pinaceae. Over five years ago research was begun at the University of Calgary to investigate this apparent anomaly.

There are a number of reasons why GA_3 might not be effective in the Pinaceae. Firstly, it might not be the appropriate "flowering GA". Secondly, many, if not all species within the Pinaceae have finite periods during which sexual differentiation of pre-existing undifferentiated lateral primordia takes place (Owens, 1969; Owens and Molder, 1975), unlike species in the Cupressaceae and Taxodiaceae where differentiation of previously vegetative meristems into strobili appears to occur anytime when active growth (i.e., new meristem formation) is taking place (Owens and Pharis, 1967, 1971 and [6]). Thirdly, other hormonal co-factors (auxins, cytokinins?) might be limiting.

It thus seemed logical to investigate the possibility of a correlation between the qualitative or quantitative spectrum of endogenous GAs and treatments or situations where a high level of flowering might be expected, said investigations to take place during the period when sexual differentiation was occurring.

At the University of Calgary, Drs. J.L. Glenn and C.G. Kuo embarked on experiments with Arizona cypress, whereby environmental factors such as daylength, or cultural treatments such as the use of growth retardants or nitrogen deficiency were used to modify the flowering response prior to examining the endogenous GA spectrum. As well, co-operative experiments began with Drs. L. Ebell, Canadian Forestry Service; J. Rediske, S. Ross and S. Cade, Weyerhaeuser Co., J. van Buijtenen, Texas Forest Service, and G.B. Sweet, New Zealand Forest Service to examine the endogenous GAs of juvenile and sexually mature trees, "good" and "poor" flowering clones, and also trees to which cultural treatments such as non-destructive girdling, drought, and nitrate N and ammonium N fertilization had been administered.

Daylength was used by Glenn (1973) as an adjunct treatment to exogenous GA_3 and he noted, for Arizona cypress grown at 4, 8, 12, 16 and 24 h daylengths, that seedlings under the longer photoperiods required less exogenous GA_3 to initiate flowering, flowered sooner, and produced a greater flowering response to a given amount of GA_3. An examination of the endogenous GA spectrum of control plants (i.e., received no exogenous GA_3) showed a massive increase in endogenous GA_3[7] and other GAs of a more polar nature as length of day increased (Glenn, 1973). Kuo (1973) and Kuo and Pharis (1975) used the growth retardant AMO-1618 (2'-isopropyl-4'-(trimethyl-ammonium chloride)-5'-methylphenyl piperidine carboxylate) to reduce endogenous GA levels in Arizona cypress and subsequently noted a positive correlation between the number of strobili induced by increasing amounts of exogenous GA_3 and the amount of GAs extractable from the seedlings. Conversely, there was a negative correlation between the number of strobili produced by a given amount of exogenous GA_3 and the concentration of growth retardant applied. Thus, exogenous GA_3 would appear, in both these cases, to supplement endogenous levels in effecting precocious flowering for this species of Cupressaceae.

Kuo (1973) also induced precocious flowering in Arizona cypress through the use of nitrogen deficiency (as had been done earlier for sugi by Lyr and Hoffmann, 1964), and subsequently examined the spectrum of endogenous GAs. Unlike the correlations noted above, they found a shift in endogenous GAs from those of a polar nature to GAs of a relatively less-polar nature in the seedlings induced to flower with the N-deficiency treatment. Hashizume (1969) has analyzed tissue for both GAs and auxins from girdled trees of sugi which flowered and compared them with non-girdled controls. He noted lowered GA levels with girdling, but increased levels of auxins, the latter occurring also after exogenous GA_3 application. However, the methods used by Hashizume (1969) to separate the GAs prior to bioassay allow the less polar GAs to co-chromatograph with an inhibitor B fraction and the latter was also increased significantly by girdling. Thus, Hashizume (1969) might not have been able to detect changes in the less-polar GAs since the bioassay measures a "balance" between promotive and inhibitory substances in a plant extract.

Good and poor flowering clones of slash pine (*Pinus elliottii* Engelm.) and Monterey pine were also analyzed for endogenous GAs. Preliminary work on samples taken at one date during the period when sexual differentiation would be expected to take place showed vastly increased levels of all GAs for "good flowering" clones of slash pine[8], but the reverse was true for Monterey pine[5]. This apparent contradiction may be resolved by future analysis of tissue from subsequent sample dates[8], although it is quite possible that "poor flowering"[5] may be a result of many causes other than a "deficiency" of certain GAs.

Results from the analysis of tissue from juvenile and sexually mature trees of Douglas fir was somewhat ambiguous (perhaps because tissue was sampled during a poor flowering year?), GA_3[7] being high in both (but highest in mature) and the less-polar GAs being somewhat higher in juvenile than in sexually mature trees[9]. Girdling increased the levels of all GAs, especially the less-polar forms[9], but again, due to a poor flowering year, there was no significant increase in flowering due to girdling. However, seedlings of Douglas fir subjected to water stress (a treatment known to promote flowering for this species (Ebell, 1967)) showed significant decreases in endogenous GA_3[7] and an increase in less-polar GAs[9]. A similar situation was noted for sexually mature Douglas fir trees given nitrate N, ammonium N, or no nitrogen (control)[9]. Here, as noted previously by Ebell and McMullan (1970) and Ebell (1972), nitrate N-treated trees flowered profusely, ammonium N-treated trees were intermediate, and control trees did not flower at all. These trends in flowering were paralleled by increases in endogenous levels of the less-polar GAs, the nitrate N-treated trees having much greater amounts of less-polar GAs than ammonium N-treated or control trees. Conversely, the control trees had the highest levels of GA_3[7] than other polar GAs[9].

On the basis of the above correlations, a decision was made in 1973 to initiate a study whereby certain of those less-polar GAs which were available in large enough amounts would be tested on previously non-flowering 4-year-old grafted ramets of adult-phase Douglas fir. The results were most promising (Ross and Pharis, 1976), flowering being significantly promoted by exogenous applications of GA_4 and GA_7 (as a mixture: ca. 70% A_7 : 30% A_4), applied alone and in combination with GA_5 and GA_9. Applications were made biweekly between late March and late June (near Jefferson, Oregon), 400 µg (at each application date) of $GA_{4/7}$ giving a 5-fold increase in ovulate and 3-fold increase in staminate strobilus production over untreated controls. Non-destructive girdling, although ineffective itself in promoting flowering, enhanced the promotive effect of GA treatments.

Encouraged by these results, we initiated a cooperative experiment in 1974 with the Canadian Forestry Service and Pacific Logging using 6-year-old half-sib seedlings of Douglas fir at Saanichton, British Columbia. The results were spectacular (Fig. 1), the best treatment (200 µg each of $GA_{4/7}$ mixture + GA_9 + 5 µg of NAA applied biweekly from April through June) producing over 7 female strobili per branch (Pharis et al., 1976). Again, non-destructive girdling was used as an adjunct treatment. Concurrently, applications of less-polar GAs to 6-year-old seedlings of lodgepole pine (Pharis et al., 1975) pro-

STRUCTURES OF GIBBERELLINS

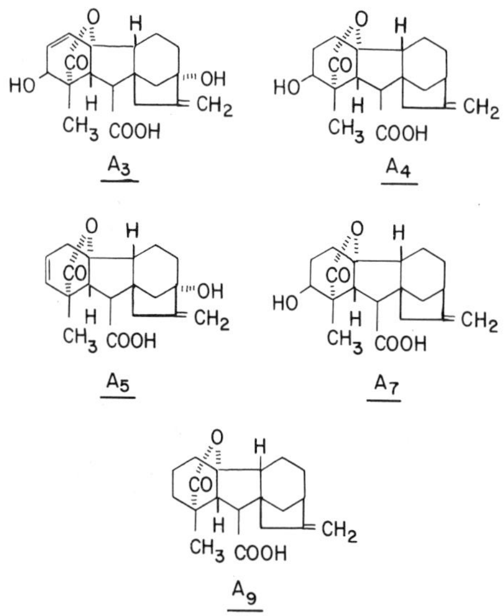

Fig. 1. Structures of five gibberellins (GAs) known to promote flowering in conifers. All are highly effective in certain species within the Cupressaceae and Taxodiaceae, but only GA_4 and GA_7 (as a mixture) and GA_9 are highly effective in promoting flowering in two species of Pinaceae, Douglas fir and lodgepole pine. GA_5 is only nominally effective by itself, but in combination with an auxin, or with other GAs is quite active in Douglas fir. GA_3, the least expensive and most readily available, shows moderate flowering activity on saplings of Japanese larch and grafted ramets of loblolly pine and will increase the flowering of Douglas fir when applied together with a $GA_{4/7}$ mixture. Otherwise, however, GA_3 has been essentially ineffective in the Pinaceae. Interconversion occurs between the GAs, and GA_3 is a known product of GA_7, which in turn can be made from GA_4 by the fungus, *Gibberella fujikuroi* (see reviews by West, 1973; MacMillan, 1974). All five GAs are native to higher plants and GA_3, GA_4, GA_7 and GA_9 are also produced by the fungus. GA_3 and a $GA_{4/7}$ mixture are available commercially

duced significant flowering and Ross (1976a, b) obtained significant flowering with $GA_{4/7}$ on 2-year-old grafted ramets of Douglas fir (4-year-old seedlings showed a small, but non-significant promotion) and 4-year-old grafted ramets of loblolly pine (*Pinus taeda* L.). In the latter case, the GA_3 was also effective, bringing to mind the past work by the Japanese workers (see earlier paragraphs in this section).

It would thus appear that judicious application of certain of the less-polar GAs (most notably the $GA_4/_7$ mixture and combinations of other GAs with it) can promote flowering in Douglas fir and lodgepole pine seedlings as well as in non-flowering grafts of Douglas fir and loblolly pine (Fig. 2). Application during that period of time when sexual differentiation is occurring appears to be critical, the treatment (in Douglas fir at least) causing differentiation of previously initiated

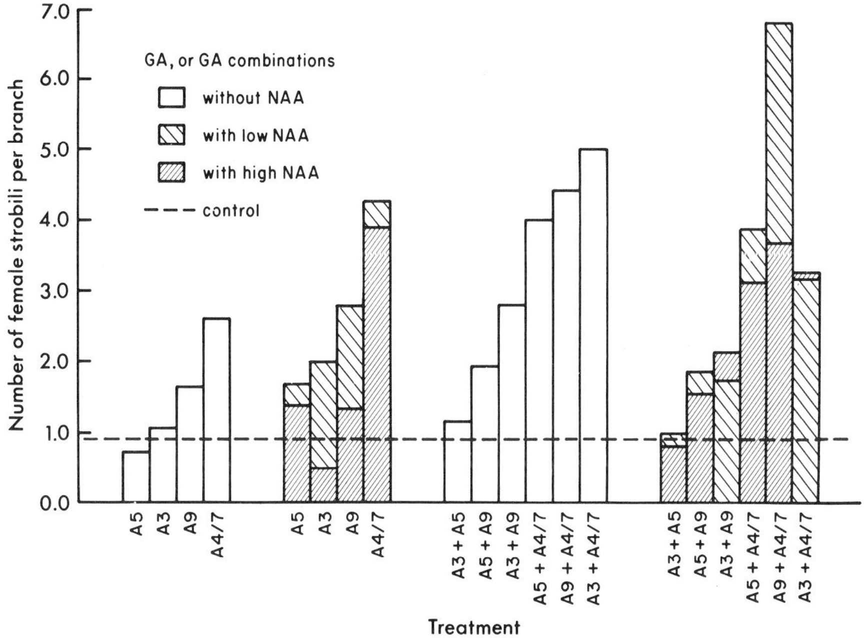

Fig. 2. Efficacy of different gibberellins (GAs) and their combinations on the promotion of flowering in 6-year-old seedlings of Douglas fir. GAs were applied biweekly in 80% ethanol (10 µl) at 200 µg each application over a 3 month period from April to June, 1973. Some treatments received the auxin, naphthaleneacetic acid (NAA) at 5 µg (low) or 25 µg (high) on each application date. All branches (including controls) had been non-destructively girdled with overlapping halfmoon girdles a week prior to the first hormone treatment to enhance the GA effect. The experiment was run with the cooperation of the Canadian Forestry Service and Pacific Logging in Pacific Logging's Saanichton seed orchard on Vancouver Island. (From Pharis, Ross and McMullan, unpublished data)

lateral primordia which otherwise would have become vegetative lateral buds (Pharis et al., 1976). Rather large amounts of GAs appear to be required and a topical application (i.e., injected behind the bud into woody tissue in 10 µl of 80% ethanol initially; then, as the new shoot "hardens", applied along the stem in 80% ethanol) has been used thus far. However, Dr. S. Ross of Weyerhaeuser [10] has indicated that preliminary experiments with Douglas fir using foliar sprays and a "continuous drip" into the main stem show promise.

One of the reasons why such large amounts of these less-polar GAs may be required is their rapid metabolism (Wample et al., 1975) into other more polar compounds, including GAs (Kamienska et al., 1974, 1976b; Kamienska and Pharis, 1975; Wample et al., 1975). It would appear then that the correlation between high levels of endogenous less-polar GAs and cultural treatments which promoted flowering (Kuo, 1973 and [9]) is a valid one since exogenous application of certain less-polar GAs is a highly effective treatment. Cultural treatments (and natural traumas?) may thus cause a promotion of flowering in conifers by increasing GA synthesis, interfering with GA interconversion, or both, the end result being a build-up of the less-polar GAs which would otherwise normally be rapidly converted (Wample et al., 1975) to more polar GAs.

Sexuality

Manipulation of sex for induced strobili is not only of interest to the tree breeder attempting to obtain both pollen and receptive ovulate strobili at an early age, but also to the seed orchard manager who, in general, wishes to skew flowering very much toward femaleness. It has been known for some time that for species in the Cupressaceae and Taxodiaceae which have been induced to flower with GA_3, daylength is a controlling factor in sexuality (Hashizume, 1959; Migita, 1963; Pharis and Morf, 1967, 1970), long day increasing maleness and short days femaleness. Apparently, for some species of Pinus, a similar situation exists (Larson, 1961; Longman, 1961; Giertych, 1967), although here the information was obtained on natural flowering at various latitudes without the use of GAs.

Although sex of the strobilus can be manipulated to a certain extent by absolute length of day in nature, photoperiod per se may not be the causative factor since both male and female differentiation can take place under the same photoperiod (Owens and Pharis, 1971). Rather, the major factor for western red cedar (*Thuja plicata* Donn) appears to be whether the daylength is "increasing" or "decreasing". If the former, a greater proportion of staminate strobili are differentiated, if the latter ovulate strobilus differentiation predominates (Owens and Pharis, 1971).

While such "photoperiodic" mechanisms may be operative for the Cupressaceae, Taxodiaceae and certain species of Pinus, the role of photoperiod or "lengthening or shortening" day in determining sexuality of strobili in certain species such as Douglas fir (where pre-existing primordia differentiate sexually into both males and females while days are lengthening) or loblolly pine (where certain ecotypes will produce both males and females on the last of three flushes in late August[10]) has yet to be ascertained. Even if daylength tells us that other factors may override such a control, years in which one sex or the other predominate are not uncommon. Nor, is it uncommon to see individual trees with a propensity toward maleness or femaleness. A similar skewing of sex ratios may also occur during the precocious flowering of very young seedlings (Ross, 1976b).

It is attractive to speculate that these "other factors" may involve plant hormones such as auxins and gibberellins, or perhaps a "balance" between different classes of hormones. In nature, ovulate strobili usually occur at the top of the tree, or on the outer part of branches, and males at the bottom or on the inner part of branches. This "gradient of sexuality" can readily be seen for western red cedar seedlings which have been induced to flower with GA_3 (Pharis and Morf, 1968). Heslop-Harrison (1963) noted that sex of the flower in many plants may be controlled by auxin gradients originating in the branch apex during the course of normal differentiation. Indeed, the apparent sex reversal in western red cedar (Pharis and Morf, 1968) of ovulate strobili (first initiated on the middle branches) to staminate strobili as the seedling continued to grow (and the position of the ovulate strobili was relegated to the "inside" of a lower branch) also fits well with a theory of "gradients originating from the apex" determining the sex of these GA_3-induced strobili. It should, of course, be recognized that such a gradient could mean either the presence, absence, or balance of various substances.

The exogenous application of auxins will increase the proportion of ovulate strobili produced by some Pinus species and also enhance femaleness in certain species in the Cupressaceae and Taxodiaceae (Saito,

1957; Hashizume, 1961, 1966, 1967, 1969, 1973) as will Ethrel given with GA_3 under certain photoperiod sequences (Bonnet-Masimbert, 1970). However, high levels (100 µg/plant applied biweekly from July to mid-September) of auxin reduced the number of GA-induced ovulate strobili by 1/3 on lodgepole pine seedlings relative to low (10 µg/plant) levels (Wample and Pharis, 1975). In Douglas fir, NAA tended to enhance maleness when applied alone (Ross, 1976b) or together with GAs (Pharis et al., 1976; Ross, 1976b) although low levels of NAA did promote additional ovulate strobilus production in certain of the very highly effective GA combinations (Pharis et al., 1976). In Arizona cypress (Bonnet-Masimbert, 1970) Ethrel, when given with GA_3 under photoperiods favoring maleness, increased the ratio of male : female strobili, but when given under photoperiods favoring femaleness, enhanced the number of female strobili produced (there were no males). Further, use on loblolly pine grafted ramets of 2,3,5-triiodobenzoic acid (TIBA) as a foliar spray in aqueous solution (TIBA is a known inhibitor of polar auxin transport (Niedergang-Kamien and Skoog, 1956)) enhanced ovulate strobilus production (Ross, 1976a). However, in Ross' study with loblolly pine very few staminate strobili were produced by any treatment.

Finally, an analysis of endogenous auxin levels in strobili of sugi and certain Pinus species (Hashizume, 1960, 1969; Giertych and Forward, 1966) showed them to be highest in ovulate strobili and lowest in staminate strobili.

Thus, while sexuality in nature is often correlated with situations indicative of gradients originating from the apex, and analysis of endogenous auxins of the strobili themselves point to auxin being associated with femaleness, exogenous applications of auxins have, depending perhaps upon species and the adjunct treatment, promoted both maleness and femaleness. Where auxin has consistently promoted maleness (ignoring for the time-being TIBA as an "anti-auxin") GAs have been an adjunct treatment. It is thus possible that a balance between the two hormones is the most critical factor, high auxin : GA ratios promoting femaleness, moderate auxin : GA ratios promoting maleness. The rather disquieting facts that 1. low or moderate levels or exogenous GA_3 will favor maleness and higher concentrations favor femaleness in sugi and Arizona cypress (Hashizume, 1959; Pharis and Morf, 1970), and 2. the transition of recently-formed microsporangia to megasporangia with GA_3 (Hashizume, 1960) could be explained by a GA_3-enhanced production of endogenous auxin as noted by Hashizume (1961). Given that daylength can control endogenous GA levels in Arizona cypress (Glenn, 1973), one could turn full circle and speculate that daylength effects on sexuality are exerted via endogenous GA concentration (i.e., GA : auxin balance) and further that high GA concentration may also increase endogenous auxin levels, thus changing the GA : auxin balance. Such a system, confusing as it may seem, might explain the quite contradictory results obtained through the exogenous application of auxin.

Several studies have shown that form of GA may influence sex of strobili (Ross and Pharis, 1976; Ross, 1976b and [10]), GA_5 preferentially promoting maleness. The reasons for this effect are unclear, although it should be noted that GA_5 shows widely divergent effects on flowering in Arizona cypress (Table 1) depending upon carrier and mode of application and that a less-polar GA-like substance (thought first to be GA_5, but is probably GA_{20}) extractable from excised pea internodes will not diffuse out, whereas a very polar GA-like substance also extractable from the internodes will (Jones and Lang, 1968). This could indicate that the GA_5 (which is injected into the woody shoot of Douglas fir behind the bud) moved only slowly into the new shoot, thus exerting its "flowering-promoting effect" primarily on the basal portion (i.e., male portion) of the shoot. Thus it is interesting to note that when GA_5 was applied topically in 90% dimethyl sulfoxide instead of 80% ethanol, staminate and ovulate strobili were equally enhanced in 8-year-old Douglas fir grafts[10]. However, Jones and Lang (1968) note no restriction of diffusion of exogenous GA_5 when applied to the excised pea internodes, and further work with radioactive GA_5 will be necessary to resolve this problem in Douglas fir.

Development to Anthesis

It is known, at least for Arizona cypress, that discontinuance of exogenous GA_3 application to young seedlings will result in abortion of many of the previously induced strobili (Pharis and Morf, 1967). Also, growth of callus derived from the somatic diploid axis of the staminate strobilus of mourning cypress (*Cupressus funebris* Endl.) requires exogenous GA_3 for normal growth (this is not the case however, for callus of similar origin from *Libocedrus decurrens* (Straus and Epp, 1960)). Thus, continued growth and development of conifer strobili prior to anthesis may require GA.

For at least certain conifer species of temperate zone origin, further development of the strobilus toward shedding of the pollen and female receptivity is influenced both by daylength and low temperature. That is, strobili of several Cupressus species[6] and western redcedar (Pharis and Morf, 1968, 1972; Pharis et al., 1969a) will not proceed toward a stage of anthesis-like development until a sequence of long day → short day → long day and/or low temperature is given. Low temperature appears to be the major factor and short day only a nominal factor. If this requirement were expressed in terms of "percentage of strobili proceeding toward anthesis", values of approximately 10% could be assigned to "short day" and 90% to "low temperature", the combination producing complete development. With regard to low temperature, staminate strobili of western redcedar require fewer cold days than do ovulate strobili (Pharis and Morf, 1972), an observation of interest from a "teleological" standpoint.

These observations are not unexpected since strobili of temperate-zone conifers normally overwinter prior to anthesis. It is however of import to geneticists and seed orchard managers who wish to work with temperate-zone species from high latitudes or with high elevation ecotypes since unwise location of the seed or breeding orchard may preclude normal development if winter temperatures are too mild. Unpublished research[6] at the University of Calgary indicates a rather broad range of temperature requirements for anthesis-like development of species within the Cupressaceae, varying all the way from no low temperature requirement for Mexican cypress (*Cupressus lusitanica* Mill.) of Central America origin, a nominal requirement for Arizona cypress (southern California origin) and Mendocino cypress (*Cupressus pygmaeae* (Lemm.) Sarg. of northern California origin) to a lenghty cold requirement (5 to 7 weeks) for western redcedar (British Columbia origin; Pharis and Morf, 1972). While such a requirement has only been demonstrated for the above-mentioned species, it is likely that all temperate-zone conifers possess it in varying degree. Analagous situations certainly exist for many other woody species (see reviews by Vegis, 1964; Wareing and Saunders, 1971).

Thus, the tree breeder should be aware that GA-induced strobili may require a continued supply of the hormone during early growth and development <u>and</u> that anthesis-like development of strobili, for temperate-zone species or high elevation ecotypes, will require a period of low temperature (short day, although helpful is not necessary (Pharis and Morf, 1972)).

Development after Anthesis

Growth and maturation of the cone and seed may take from 6 to 18 months depending upon species. During this time abortion of the cone, or arrested development of the megagametophyte, may occur from a wide variety of causes (see review by Sweet, 1973). Fertilization may not occur due to improper pollen germination, poor pollen tube growth, incompatibility factors, or a combination of all of these; abortion of the cone or parthenocarpy being the end result. Certain of these problems appear to have a hormonal basis and will be discussed below.

Firstly, there are indications that hormones play roles in gymnosperms, both for pollen germination (Kaurov and Vakula, 1961) and pollen tube growth (see review by Stanley and Linskens, 1974). Gibberellin A_3 will stimulate the germination (Kaurov and Vakula, 1961) of jack pine (*Pinus banksiana* Lamb.) pollen and pollen tube growth of Pinus species can be stimulated by indole-3-acetic acid and the ether extract from *Pinus montana* (Annhaeusser, 1953 cited in Stanley and Linskens, 1974), although such an ether extract does not preclude the presence of certain of the less-polar GAs. Gibberellins will also stimulate pollen tube growth in Douglas fir (Ching and Ching, 1959) and a GA-like substance extracted from Monterey pine pollen will, when applied to germinating pollen, stimulate its tube growth (Sweet and Lewis, 1971). Conifer pollen contains auxin-like substances (Stanley and Linskens, 1974) and from pollen of knobcone pine (*Pinus attenuata* Lemm.) Kamienska et al. (1976a) have characterized GA_4, GA_7, and GA_3 by gas chromatography - mass spectrometry and GA_1, GA_2 (also an artifact of the purification system) and GA_{34} by gas-liquid radiochromatography (Kamienska et al., 1976b). Dormant pollen of knobcone pine contains large quantities of GA_4, GA_7 and other less-polar GAs and these decrease during germination (Kamienska and Pharis, 1975). By hour 15 of germination the less-polar GAs and GA-like substances have reached their lowest level and a relatively high level of GA_3 and other polar GAs is present (Kamienska and Pharis, 1975), implying interconversion during germination.

Thus, while GAs (and perhaps auxins) have been implicated in pollen tube growth, and their exogenous application to germinating pollen *in vitro* will stimulate tube growth, the significance of this to the tree breeder still remains to be determined.

Of more practical import may be the use of auxins and GAs to maintain cone and ovule development. While Sweet (1973) noted that the results are somewhat ambiguous where growth regulators have been applied to pollinated strobili, I believe that past work with unpollinated strobili lend some hope for developing useful techniques to reduce cone abortion and arrested ovule development. For example, it has been noted that exogenous application of GA_3, cytokinins, auxins and their combinations have been used to delay abscission of unpollinated strobili (Sweet, 1973). Application of GA_3 alone, or with an auxin, is even reported to maintain cone and ovule development in the absence of pollen and produce viable seed in Douglas fir (Stettler et al., 1969; Livingston, 1971) and several species of Pinus[11]. Sweet and Lewis (1971) have also postulated that substances diffusing from germinating pollen might stimulate auxin or other growth hormone production by the ovule, which in turn would maintain cone development, and it appears that endogenous GA-like substances increase in ovulate strobili after pollination (Kopcewicz, 1969).

Given the above, it is possible that analysis of growth regulator levels in problem situations may indicate a hormone deficit or imbalance which could be corrected by judicious application of the appropriate substance. Along the same lines, preliminary analysis[12] of seed yield where GAs were used to enhance flowering on 6-year-old sibs of Douglas fir (Pharis et al., 1976) indicates that the significant increase in cones per branch was accompanied by an increase in both total and filled seeds per cone.

Conclusions

From the above, I believe that it is possible to recommend a practical approach whereby use of plant hormones can assist the tree breeder and seed orchard manager in promoting early flowering, regularizing cone crops and enhancing seed yield. Most of these approaches will be experimental in nature and may have to follow research designed to determine the levels or balance of certain of the plant growth regulators during critical periods of differentiation and development. Even so, I believe that an empirical approach, based on our current level of knowledge, may well be warranted.

Firstly, genetics work with members of the Cupressaceae and Taxodiaceae can perhaps best proceed by using applications of GA_3 to the seedling or orchard tree - both precocious and enhanced flowering should result. Mode of application can involve a periodic aqueous (using a surfactant) or ethanolic (80%) spray, injection into the main stem in ethanolic solution (5 to 80%), or as a root drench in aqueous solution. Depending upon the photoperiod under which treatment is initiated, sexuality of the induced strobili should be readily manipulable, either by varying GA_3 concentration or by applying auxin or Ethrel in addition to GA_3. In very young seedlings, or where environmental factors mitigate against heavy flowering, levels of GA_3 may have to be increased and/or duration of treatment lengthened.

Secondly, for species within the Pinaceae, the topical application of rather large and frequent dosages (300 to 600 µg applied weekly or biweekly) of $GA_{4/7}$ in 60 to 80% ethanolic solution to buds and shoots during that period when sexual differentiation takes place will, in my opinion, stand a good chance of promoting early or enhanced flowering _if_ additional cultural treatments known to be even moderately successful in enhancing flowering are used in conjunction with the $GA_{4/7}$ application. Here, depending upon species, non-destructive girdling (known to synergistically enhance flowering in the presence of GAs (Ross and Pharis, 1976 and [13])), nitrate fertilization (but _not_ urea), bagging of branches with clear plastic to increase temperature during differentiation, moderate water stress, etc., could all be useful adjunct treatments. Application of modest amounts of auxin (10 to 20 µg NAA/bud/application) together with the $GA_{4/7}$ would appear to hold promise as well.

Thirdly, where the choice still exists, breeding and seed orchards should be located in dry, sunny locations, ideally ones where irrigation can be used judiciously during conelet development, but withheld as necessary during the period of strobilus differentiation. Photoperiod during the differentiation period should be kept in mind with regard to sexuality when species or clones from different latitudes are to be moved into the orchard. The location of the orchard must, of course, take into account any low temperature requirement necessary for normal strobilus development toward anthesis.

Fourthly, although the research has yet to be accomplished, I would speculate that careful application of certain of the less-polar GAs, perhaps together with small amounts of auxins, may be useful in preventing cone abortion where weather and other factors are unfavorable for continued cone development.

While sexuality may not be manipulated in the Pinaceae as readily as in the Cupressaceae and Taxodiaceae by timing of application of the GA treatments, the possibility of total shading or supplemental lighting should be kept in mind as adjunct treatments to shift the sex of induced strobili one way or the other (at least for those species that use photoperiod as a determining factor). Exogenous application of auxins may hold some promise in manipulating sex, but from the large amount of variation noted in experiments cited above, the breeder or orchard manager will have to be very careful in amounts, not only of auxin but perhaps also of GAs.

Finally, successful manipulation of flowering through the use of plant hormones does not appear to be a simple matter subject to ready solution by use of only one hormone or technique. Rather, each species and site will require careful analysis with regard to timing, mode and duration of application, amounts of hormone, and appropriate adjunct cultural treatment (if any). It appears, at least to me, that recent successes in the Pinaceae with application of certain of the GAs, together with auxins, have depended upon many other factors being favorable for sexual differentiation. In very young trees, or during "poor differentiation years", factors other than hormone concentrations may become limiting. Care to minimize these other factors could be the key to successful manipulation of early flowering and regularization of cone crops through the use of plant hormone application.

Acknowledgements. Much of the work reported herein was accomplished through cooperative experiments with Mrs. E. McMullen, Messrs. W. Morf and R. Wample, Drs. H. Aoki, S. Cade, R. Durley, L. Ebell, J. Glenn, C. Heaman, A. Kamienska, C. Kuo, N. Murofushi, G. Puritch, J. Rediske, S. Ross and J. van Buijtenen. Their contributions are valued most highly. Financial support for the projects has come from the Canadian Forestry Service (Environment Canada), National Research Council of Canada, N.A.T.O., and the Weyerhaeuser Foundation.

Postscript. It has just been learned from Drs. S. Ross and J. Oh, Weyerhaeusen Co., Centralia, Wash., USA, that GA_7 appears to be the active component of the $GA_{4/7}$ mixture with regard to the promotion of flowering in Douglas fir grafts. Further, Dr. Ross has now obtained highly significant flowering on 4-year-old seedlings of Douglas fir using a $GA_{4/7}$ mixture (70% GA_7) in conjunction with certain cultural treatments (girdling, water stress, or nitrate N fertilization).

Footnotes

[1]Pers. comm. H. Wellendorf, Arboretum, Royal Veterinary and Agricultural University, Hørsholm, Denmark.

[2]Unpubl. res. results. A. Dunberg, Royal College of Forestry, University of Umeå, Sweden.

[3]Pers. comm. A. Fletcher, Forestry Commission, Roslin, Midlothian, Scotland, U.K.

[4]Pers. comm. J.N. Owens, Biology Dept., University of Victoria, Victoria, B.C., Canada.

[5]Unpubl. res. results. R.P. Pharis and G.B. Sweet.

[6]Unpubl. res. results. R.P. Pharis and W. Morf.

[7]Unpubl. res. results. R.C. Durley and R.P. Pharis.

[8]Unpubl. res. results. R.P. Pharis and J. van Buijtenen.

[9]Unpubl. res. results. R.P. Pharis, J. Rediske, S. Cade, S. Ross and L. Ebell.

[10]Pers. comm. S. Ross, Weyerhaeuser Co., Centralia, Washington, USA.

[11]Pers. comm. S. Krugman, U.S. Forest Service, Washington, D.C., USA.

[12]Unpubl. res. results. S. Ross, C. Heaman and R.P. Pharis.

[13]Unpubl. res. results. R.W. Skadsen, S. Ross and S. Cade, Weyerhaeuser Co., Centralia, Washington, USA.

References

Annhaeusser, H.: Keimung und Schlauchwachstum des Gymnospermenpollens unter besonderer Berücksichtigung des Wuchsstoffproblems. Beitr. Biol. Pflanz. 29, 297-338 (1974). 1953 (cited in R.G. Stanley, H.F. Linskens: Pollen, Biology, Biochemistry, Management. Berlin, Heidelberg, New York: Springer 1974).

Bonnet-Masimbert, M.: Articifical induction of male and female flowers on young seedlings of *Cupressus arizonica* (Greene) and *Chamaecyparis lawsoniana* (Parl.)(ed. Finnish Forest Res. Inst.), Helsinki: IUFRO Section 22 Working Group on Sexual Reproduction of Forest Trees, 1970.

Ching, K.K., Ching, T.M.: Extracting Douglas-fir pollen and effects of gibberellic acid on its germination. Forest Sci. 5, 74-80 (1959).

Doorenbos, J.: Juvenile and adult phases in woody plants. In: Encyclopedia Plant Physiology (ed. W. Ruhland), Vol. XV, pp. 1222-1235. Berlin, Heidelberg, New York: Springer 1965.

Ebell, L.F.: Cone production induced by drought in potted Douglas-fir. Can. Dept. Forestry Bi-Mon. Res. Notes 23, 26-27 (1967).

Ebell, L.F.: Girdling: its effect on carbohydrate status and on reproductive bud and cone development of Douglas-fir. Can. J. Botany 45, 453-466 (1971).

Ebell, L.F.: Cone induction response of Douglas-fir to form a nitrogen fertilizer and time of treatment. Can. J. Forest Res. 2, 317-326 (1972).

Ebell, L.F., McMullan, E.E.: Nitrogenous substances associated with differential cone production response of Douglas-fir to ammonium and nitrate fertilization. Can. J. Botany 48, 2163-2177 (1970).

Giertych, M.: Analogy of the difference between male and female strobilus in Pinus to the differences between long- and short-day plants. Can. J. Botany 45, 1907-1910 (1967).

Giertych, M., Forward, D.F.: Growth regulation changes in relation to growth and development of *Pinus resinosa* Ait. Can. J. Botany 44, 717-738 (1966).

Glenn, J.L.: Investigations concerning the physiology of gibberellin-induced growth and sexual differentiation phenomena in Arizona cypress (*Cupressus arizonica*, Yreene) seedlings. Unpubl. Ph.D. Thesis, Calgary, Univ. of Calgary, 1973.

Hashizume, H.: The effect of gibberellin upon flower formation and sex transition to female in *Chamaecyparis obtusa* and *C. lawsoniana*. J. Jap. Forest Soc. 41, 458-463 (1959).

Hashizume, H.: The effect of gibberellin upon sex differentiation in *Cryptomeria japonica* strobili. J. Jap. Forest Soc. 42, 176-180 (1960).

Hashizume, H.: The effect of gibberellin on sex differentiation in *Cryptomeria japonica* strobili. II. Effect of auxins and urea on gibberellin-induced sex transition to female in strobili. J. Jap. Forest Soc. 43, 47-49 (1961).

Hashizume, H.: Initiation and development of flower buds in *Chamaecyparis obtusa*. J. Jap. Forest Soc. 45, 135-151 (1963).

Hashizume, H.: Chemical regulation of flower-bud formation and flower-sex differentiation in conifers. Chem. Regul. Plants 1, 148-156 (1966).

Hashizume, H.: Experimental induction of female flowers in young Japanese larch (*Larix leptolepis*, Gordon). J. Jap. Forest Soc. 49, 405-408 (1967).

Hashizume, H.: Auxins and gibberellin-like substances existing in the shoots of conifers and their roles in flower bud formation and flower sex differentiation. Bull. Tottori Univ. Forestry 4, 1-46 (1969).

Hashizume, H.: Studies on flower bud formation, flower sex differentiation and their control in conifers. Bull. Tottori Univ. Forestry 7, 1-139 (1973).

Heslop-Harrison, J.: Plant growth substances. In: Vistas in Botany III. (ed. W.B. Turrill), pp. 104-194. New York: Pergamon Press 1963.

Holmsgaard, E., Olsen, H.C.: Experimental induction of flowering in beech. Det forstlige Forsøgsuoesen i Danmark 232, 1-17 (1966).

Jackson, D.I., Sweet, G.B.: Flower initiation in temperature woody plants. Hort. Abstr. 42, 9-24 (1972).

Jones, R.L., Lang, A.: Extractable and diffusible gibberellins from light- and dark-growth pea seedlings. Plant Physiol. 43, 629-634 (1968).

Kamienska, A., Durley, R.C., Pharis, R.P.: Endogenous gibberellins of pine pollen. I. Characterization of gibberellins A_3, A_4 and A_7 by gas chromatography-mass spectrometry. Phytochemistry 15, 421-424 (1976a).

Kamienska, A., Durley, R.C., Pharis, R.P.: Endogenous gibberellins pine pollen. III. Interconversion of (^{3}H)-gibberellin A_4 to gibberellins A_1, A_2 and A_{34} by germinating pollen of *Pinus attenuata*. Plant Physiol., in press (1976b).

Kamienska, A., Pharis, R.P.: Endogenous gibberellins of Pinus pollen. II. Changes during germination of *P. attenuata*, *P. coulteri* and *P. ponderosa*. Plant Physiol. 56, 655-659 (1975).

Kamienska, A., Pharis, R.P., Wample, R.L., Kuo, C.C., Durley, R.C.: Occurrence and metabolism of gibberellins in conifers. In: Plant Growth Substances - 1973 (ed. S. Tamura), pp. 305-313. Tokyo: Hirokawa Publishing Co. 1974.

Kato, Y.: Stimulation of differentiation of flower buds in sugi (*Cryptomeria japonica*). J. Jap. Forest Soc. 41, 138-141 (1959).

Kato, Y., Fukuhara, N.: Stimulation of differentiation of flower buds in conifers by gibberellin. Trans. Meet. Soc. for Stud. Gibb. 4, 36-40 (1960).

Kato, Y., Miyake, I., Ishikawa, H.: Initiation of flower bud by gibberellin in *Cryptomeria japonica*. J. Jap. Forest Soc. 40, 35-36 (1958).

Kaurov, I.A., Vakula, V.S.: Vlijanie gibberellina na prorastanie pyl'cy drevenykh rastenij. Botan. Zh. 46, 1125-1133 (1961).

Klebs, G.: Über das Verhalten der Außenwelt zur Entwicklung der Pflanze. Sitz. der Akad. Wiss. Heidelberg, Ser. B, No. 5 (1913) (from A. Lang, 1965). In: Encyclopedia of Plant Physiology.

Kopcewicz, J.: The dynamics of growth substances during the development of the pine inflorescence. II. Female inflorescences. Rocz. Nauk Roln. 95, 231-238 (1969).

Kuo, C.C.: Growth retardation and nutritional stress in relation to vegetative growth and reproductive differentiation of seedlings of *Cupressus arizonica* Greene. Unpubl. Ph.D. Thesis, Calgary, The University of Calgary, 1973.

Kuo, C.C., Pharis, R.P.: The effects of growth retardants, AMO-1618 and B-995, on growth and gibberellin content of vegetative tissue of Arizona cypress (*Cupressus arizonica*) seedlings. Physiol. Plantarum 34, 288-292 (1975).

Lang, A.: Physiology of flower initiation. In: Encyclopedia of Plant Physiology, Vol. XV, pt. 1, pp. 1380-1536. Berlin, Heidelberg, New York: Springer 1965.

Larson, P.R.: Influence of date of flushing on flowering in *Pinus banksiana*. Nature 192, 82-83 (1961).

Livingston, G.K.: Experimental studies on the induction of haploid parthenogenesis in Douglas-fir and the effects of radiation on the germination and growth of Douglas-fir pollen. Unpubl. Ph.D. Thesis. Seattle: Univ. of Washington, 1971.

Longman, K.A.: Five factors affecting flowering initiation in certain conifers. Proc. Linn. Soc. London 72, 124-127 (1961).

Lyr, V.H., Hoffmann, G.: Über den Einfluß der Mineralsalzernährung auf die Frühfruktifikation von *Cryptomeria japonica* (L.F.D.) Don. Flora 154, 189-208 (1964).

MacMillan, J.: Recent aspects of the chemistry and biosynthesis of the gibberellins. In: The Chemistry and Biochemistry of Plant Hormones (ed. V.C. Runeckles, E. Sondheimer, D.C. Walton), pp. 1-20. London, New York: Academic Press 1974.

Michniewicz, M.: The dynamics of gibberellin-like substances and growth inhibitors in ontogeny of conifers. Wiss. Z. Univ. Rostock 416, 577-583 (1967).

Migita, K.: Flow bud formation of *Cryptomeria japonica* through the action of gibberellic acid in relation to growing conditions. III. Effect of photoperiodic and fertilizing treatment. J. Jap. Forest Soc. 45, 393-396 (1963).

Niedergang-Kamien, E., Skoog, F.: Studies on polarity and auxin transport in plants. I. Modification of polarity and auxin transport by triiodobenzoic acid. Physiol. Plantarum 9, 60-73 (1956).

Ohba, K., Okada, Y., Murai, M.: Advancing the generation of sugi, *Cryptomeria japonica* D. Don by greenhouse cultivation. J. Jap. Forest Soc. 53, 133-137 (1971).

Owens, J.N.: The relative importance of initiation and early development in cone production in Douglas-fir. Can. J. Botany 47, 1039-1049 (1969).

Owens, J.N., Molder, M.: Development of long-shoot terminal buds of *Pinus contorta* ssp. *contorta*. In: Management of Lodgepole Pine Ecosystems (ed. D.M. Baumgartner), Vol. I, Proceedings, pp. 86-105. Pullman, Wash.: Wash. State University, 1975.

Owens, J.N., Pharis, R.P.: Initiation and ontogeny of the microsporangiate cone in *Cupressus arizonica* in response to gibberellin. Am. J. Botany 54, 1260-1272 (1967).

Owens, J.N., Pharis, R.P.: Initiation and development of western red cedar cones in response to gibberellin induction and under natural conditions. Can. J. Botany 49, 1165-1175 (1971).

Pharis, R.P., Crozier, A., Reid, D.M., Kuo, C.C.: Activity of gibberellin A_{24} in 10 plant bioassays. Can. J. Botany 47, 815-820 (1969b).

Pharis, R.P., Kuo, C.C.: Physiology of gibberellins in conifers. Can. J. Forestry Res., in press (1975).

Pharis, R.P., Morf, W.: Experiments on the precocious flowering of western red cedar and four species of Cupressus with gibberellin A_3 and $A_{4/7}$ mixture. Can. J. Botany 65, 1519-1524 (1967).

Pharis, R.P., Morf, W.: Physiology of gibberellin-induced flowering in conifers. In: Biochem. and Physiology of Plant Growth Substances (ed. F. Wightman, G. Setterfield), pp. 1341-1356. Ottawa: Runge Press 1968.

Pharis, R.P., Morf, W.: Sexuality in conifers: effects of photoperiod and gibberellin concentration on the sex of gibberellin-induced strobili of western red cedar (*Thuja plicata* Donn.). Zesz. Nauk. Uniw. W. Kopernika, Torun. 23, Biol. 13, 85-89 (1970).

Pharis, R.P., Morf, W.: Short day and cold as causative factors in the anthesis-like development of strobili of western red cedar (*Thuja plicata*). Can. J. Botany 50, 2683-2685 (1972).

Pharis, R.P., Morf, W., Owens, J.N.: Development of the gibberellin-induced ovulate strobilus of western red cedar: quantitative requirement for long day → short day → long day. Can. J. Botany 47, 415-420 (1969a).

Pharis, R.P., Owens, J.N.: Hormonal induction of flowering in conifers. Yale Sci. Mag. 41, 10-19 (1966).

Pharis, R.P., Ross, S.D., Wample, R.L., Owens, J.N.: Promotion of flowering in conifers of the Pinaceae by certain of the gibberellins. In: Symposium on Juvenility in Woody Perennials (ed. R.E. Zimmermann. College Park, Md.: Acta Horticulturae, 56, in press (1976).

Pharis, R.P., Ross, S.D.: Promotion of flowering in the Pinaceae by gibberellins. II. Seedlings of Douglas-fir. Physiol. Plantarum, in press (1975).

Pharis, R.P., Ruddat, M., Phillips, C.C., Heftmann, E.: Precocious flowering of Arizona cypress with gibberellin. Can. J. Botany 43, 923-927 (1965).

Pharis, R.P., Wample, R.L., Kamienska, A.: Growth, development, and sexual differentiation in Pinus, with emphasis on the role of the plant hormone, gibberellin. In: Management of Lodgepole Pine Ecosystems (ed. D.M. Baumgartner), Vol. I, pp. 106-134. Pullman, Wash.: Wash. State University, 1975.

Puritch, G.S.: Cone production in conifers. A review of the literature and evaluation of research needs (with an economic analysis by A.H. Vyse). Pac. Forestry Res. Centre, Victoria, B.C.: Can. For. Service. Information Rep. BC-X-65 1972.

Robinson, L.W., Wareing, P.F.: Experiments on the juvenile-adult phase change in some woody species. New Phytol. 68, 67-78 (1969).

Ross, S.D.: Promotion of flowering in the Pinaceae by gibberellins. IV. Sexually mature, non-flowering mature grafts of loblolly pine. Physiol. Plantarum, accepted for publication (1976a).

Ross, S.D.: Promotion of flowering in Pinaceae by gibberellins. V. Four-year-old seedlings and two-year-old grafts of Douglas fir. In: Symposium on Juvenility in Woody Perennials (ed. R.H. Zimmerman). College Park, Md.: Acta Horticulturae, 56, in press (1976b).

Ross, S.D., Pharis, R.P.: Promotion of flowering in the Pinaceae by gibberellins. I. Sexually mature, non-flowering grafts of Douglas-fir. Physiol. Plantarum 36, 182-186 (1975).

Saito, Y.: Artificial control of sex differentiation in Japanese red pine and black pine strobili. J. Fac. Agr. Tottori Univ. 3, 1-29 (1957).

Saito, Y., Hashizume, H.: The effect of gibberellin application in flower bud differentiation and growth of *Metasequoia glyptostroboides* Hu et Cheng. Abstr. 69th Meet. J. For. Soc., p. 306 (1959).

Shidei, T., Akai, T., Ichikawa, S.: Flower-buds formation in sugi (*Cryptomeria japonica*) and metasequoia (*M. glyptostroboides*) by gibberellic and treatment. J. Jap. Forest Soc. 42, 363-368 (1960).

Shidei, T., Yoshikawa, K.: Stimulation of flower buds by gibberellin treatment in *Pinus* and Taxodiaceae, pp. 22-24. Trans. 6th Meet. Soc. Study Gibberellins 1963.

Stanley, R.G., Linskens, H.F.: Pollen, biology, biochemistry, management. Berlin, Heidelberg, New York: Springer 1974.

Stettler, R.K., Bawa, K.S., Livingston, G.K.: Experimental induction of haploid parthenogenesis in forest trees. In: Induced Mutations in Plants, pp. 611-617. Vienna: Intern. At. Energy Agency (STl/PUB/231), 1969.

Straus, J., Epp, R.R.: Response of *Cupressus funebris* tissue cultures to gibberellins. Science 131, 1806-1807 (1960).

Sweet, G.B.: Shedding of reproductive structures in forest trees. In: Shedding of Plant Parts (ed. T.T. Kozlowski), pp. 341-382. New York, London: Academic Press 1973.

Sweet, G.B., Lewis, P.N.: Plant growth substances in pollen of *Pinus radiata* at different levels of germination. New Zealand J. Botany 9, 146-156 (1971).

Vegis, A.: Dormancy in higher plants. Ann. Rev. Plant Physiol. 15, 185-224 (1964).

Wample, R.L., Durley, R.C., Pharis, R.P.: Metabolism of gibberellin A_4 by vegetative shoots of Douglas fir at 3 stages of ontogeny. Physiol. Plantarum 35, 273-278 (1975).

Wareing, P.F.: Problems of juvenility and flowering in trees. J. Linnean Soc. Lond. 56, 282-289 (1959).

Wareing, P.F., Saunders, P.F.: Hormones and dormancy. Ann. Rev. Plant Physiol. 22, 261-288 (1971).

West, C.A.: Biosynthesis of gibberellins. In: Control of Biosynthesis in Higher Plants (ed. B.V. Milborrow), pp. 143-177. London, New York: Academic Press 1973.

Zimmerman, R.H.: Juvenility and flowering in woody plants: a review. Hort. Science 7, 447-455 (1972).

Subject Index

R. FRANKEL, E. GALUN
Pollination Mechanisms in Plants and their Utilisation in Plant Breeding Procedures
(Monographs on Theoretical and Applied Genetics, Vol. 2)
In preparation
From the Contents: Implications of Pollination Mechanisms in Plant Breeding and Crop Production. - Reproduction in Higher Plants. - Ecology and Dynamics of Pollination.

J. SYBENGA
Meiotic Configurations
A Source of Information for Estimating Genetic Parameters.
65 figures. 64 tables. X, 251 pages, 1975
(Monographs on Theoretical and Applied Genetics, Vol. 1)
Contents: Introduction. - The analysis of crossing-over. - The analysis of chromosome pairing. - The analysis of distribution: centromere coorientation.

R.G. STANLEY, H.F. LINSKENS
Pollen
Biology, Biochemistry, Management
64 figures. 66 tables. VIII, 307 pages. 1974
Contents: Biology: Development. Wall Formation. Dehiscence, Size and Distribution. - Management: Collection and Uses. Storage. Viability Tests. Nutritive Role. - Biochemistry: General Chemistry. Carbohydrates and Cell Walls. Organic Acids, Lipids and Sterols. Amino Acids and Proteins. Pollinosis. Nucleic Acids. Enzymes and Cofactors. Pollen Pigments. Growth Regulators.

M.H. ZIMMERMANN, C.L. BROWN
Trees
Structure and Function. Springer Study Edition.
With a chapter on irreversible thermodynamics of transport phenomena by M.T. Tyree
2nd printing. 134 figures. XIII, 336 pages. 1974
Contents: Primary Growth. - Secondary Growth. - Growth and Form. - Transport in the Xylem. - Transport in the Phloem. - The Steady State Thermodynamics of Translocation in Plants. - Storage, Mobilization and Circulation of Assimilates.

K. STERN, L. ROCHE
Genetics of Forest Ecosystems
70 figures. X, 330 pages. 1974
(Ecological Studies, Vol. 6)
Contents: The Ecological Niche. - Adaptations. - Genetic Systems. - Adaptive Strategies. - Forest Ecosystems. - How Man Affects Forest Ecosystems.

Springer-Verlag Berlin Heidelberg New York